GENERAL THEORY OF SUPERCONDUCTIVITY

GENERAL THEORY OF SUPERCONDUCTIVITY

FU-SUI LIU AND YUMIN HOU

Nova Science Publishers, Inc.
New York

NOTICE TO THE READER

Library of Congress Cataloging-in-Publication Data
Liu, Fu-sui.
General theory of superconductivity / Fu-sui Liu and Yumin Hou.
p. cm.
Includes bibliographical references and index.
ISBN-13: 978-1-60021-803-3 (hardcover)
ISBN-10: 1-60021-803-2 (hardcover)
1. Superconductivity. 2. High temperature superconductivity. I. Hou, Yumin. II. Title.
QC611.92.L58 2006
537.6'23–dc22

2007029303

Published by Nova Science Publishers, Inc. ✤ New York

Contents

Preface

The research field of the high-T_c superconductivity will approach its twenty-first birthday. The first publication in this field appeared in September, 1986. In the past twenty years a wealth of experience and knowledge in both experiment and theory on the high-T_c and low-T_c superconductivity was accumulated. It is the time to make a complete summary and upgrade the understanding of the superconductivity and the pseudogap. "General Theory of Superconductivity" is born in this background.

Many books about theory of superconductivity are available now, which may be called special theory of superconductivity. There are some main differences between the general and the special theory of superconductivity. First, the special theory only studies the conventional superconductor, but the general theory studies both the conventional and unconventional superconductor. High-T_c cuprates are one kind of unconventional superconductors. The essential difference between the conventional superconductor and the high-T_c cuprates is the coherence length of the Cooper pair ξ_o. For the conventional superconductor $\xi_o \sim 10^3 - 10^4$ $\mathring{A}$, but for the high-T_c cuprates $\xi_o \sim 20 - 30$ $\mathring{A}$. Second, the special theory only studies the superconducting state and only gives the equation to determine the pairing temperature T^*, at which the gap begins to appear, but the general theory studies both the superconducting and the pseudogap state and gives different equations to determine T^* and superconducting transition temperature T_c. Third, the special theory mainly discusses the phonon-mediated superconductivity, but the general theory discusses both the phonon-mediated and the two local spin-mediated superconductivity. The latter is the mechanism of the superconductivity and pseudogap in the high-T_c cuprates, and can quantitatively explain many experimental facts. Fourth, the special theory only studies macroscopic superconductivity, but the general theory studies both macroscopic and microscopic superconductivity. Fifth, the general theory introduces some new terms and gives many predictions. In addition, the general theory can give some further discussions on mechanism of superfluidity in ^{4}He and ^{3}He.

To popularize the general theory, we are motivated to write this book as a textbook and expect that the research work will progress on the new level of the general theory. Easy to understand is one of our purposes. So this book interprets the general theory in simple language and graduate students can well know the methods to determine T^*, T_c, gap and pseudogap by using the program in chapter 5. This book concerns the physical pictures of theoretical results, the explanations of experiments and physical predictions besides pure theory. Therefore, the title of this book can be also called "General Physics of Superconductivity".

The whole book consists of seven chapters and one appendix. Chapter 1 introduces the basic phenomena and expounds the phenomenological theories. Chapters 2 and 3 give the theoretical explanations of concepts in chapter 1. Chapter 2 introduces BCS pairing theory of both conventional and unconventional superconductors. The gap and T^* equations are established in this chapter. Chapter 3 proposes a mechanism of the superconductivity and pseudogap in the high-T_c cuprates, and gives the T_c equation. Chapter 4 applies the microscopic theories in chapters 2 and 3 to explain many experiments in the high-T_c cuprates quantitatively. Since many experiments can be explained by the two local spin-mediated mechanism quantitatively in terms of unified values of related parameters, we believe that this mechanism is correct for the high-T_c cuprates. Some readers might be doubtful about the correctness of the new concepts in chapter 1. However, the theoretical grounds of these new concepts are given in chapters 2, 3, and 4. All the doubts about the new concepts will be removed after reading chapters 2, 3, and 4. Chapter 5 gives the numerical method to solve the non-linear integral equation used in chapter 4. Chapter 6 discusses many systems, in which the mechanism of superconductivity and pseudogap might be the same as that of the high-T_c cuprates. This chapter also discusses the mechanism of superfluidity in ^{3}He and ^{4}He. Chapter 7 expounds the extended Abrikosov pseudo-fermion representation of local spin operator, which is used in chapter 3. Since chapter 7 might be a little hard to understand, an appendix is attached to expound the basis of the pseudo-fermion representation in detail.

Authors
March 15, 2007.

Chapter 1

Introduction

1.1. Basic Phenomena and Concepts

1.1.1. Electric Properties (1911)

If direct current resistance $R = 0$, when $T \leq T_c$ [1] (1911), then the T_c is called superconducting transition temperature, or critical temperature. It is also called temperature of long-range phase coherence of Cooper pairs in general theory of superconductivity. The state at $T \leq T_c$ is called superconducting state. (For the exact statement on T_c of conventional superconductors see the sixth paragraph.) Bardeen, Cooper, and Schrieffer (BCS) proposed their epoch-making pairing theory of superconductivity [2], which forms the subject of chapter 2. In BCS theory, it was shown that even a weak net attractive interaction between electrons can cause an instability of the ordinary Fermi-sea ground state of the electron gas with respect to the formation of bound pairs of electrons occupying states with equal and opposite momentum. These so-called Cooper pairs have spatial extension or average size or coherence length ξ_o. One might well ask what really is the essential universal characteristic of the (macroscopic) superconducting state. The answer is that the existence of Cooper pairs is only a necessary condition. To have (macroscopic) superconductivity the Cooper pairs have to maintain long-range phase coherence over macroscopic distances. When the superconducting state is at rest, the long-range phase coherence contains two types. One is long-range unified phase of all Cooper pairs which comes from strong overlap of Cooper pairs. Another one is long-range phase difference locking which comes from separation of Cooper pairs. Note that when the superconducting state is at rest, the phase difference locked is actually zero (See section 1.1.5.). Microscopic superconductivity and supercurrent can occur without the long-range phase coherence. For the microscopic superconductivity and supercurrent see sections 2.6.2, 2.7.4 and 2.8. In this book "superconductivity" connotes "macroscopic superconductivity".

There are two kinds of superconductors: conventional and unconventional superconductors. A great part of low-T_c superconductors belong to the conventional superconductors. The whole Josephson tunneling junction [3], the granular superconductor [4], and the high-T_c cuprates [5] belong to the unconventional superconductors. The basic difference between conventional and unconventional superconductors is as follows. In conventional superconductors the size of a Cooper pair is large (about 10^3 - 10^4 $\mathring{A}$). Therefore, the Cooper pairs are easy to strongly overlap. In unconventional superconductors there is, at least, one region in which there is no the Cooper pairs. In the high-T_c cuprates the size of a Cooper pair is small (about 25 Å). So, the Cooper pairs are separated into many single Cooper pairs (units), and there are no Cooper pairs in many regions in the high-T_c cuprates. For the whole Josephson tunneling junction the Cooper pairs are separated into two superconducting units disconnected with each other, i.e., two superconducting electrodes, by the middle insulating layer, in which there are no the Cooper pairs. For the granular superconductors the Cooper pairs are separated into many superconducting units disconnected with each other, i.e., superconducting grains. So, the granular superconductor also contains many regions, in which there are no Cooper pairs. Note that the Cooper pairs in the superconducting units separated are of the long-range phase coherence in the whole Josephson tunneling junction and the granular superconductors. The superconducting unit may be a superconducting film for a Josephson tunneling junction, or a superconducting grain for a granular superconductor. The unit of the high-T_c cuprates contains only a single Cooper pair. There is Josephson tunneling coupling between the two nearest-neighbor superconducting units in the whole Josephson tunneling junction and the granular superconductor, and between the two nearest-neighbor units in the high-T_c cuprates (See section 3.3.).

When the Cooper pairs begin to appear at T^* (pairing temperature) in the units, the high-T_c cuprates is still in normal state because the zero-point fluctuation of the phase difference between any two nearest-neighbor units, and the thermal disturbance can destroy the phase difference locking, and thus there is no long-range phase difference locking. Only at a lower temperature T_c ($< T^*$), when the Josephson tunneling coupling energy (See section 3.3.) can overcome both the thermal disturbance energy and the zero-point fluctuation energy, there is the long-range phase difference locking over macroscopic distances, and thus the superconductivity occurs in the high-T_c cuprates. Note that the long-range phase difference locking is also a type of the long-range phase coherence besides the unified phase of all Cooper pairs. The Cooper pairs at $T_c < T < T^*$ may be called non-supercurrent Cooper pairs or Cooper pairs without the long-range phase coherence. The Cooper pairs at $T < T_c$ consist of two sorts (See section 3.3.). One sort is the supercurrent Cooper pairs or Cooper pairs with the long-range phase coherence, and the other one is non-supercurrent Cooper pairs or Cooper pairs without the long-range phase coherence.

When the superconductivity begins to appear at T_c' in the superconducting units of Josephson tunneling junction, the whole Josephson tunneling junction is still in normal state. Only at a lower temperature T_c ($< T_c$'), when the Josephson tunneling coupling can overcome both the thermal disturbance and the zero-point fluctuation of phase difference on the two sides of the insulating barrier layer, there is the phase difference locking between the two superconducting units on the two sides of the insulating barrier layer. The T_c is called superconducting transition temperature of the whole Josephson tunneling junction. The Josephson tunneling junction belongs to one sort of the weakly coupled systems. The analyses for other weakly coupled systems are similar to the whole Josephson tunneling junction.

When the superconductivity begins to appear at T_c' in the superconducting units of granular superconductor, the granular superconductor is still in normal state. Only at a lower temperature T_c ($< T_c$'), when the Josephson tunneling coupling can overcome both the thermal disturbance and the zero-point fluctuation of phase coherence, there is the phase difference locking between any two nearest-neighbor superconducting units , i.e., the long-range phase difference locking. The T_c is called superconducting transition temperature of granular superconductor.

Let us discuss exactly the characteristic temperatures of the conventional superconductors, such as tin and lead. In our physical views, the conventional superconductors have three characteristic temperatures: T^*, $T_{c,1}$, and T_c. T^* is a temperature at which the Cooper pairs without the long-range phase coherence begin to appear. $T_{c,1}$ is a temperature at which the Cooper pairs with the long-range phase difference locking (one type of the long-range phase coherence) begin to appear. T_c is a temperature at which the Cooper pairs with the unified phase (another type of the long-range phase coherence) begin to appear. The reasons are as follows. In conventional superconductors the average size of a Cooper pair ξ_o is at the order of $10^3 \sim 10^4$ $\mathring{A}$. Assume $\xi_o = 10^4$ $\mathring{A}$ and electron number density is $n = 10^{22}/cm^3$. (Note that the ξ_o in the high-T_c cuprates is only 20 - 30 $\mathring{A}$.) At T^* the Cooper pairs begin to appear. Because the number of Cooper pairs is very small at $T \approx T^*$, the distance between any two centers of Cooper pairs is larger than ξ_o. When the temperature lowers, the number of Cooper pairs increases. At $T_{c,1}$ ($< T^*$) the number density of Cooper pairs is $2 \times 10^{12}/cm^3$, i.e., there are 2 Cooper pairs in $(10^{-4}\ cm)^3$ volume. In this case the Josephson coupling energy overcomes both the zero-point fluctuation energy of phase coherence of two nearest-neighbor Cooper pairs and thermal disturbance energy (See section 3.3.), and thus the long-range phase difference locking between Cooper pairs is established. At $T_{c,1}$ the supercurrent Cooper pairs begin to appear. When the temperature lowers from $T_{c,1}$, the number of Cooper pairs continues to increase. At T_c ($< T_{c,1}$) the number density of Cooper pairs is $10^{13}/cm^3$, i.e., there are 10 Cooper pairs in $(10^{-4}\ cm)^3$ volume. In this case the Cooper pairs begin to strongly overlap, and thus the unified phase of all Cooper pairs is formed. From the measurements of resistance and specific heat jump the T^* and $T_{c,1}$ are in a very narrow temperature interval 10^{-3} K. Therefore, in physics, we deduce that $T_{c,1}$ and T_c are in a very narrow temperature interval 10^{-3} K as well. Because $T^* \approx T_{c,1} \approx T_c$ we can roughly say that the specific heat jumps at T_c instead at T^* and $T_{c,1}$, and the zero resistances occur at T_c instead at $T_{c,1}$. At present, there is only conceptual sense instead of observation sense to distinguish three temperatures for the conventional superconductors.

We interpret the alternating current resistance Z in superconducting state by the following table.

$$Z \begin{cases} = 0, & \text{when} \quad T = 0 \quad \text{and} \quad \begin{cases} \hbar\omega \leq 3.5 k_B T_c & \text{for} \quad \text{Sn}; \\ \hbar\omega \leq 7 k_B T_c & \text{for} \quad \text{YBa}_2\text{Cu}_3\text{O}_{7-\delta}; \\ \hbar\omega \leq 4.1 k_B T_c & \text{for} \quad \text{Pb}; \end{cases} \\ > 0, & \text{when} \quad T = 0 \quad \text{and} \quad \begin{cases} \hbar\omega > 3.5 k_B T_c & \text{for} \quad \text{Sn}; \\ \hbar\omega > 7 k_B T_c & \text{for} \quad \text{YBa}_2\text{Cu}_3\text{O}_{7-\delta}; \\ \hbar\omega > 4.1 k_B T_c & \text{for} \quad \text{Pb}; \end{cases} \\ > 0, & \text{when} \quad 0 < T < T_c \quad \text{and} \quad \hbar\omega > 0; \\ \approx Z_n, & \text{when} \quad 0 < T < T_c \quad \text{and} \quad \hbar\omega \gg k_B T_c. \end{cases}$$

Here ω is the frequency of alternating current, and Z_n is the alternating current resistance in normal state. For the high-T_c cuprates the condition of $Z = 0$ at $T = 0$ is that they are in superconducting state. If they are in pseudogap state (See section 1.1.6.), then always $Z > 0$ at any temperature including $T = 0$. From the alternating current resistance at $T = 0$

(very low temperature) in the above table we know that there is a threshold energy for absorption of electromagnetic wave. The existence of the threshold energy means that there is an energy gap in the single particle excitation spectrum. BCS pairing theory gives that the threshold energy is equal to doubled energy gap of superconductor (See chapter 2.), and the energy gap leads to formation of Cooper pairs. The gap or Cooper pairs alone can not lead to superconductivity because the Cooper pairs may be separated into two sorts: supercurrent and non-supercurrent Cooper pairs, i.e., Cooper pairs with and without the long-range phase coherence.

1.1.2. Diamagnetism (Meissner effect, 1933)

Inside superconductors in superconducting state the magnetic induction $B = 0$, and the $B = 0$ is history-independent. This phenomenon is called Meissner effect [6]. Meissner effect is a perfect diamagnetism. This perfect diamagnetism occurs at magnetic field smaller than the thermodynamic critical field H_c for type-I superconductor (See Fig. 1.5.), and at magnetic field smaller than the lower critical magnetic field H_{c1} for type-II superconductor (See Fig. 1.5.). From Meissner effect we realize that superconductors in superconducting state are different from the perfect conductors since the latter tends to trap magnetic flux in, and leads to $B = 0$ history-dependent. The origin of Meissner effect is as follows. The magnetic field parallel to the surface of superconductor decays exponentially, with a penetration depth λ_L, in the interior. $\lambda_L(T = 0$ K) ranges from 500 to 10 000 $\mathring{A}$, depending on the material. From Maxwell fourth equation, we know that accompanying this parallel field is a surface current density. This surface current density screens the magnetic field from the interior of superconductor.

1.1.3. Electronic Specific Heat

In the conventional superconductors, such as tin and lead, when applied magnetic field is zero, the electronic specific heat, C_e, as a function of temperature is shown in Fig. 1.1. Usually people say that the specific heat jump occurs at T_c. But in the view of general theory, for the conventional superconductors the specific heat jump occurs at T^* and $T_{c,1}$. At T^* the Cooper pairs without the long-range phase coherence begin to condense into the chemical potential. This is a second order phase transition. At $T_{c,1}$ the Cooper pairs with the long-range phase difference locking, i.e., the supercurrent Cooper pairs, begin to appear. This is also a second order phase transition. At T_c there is no new phase transition. That the transition width of the jump is only 10^{-3} K [7,8] means that T^* and $T_{c,1}$ are in a very narrow temperature interval. So $T_c \approx T_{c,1} \approx T^*$ and we can roughly say that the specific heat jumps at T_c for conventional superconductors.

The measurement showed that the electronic specific heat well below T_c was dominated by an exponential dependence so that [9,10]

$$C_{es} = \gamma T_c d e^{-bT_c/T}, \tag{1.1}$$

where the normal state electronic specific heat is $C_{en} = \gamma T$, and d and b are numerical constants. $b = 1.5$. Such an exponential dependence implies a minimum excitation energy per quasiparticle of $1.5 k_B T_c$ for some materials, i.e., the existence of gap for quasiparticle

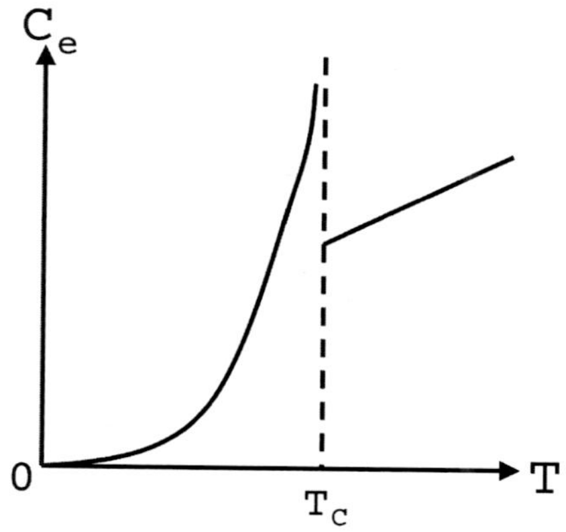

Figure 1.1. The usual schematic plot of C_e vs. T for conventional superconductors, such as tin. The transition width at T_c is only 10^{-3} K. At $T > T_c$ $C_e = C_{en}$. At $T \leq T_c$ $C_e = C_{es}$.

excitations. Note that the existence of gap does not mean superconductivity, in general. The appearance of superconductivity needs the long-range phase coherence of the Cooper pairs.

Let us interpret the second order phase transition in the high-T_c cuprates, which are one sort of the unconventional superconductors, in detail in this paragraph. At $T = T^*$ ($> T_c$) the Cooper pairs begin to appear. Although the Cooper pairs at $T_c < T < T^*$ have no long-range phase coherence, they condense into the chemical potential, which is the Bose-Einstein condensation. Therefore, the high-T_c cuprates have a second order phase transition at $T = T^*$. The magnitude of the specific heat jump at $T = T^*$ is given in section 2.5.4. The order parameter of the second order phase transition at $T \leq T^*$ is the amplitude of the wavefunction of Cooper pairs. The Cooper pairs at $T_c < T < T^*$ have no long-range phase coherence, and thus they are not the supercurrent Cooper pairs. The section 3.3 of this book demonstrates that at $T = T_c$ ($< T^*$) the supercurrent Cooper pairs begin to appear. This is also a second order phase transition. The order parameter of this second order phase transition is related to the number density of supercurrent Cooper pairs. The number density of supercurrent Cooper pairs is a complex function of the wavefunction of Cooper pair (See section 3.3.). At $T < T_c$ the high-T_c cuprates contain two sorts of Cooper pairs: supercurrent and non-supercurrent (See section 3.3.). Note that for the high-T_c cuprates the widths of specific heat jumps at T^* and T_c are wide because the gap has node and is anisotropic (See section 3.2.2.).

1.1.4. Isotope Effect (1950)

The phenomenon is that for element superconductors, $T_c \propto M^{-\alpha}$, where M is the atom mass and isotope exponent $\alpha \simeq 0.5$ [11]. The larger the M is, the lower the T_c. From this isotope effect people infer the importance of electron-phonon interaction for the superconductivity in these element superconductors. The isotope effect was found in some superconducting alloys and compounds as well. Actually, the electron-phonon interaction is important for a great part of superconductors. Note that the superconductivity caused by another mechanism other than the electron-phonon interaction can also have the isotope effect, which is shown for the T_c cuprates in section 4.6.

1.1.5. Josephson Effect (1962)

Josephson made from microscopic theory a prediction that a zero voltage supercurrent

$$I_s = I_c \sin(\phi_1 - \phi_2) \tag{1.2}$$

should flow between two superconducting electrodes in superconducting states, separated by a thin insulating barrier layer [3]. Here, $(\phi_1 - \phi_2)$ is the difference in the phases of the Cooper pairs, responsible for superconductivity, in the two electrodes, and I_c is the maximum supercurrent that the junction can support. The structure consisting of one thin insulating barrier layer and two superconducting electrodes on the two sides of the barrier layer is called Josephson tunneling junction or S-I-S junction. The supercurrent Cooper pair tunneling costs no energy when the two superconducting electrodes have the same chemical potential ($\approx$ Fermi level). When the dc supercurrent I_s is zero, i.e., the superconducting state of Josephson junction is at rest, the phase difference $(\phi_1 - \phi_2)$ is zero as well. When we increase the supercurrent, the phase difference $(\phi_1 - \phi_2)$ between the two superconducting electrodes adjusts to make the Josephson supercurrent $I_c \sin(\phi_1 - \phi_2)$ just the experimental value I_s. The increase of phase difference continues for increasing I_s until the point is reached where $I_s = I_c$, or $(\phi_1 - \phi_2) = \pi/2$. Then the current jumps from zero-voltage line to the curve of single-particle on the coordinate system of current vs. voltage.

Josephson further predicted that the I_s oscillates as a function of time if there is a voltage drop, V, on the S-I-S junction,

$$I_s = I_c \sin\left((\phi_1 - \phi_2)_\circ + \frac{2|e|V}{\hbar} t \right), \tag{1.3}$$

where e is the charge of free electron, and $(\phi_1 - \phi_2)_\circ$ is the phase difference at $V = 0$.

These predictions, known as dc and ac Josephson effect, respectively, have been confirmed by a large number of experiments.

The physical mechanism of the dc Josephson effect in S-I-S junction is as follows. Assume that the values of the superconducting transition temperature T_c' of the two superconducting electrodes are the same. At $T \leq T_c'$ the two superconducting electrodes are in superconducting states separately, and thus the supercurrent can flow in them separately. The whole Josephson tunneling junction is still in normal state at $T \cong T_c'$ because the Cooper pairs in two superconducting electrodes have not any long-range phase coherence yet. As the temperature goes down and reaches T_c, at which the Josephson coupling energy begins to overcome the zero-point fluctuation energy and the thermal disturbance energy [12], there is the phase difference locking across the insulating barrier layer, and thus the whole Josephson tunneling junction is in superconducting state. From Eq. (1) we know that the long-range phase coherence of Cooper pairs in the two electrodes, i.e., for example, ϕ_1 and ϕ_2 are constants for conventional superconductor, and the phase difference locking across the insulating barrier layer, i.e., constant phase difference $(\phi_1 - \phi_2)$ does not be destroyed by the zero-point fluctuation energy and thermal disturbance energy, are the two necessary and sufficient conditions for the dc Josephson effect. Note that for the usable Josephson

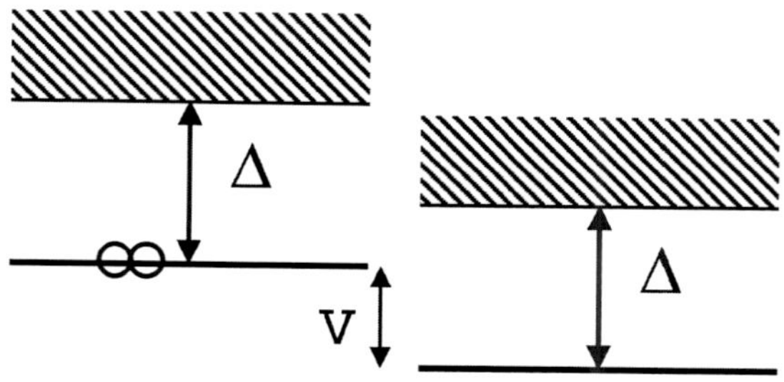

Figure 1.2. The schematic energy level diagram for the two superconducting electrodes. The shading denotes the energy levels occupied by quasiparticles. The two circles denote a Cooper pair. The real line denotes the chemical potential. Note that the Fermi energy level is nearly equal to the chemical potential.

tunneling junction the necessary value of Josephson coupling energy to overcome the zero-point fluctuation energy is actually very small (10^{-10} eV [12]). Therefore, T_c is nearly equal to T_c' for the usable Josephson tunneling junction.

Let us propose a mechanism of the ac Josephson effect. Assume that the two superconducting electrodes of the Josephson tunneling junction consist of the same superconductors. The schematic energy level diagram for the two superconducting electrodes is shown Fig. 1.2. All the Cooper pairs occupy the same energy level (Fermi energy level). The Fermi energy level can contain many Cooper pairs because the Pauli principle can not apply to Cooper pairs with spins 0 and 1. The oscillatory current in Eq. (1.3) comes from the tunneling of Cooper pairs between the Fermi energy level of one superconducting electrode and the Fermi energy level of the other [13]. For example, the Cooper pair on the left Fermi energy level tunnels to the right Fermi energy level. This tunneling of the Cooper electron pair does not conserve energy but takes an energy $2|e|V$ due to the voltage V. This Cooper pair must tunnel back, since it can not complete a transition where energy is not conserved. So long as the voltage V exists, the Cooper pair will tunnel to and from between the two superconducting electrodes [13]. Naturally, the frequency ω coming and going should satisfy the energy conservation,

$$\hbar\omega = 2|e|V. \tag{1.4}$$

From this relation, Eq. (1.4), we can understand Eq. (1.3).

Here we give an another similar mechanism for the ac Josephson effect. We assume that the Cooper electron pair on the left Fermi energy level tunnels to the right Fermi energy level. This tunneling of the Cooper pair does not conserve energy but takes an energy $2|e|V$. To keep the energy conservation the tunneling Cooper pair has no choice but to emit a photon with energy $\hbar\omega = 2|e|V$, and to return to the left Fermi energy level. The Cooper pairs move to and from between the left and the right, and thus the oscillatory current and the external radiation are formed in the S-I-S junction.

According to our mechanism, it is not necessary for Cooper pairs to have the long-range phase coherence and the phase difference locking across the barrier layer in the ac

Josephson effect. The only necessary condition for the ac Josephson effect is that there are Cooper pairs condensed into the Fermi energy levels in the two superconducting electrodes. There are five points which should be noted. First, to have a non-zero total oscillatory current in the S-I-S junction, the $(\phi_1 - \phi_2)_\circ$ in Eq. (1.3) should be a constant for all tunneling Cooper pairs, i.e., the Cooper pairs in the two superconducting electrodes should have the long-range phase coherence and the phase difference across the barrier layer should be locked. Second, if $V > 2\Delta(T)/|e|$, then there is single particle tunneling besides Cooper pair tunneling [13]. Here $\Delta(T)$ is the gap at T. Third, the function of the I in S-I-S junction is only a weak link. The S-I-S junction belongs to weakly coupled system. The ac Josephson effect was found as well in many other weakly coupled systems [14, 15], such as S-N-S, S-c-S, and point contact (in superconducting state) junctions [15]. Here N and c represent normal metal and constriction, respectively. Fourth, the ac Josephson effect occurs in the simple break junction [16] as well because the break junction belongs to weakly coupled system. Fifth, the macroscopic occupation of the zero momentum state is called Bose-Einstein condensation. According to our mechanism of the ac Josephson effect, the only necessary condition to yield the bias-induced oscillatory current in weakly coupled systems is the condensation of particles into a single level. Therefore, to observe the ac Josephson effect the two electrodes of the weakly coupled system may be, in principle, the materials in which there is Bose-Einstein condensation.

The readers might hope us to give a microscopic derivation to prove our mechanism of the ac Josephson effect. We will give this microscopic derivation in sections 2.7.3 and 2.7.4. Also see Ref. [13]. Note that the ϕ in Mahan's derivation can not be understood as the phase of the Cooper pair. The ϕ is the phase of the tunneling matrix elements $T_{\mathbf{kq}}T_{-\mathbf{k},-\mathbf{q}}$ [13]. This ϕ is taken as zero in many papers [17].

1.1.6. Pseudogap State

Pseudogap state is a state in which there are Cooper pairs without the long-range phase coherence. For a great part of the low-T_c superconductors, the gap and the superconductivity appear nearly at the same temperature T_c, and thus they belong to the conventional superconductors. (The reason is given in sections 3.3 and 1.1.1.) In the unconventional superconductors, for example, the high-T_c cuprates [5], the gap and the Cooper pair without the long-range phase coherence appear at T^*, but the superconductivity appears at T_c [18,19]. It is very interesting to note that for the underdoped $Bi_2Sr_2CaCu_2O_{8+\delta}$ (Bi2212) with $T_c = 10$ K, the T^* is larger than 300 K [20], i.e., in some cases the value of T^* can be much larger than the value of T_c.

The gap at $T_c < T < T^*$ is called the pseudogap. The state at $T_c < T < T^*$ is called the pseudogap state. The dc resistance is not zero in the pseudogap state. Chapter 2 shows that the Cooper pairs condense into the chemical potential for the pseudogap state and section 3.3 gives the following results for the high-T_c cuprates. First, there is no the phase difference locking between nearest-neighbor Cooper pairs, and thus the number density of the supercurrent Cooper pairs is zero in the pseudogap state. Second, at $T < T_c$ there is the phase difference locking between any two nearest-neighbor Cooper pairs, and thus the number density of the supercurrent Cooper pairs is not equal to zero. Note that some non-supercurrent Cooper pairs still exist at $T < T_c$.

The unconventional superconductors can have three states, normal, pseudogap and superconducting states, corresponding to different temperatures. The usual statement is that the conventional superconductors can only have two states, normal and superconducting states, corresponding to different temperatures. However, in principle, the conventional superconductors can also have three states as that of the unconventional superconductors because the former has T^* and $T_{c,1}$. Exactly speaking, there are normal state at $T > T^*$, pseudogap state at $T_{c,1} < T < T^*$, and superconducting state at $T < T_{c,1}$ for the conventional superconductors. Section 4.2.2 points out that a pseudogap body is also a possible material, which can be in either normal or pseudogap states, corresponding to different temperatures.

We predict that a novel ac effect, which is like the ac Josephson effect, can occur in a series of weakly coupled systems in the pseudogap states, such as P-I-P, P-N-P, P-c-P junctions, point contact, or break junction. Here, P, I, N, and c represent material in pseudogap state, thin insulating barrier layer, normal metal, and constriction, respectively. The reason is that the ac Josephson effect does not need that the Cooper pairs in the two electrodes on the two sides of the weak link are in the states with the long-range phase coherence. Of course, in the case of the above weakly coupled systems the $(\phi_1 - \phi_2)_\circ$ in Eq. (1.3) is different for every tunneling Cooper pair, and thus the total oscillatory current coming from all tunneling Cooper pairs is zero. However, we can still find the external radiation. The frequency of the external radiation, ω, satisfies $\hbar\omega = 2|e|V$, where V is the voltage applied on the above weakly coupled systems. Section 4.2.3 gives some candidate materials to get the pseudogap state, and points out that room temperature electronics of superconductors in the pseudogap state will be established.

1.2. Phenomenological Theories

1.2.1. London Equations (1935)

F. and H. London proposed the following equations to describe the superconductivity, which are called London equations now [21].

$$\frac{\partial \mathbf{J}_s}{\partial t} = a\mathbf{E}, \tag{1.5}$$

$$\nabla \times \mathbf{J}_s = -\frac{a}{c}\mathbf{B}, \tag{1.6}$$

$$a = \frac{n_s e^{*2}}{m^{**}}, \tag{1.7}$$

where $\mathbf{J}_s$, n_s, e^*, m^{**}, and c are the current density, number density, effective charge, effective mass of supercurrent Cooper pairs, and light velocity, respectively. According to the BCS pairing theory in chapter 2 of this book and Ref. [2], the supercurrent Cooper pairs are Cooper pairs with the long-range phase coherence. Therefore, $m^{**} = 2m^*$ and $e^* = 2e$. m^* is the effective mass of electron in solid. e is the charge of free electron. The first London equation, Eq. (1.5), is an equation of perfect conductor (R=0). We interpret this point as follows. Assume a supercurrent Cooper pair moves in a perfect conductor. Any electric field accelerates the supercurrent Cooper pair since resistance $R = 0$. The velocity, mass, and charge of the supercurrent Cooper pair are $\mathbf{V}$, m^{**}, and e^*, respectively. Then we have

$$m^{**}\frac{\partial \mathbf{V}}{\partial t} = e^*\mathbf{E}. \tag{1.8}$$

Introducing the supercurrent density $\mathbf{J}_s = n_s e^* \mathbf{V}$, Eq. (1.8) becomes Eq. (1.5).

The second London equation gives Meissner effect. We interpret this point as follows. The second London equation, when combined with the Maxwell equation $\nabla \times \mathbf{B} = 4\pi\mathbf{J}/c$, leads to

$$\nabla \times (\nabla \times \mathbf{B}) + \lambda_L^{-2}\mathbf{B} = 0, \tag{1.9}$$

$$\lambda_L^2 = \frac{c^2 m^{**}}{4\pi n_s e^{*2}}. \tag{1.10}$$

Consider a semi-infinite superconducting slab occupying the region $x > 0$ and an applied magnetic field lying parallel to the superconductor, i.e., $\mathbf{H} \parallel \mathbf{z}$, which we may write as $\mathbf{H} = H(x)\mathbf{z}$. The $\mathbf{H}$ satisfies $\nabla \cdot \mathbf{H} = 0$. Using the vector identity

$$\nabla \times (\nabla \times \mathbf{B}) = \nabla(\nabla \cdot \mathbf{B}) - \nabla^2 \mathbf{B}, \tag{1.11}$$

Eq. (1.9) becomes

$$\frac{\partial^2 B_z(x)}{\partial x^2} = \frac{1}{\lambda_L^2} B_z(x). \tag{1.12}$$

The solution of Eq. (1.12) for superconductor in superconducting state ($x > 0$) is

$$\mathbf{B}(x) = B_z(0) e^{-x/\lambda_L}\hat{\mathbf{z}}, \tag{1.13}$$

$$B_z(0)\hat{\mathbf{z}} = \mathbf{H}. \tag{1.14}$$

Eq. (1.13) implies that a magnetic field is exponentially screened from the interior of superconductor in superconducting state with penetration depth λ_L (See Fig. 1.3.), i.e., the Meissner effect. $\lambda_L(T = 0)$ ranges from 500 to 10 000 Å, depending on the material.

A more profound motivation for the London equations is the quantum one, emphasizing use of the vector potential $\mathbf{A}$ [22]– [24]. London postulated that the superconducting state can be described by a wavefunction $\psi(\mathbf{r}) = \psi_\circ \exp[iS(\mathbf{r})]$, where the phase $S(\mathbf{r})$ is a real function. From this function we know that the canonical momentum $\mathbf{p} = \hbar\nabla S$. Noting that the canonical momentum $\mathbf{p}$ is $m^{**}\mathbf{V} + e^*\mathbf{A}/c$ [24], where $\mathbf{A}$ is the magnetic vector potential, we have

$$< \mathbf{V}_s > = \frac{\hbar}{m^{**}}\nabla S - \frac{e^*\mathbf{A}}{m^{**}c}. \tag{1.15}$$

Substituting $\mathbf{J}_s = n_s e^* < \mathbf{V}_s >$ into Eq. (1.15) yields

$$\mathbf{J}_s = \frac{n_s e^* \hbar}{m^{**}}\nabla S - \frac{a}{c}\mathbf{A}. \tag{1.16}$$

Taking the time derivative of both sides of Eq. (1.16) yields Eq. (1.5), and taking the curl of both sides of Eq. (1.16) leads to Eq. (1.6) [22]. Thus, Eq. (1.16) contains the two London equations. Eq. (1.16) is the most general London equation. Eq. (1.16) tells us that

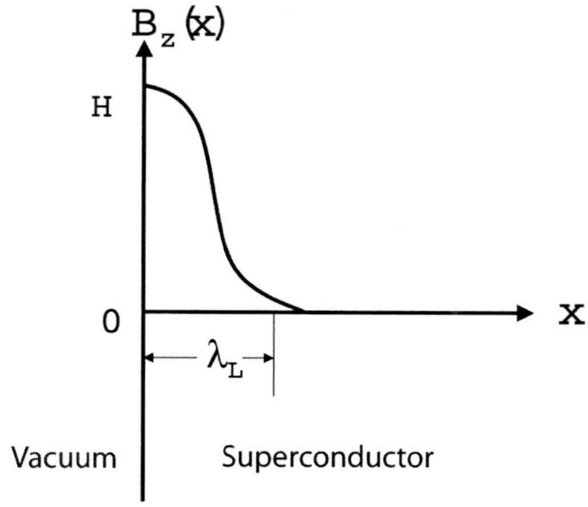

Figure 1.3. The schematic plot of $B_z(x)$ vs. x.

the $\mathbf{J}_s$ at $\mathbf{r}$ is determined by the $\mathbf{A}(\mathbf{r})$ at $\mathbf{r}$. Therefore, the relation between $\mathbf{J}_s(\mathbf{r})$ and $\mathbf{A}(\mathbf{r})$ is local. The superconductor obeying Eq. (1.16) is called London local superconductor.

The elimination of the ∇S term in Eq. (1.16) hinges upon a gauge transformation [24]. If we put $\mathbf{A} = \mathbf{A}_1 + e^*\chi/\hbar$ and $S = S_1 + e^*\chi/(c\hbar)$, where χ is a scalar function, then the values of $\mathbf{B}$ and $\mathbf{J}_s$ are unaltered. Since $\mathbf{B}$ and $\mathbf{J}_s$ are what we can measure, we can use the transformation to go from a given $\mathbf{A}$ and S to an $\mathbf{A}_1$ and S_1 which we may find more convenient. In particular, for a simply connected superconductor, we can choose a gauge function χ which makes S_1 zero. For this we just have to take $\chi = c\hbar S/e^*$, where S is our original phase function. With this choice of gauge, Eq. (1.16) becomes $\mathbf{J}_s = -a\mathbf{A}_1/c$. In a multiply connected specimen, on the other hand, the use of the gauge function $\chi = c\hbar S/e^*$ would lead to a vector potential $\mathbf{A}_1$, which was not single valued, so it is easier then to retain the phase S explicitly.

1.2.2. Pippard Nonlocal Generalization of London Equations (1950)

Assume that the supercurrent Cooper pairs have bound energy kT_c. Let Δp and v_F represent a momentum range and the Fermi velocity of supercurrent Cooper pair, respectively. Then $\Delta p \cdot v_F \approx k_B T_c$. From the uncertainty relation in quantum mechanics, $\Delta p \cdot \Delta x = \hbar$, we know that the space range or size of supercurrent Cooper pair, Δx, is

$$\Delta x = \frac{\hbar v_F}{k_B T_c}. \tag{1.17}$$

For element Sn, $v_F = 1.88 \times 10^8 cm/sec$ and $T_c = 3.72$ K [25]. Then $\Delta x = 3.84 \times 10^{-4} cm$. From this estimation we can see that the space size of a supercurrent Cooper pair is very large. All vector potentials $\mathbf{A}$ in this large space size should have influences on the supercurrent density $\mathbf{J}_s(\mathbf{r})$. Introduce a characteristic length, so-called coherence length,

$$\xi_o = g\frac{\hbar v_F}{k_B T_c}, \tag{1.18}$$

where g is a numerical constant of order unity. The coherence length is the same as the size of Cooper pair.

Pippard inferred that the relation between $\mathbf{J}_s(\mathbf{r})$ at point $\mathbf{r}$ and $\mathbf{A}$ involves contributions from $\mathbf{A}(\mathbf{r}')$ at neighboring points $\mathbf{r}'$ located in a volume with a radius of order ξ_o surrounding $\mathbf{r}$ [26]. If a superconductor has impurities, it is natural to assume that the relation between $\mathbf{J}_s(\mathbf{r})$ and $\mathbf{A}(\mathbf{r})$ will be altered a little. To account for the effect of electron scattering, Pippard introduced coherence length ξ.

$$\frac{1}{\xi} = \frac{1}{\xi_o} + \frac{1}{l}, \tag{1.19}$$

where l is the electron mean free path. The field-current relation proposed by Pippard is [26]

$$\mathbf{J}_s(\mathbf{r}) = -\frac{3n_s e^{*2}}{4\pi\xi_o m^{**}c} \int \frac{[\mathbf{A}(\mathbf{r}')\cdot\mathbf{R}]\mathbf{R}}{R^4} e^{-R/\xi} d\mathbf{r}', \tag{1.20}$$

where $\mathbf{R} = \mathbf{r} - \mathbf{r}'$. Eq. (1.20) is the Pippard nonlocal generalization of London equation. Using Eq. (1.20), Pippard found that the experimental data on both tin and aluminum can be fitted by the choice of $g = 0.15$ in Eq. (1.18) [27]. If $\lambda_L < \xi_o$, then the field has large change over the range ξ_o. In this case the relation between field and current is nonlocal, and this superconductor is called Pippard nonlocal superconductor. On the contrary, if $\lambda_L > \xi_o$, then the field has only a little change over the range ξ_o. In this case the relation between field and current is local approximately, and this superconductor is called London local superconductor.

1.2.3. Ginzburg-Landau Theory (1950)

Ginzburg and Landau introduced a macroscopic complex pseudo-wavefunction (or effective wavefunction) ψ as an order parameter to describe the second order phase transition at T_c for superconductor [28]. At first, let us discuss GL theory of the conventional superconductors. This ψ represents the wavefunction of motion of the center of mass of supercurrent Cooper pair for the conventional superconductors (See chapter 2 and Ref. [29]). The local number density of supercurrent Cooper pairs for the conventional superconductor is given by [28, 29]

$$n_s(\mathbf{r}) = |\psi(\mathbf{r})|^2. \tag{1.21}$$

Since the pseudo-wavefunction is complex, only the form $\psi\psi^*$ may enter the expansion of free energy F. We therefore write in cases of $T \approx T_c$ ($T < T_c$), applied field H $= 0$, and infinite superconductor, i.e., no boundary effect,

$$F = F_n(T) + \int d\mathbf{r} \left\{ \alpha|\psi|^2 + \frac{1}{2}\beta|\psi|^4 \right\}, \tag{1.22}$$

where $F_n(T)$ is the free energy of the normal state in the absence of a magnetic field. In Eq. (1.22) there are no terms linear and cubic in ψ. These terms contain an arbitrary phase of ψ, and should be excluded because F is physically observable. Thermodynamic stability of superconducting state requires $\beta > 0$ (otherwise the system would seek a divergently large $|\psi|$). However, $\alpha(T)$ may have either sign. If $\alpha(T) > 0$, then the minimum of Eq. (1.22) occurs at $|\psi| = 0$. If $\alpha(T) < 0$, then the minimum is at $\psi \neq 0$. Hence we can model a second order phase transition simply by arranging for the sign of $\alpha(T)$ to change at T_c, which is easily done by writing $\alpha(T) = d(T - T_c)$. Thermodynamic stability of superconducting state requires $d > 0$. It is sufficient to regard β as a constant independent of temperature. Setting $\partial F/\partial|\psi| = 0$ yields

$$|\psi| = 0, \quad T > T_c, \tag{1.23}$$

$$|\psi| = \left(-\frac{\alpha}{\beta}\right)^{\frac{1}{2}}, \quad T \approx T_c(T < T_c), \tag{1.24}$$

i.e., in this case $n_s = -\alpha/\beta$. $|\psi|$ is independent of space, which describes a homogeneous superconducting state. In cases of that the applied field is not equal to zero and/or there is a boundary effect, the F should contain kinetic term, and ψ depends on space.

$$F = F_n(T) + \int d\mathbf{r} \left\{ \alpha|\psi(\mathbf{r})|^2 + \frac{\beta}{2}|\psi(\mathbf{r})|^4 \right.$$
$$\left. + \frac{1}{2m^{**}} \left| \left[\hbar\nabla - \frac{ie^*}{c}\mathbf{A}(\mathbf{r})\right]\psi(\mathbf{r}) \right|^2 + \frac{B^2(\mathbf{r})}{8\pi} \right\}, \tag{1.25}$$

where m^{**} and e^* are the effective mass and charge of supercurrent Cooper pair. $e^* = 2e$. $m^{**} = 2m^*$. Varying $\psi^*(\mathbf{r})$ by $\delta\psi^*(\mathbf{r})$ and integrating by parts, we obtain

$$\delta F = \int d\mathbf{r}\delta\psi^*(\mathbf{r}) \left\{ \alpha\psi(\mathbf{r}) + \beta|\psi(\mathbf{r})|^2\psi(\mathbf{r}) - \frac{1}{2m^{**}} \left[\hbar\nabla - i\frac{e^*}{c}\mathbf{A}(\mathbf{r})\right]^2 \psi(\mathbf{r}) \right\}$$
$$+ \frac{1}{2m^{**}} \int\int \delta\psi^*(\mathbf{r}) d\mathbf{s} \cdot \left[\hbar\nabla - \frac{ie^*}{c}\mathbf{A}(\mathbf{r})\right]\psi(\mathbf{r}), \tag{1.26}$$

where the second integral is over the surface of the superconductor. In equilibrium, $\delta F = 0$ and both integrals should vanish for any $\delta\psi^*(\mathbf{r})$. As a result, we obtain an equation

$$-\frac{\hbar^2}{2m^{**}} \left[\nabla - \frac{ie^*}{\hbar c}\mathbf{A}(\mathbf{r})\right]^2 \psi(\mathbf{r}) + \alpha\psi(\mathbf{r}) + \beta|\psi(\mathbf{r})|^2\psi(\mathbf{r}) = 0, \tag{1.27}$$

which is called the first equation of GL equation set. The equation obtained from the surface integral in Eq. (1.26) corresponds to the boundary condition at the surface, $\mathbf{n} \cdot [\hbar\nabla - ie^*/c\mathbf{A}(\mathbf{r})]\psi(\mathbf{r}) = 0$, where $\mathbf{n}$ is unit vector normal to the surface.

Varying $\mathbf{A}(\mathbf{r})$ by $\delta\mathbf{A}(\mathbf{r})$ in the free energy, Eq. (1.25), and setting $\delta F = 0$, we obtain the Maxwell equation $\nabla \times \nabla \times \mathbf{A}(\mathbf{r}) = (4\pi/c)\mathbf{J}_s(\mathbf{r})$, and the supercurrent density $\mathbf{J}_s(\mathbf{r})$ is

$$\mathbf{J}_s(\mathbf{r}) = \frac{e^*\hbar}{i2m^{**}}[\psi^*(\mathbf{r})\nabla\psi(\mathbf{r}) - \psi(\mathbf{r})\nabla\psi^*(\mathbf{r})] - \frac{e^{*2}}{m^{**}c}|\psi(\mathbf{r})|^2\mathbf{A}(\mathbf{r}), \tag{1.28}$$

which is called the second equation of GL equation set. Using above boundary condition at the surface and the supercurrent density equation, Eq. (1.28), we can obtain $\mathbf{n} \cdot \mathbf{J}_s(\mathbf{r}) = 0$, i.e., the normal component of the supercurrent should be zero on the surface.

We discuss a case involving an inhomogeneous distribution of the order parameter, generated by the presence of a boundary, in the absence of magnetic field. Assume we have a superconducting half space occupying the region $x > 0$. We further assume that $|\psi| = 0$ at this interface $x = 0$. ψ has unified phase at $T < T_c$ for conventional superconductor, i.e., $\psi(x) = |\psi(x)|\exp(i\phi)$. In this case Eq. (1.27) becomes

$$-\frac{\hbar^2}{2m^{**}}\frac{d^2|\psi}{|}dx^2 + \alpha|\psi| + \beta|\psi|^3 = 0. \tag{1.29}$$

Defining the GL coherence length as

$$\xi_{GL}^2 = \frac{\hbar^2}{2m^{**}|\alpha|}, \tag{1.30}$$

and writing $(\beta/|\alpha|)|\psi|^2 = f^2$, we may rewrite Eq. (1.29) as

$$-\xi_{GL}^2 f'' - f + f^3 = 0. \tag{1.31}$$

Multiplying by f', we rewrite Eq. (1.31) as

$$\frac{d}{dx}\left[-\frac{\xi_{GL}^2 f'^2}{2} - \frac{1}{2}f^2 + \frac{1}{4}f^4\right] = 0. \tag{1.32}$$

The quantity in square brackets must be a constant independent of x. Far from the boundary, $f' = 0$ and $f^2 = 1$ [Eq. (1.24)]. Then Eq. (1.32) becomes

$$\xi_{GL}^2 (f')^2 = \frac{1}{2}(1 - f^2)^2. \tag{1.33}$$

The solution of Eq. (1.33) is

$$|\psi| = \left(\frac{|\alpha|}{\beta}\right)^{\frac{1}{2}} \tanh\left(\frac{x}{\sqrt{2}\xi_{GL}}\right). \tag{1.34}$$

From Eq. (1.34) we see that ξ_{GL} is approximately a measure of distance over which the order parameter is from a constant in the interior of superconductor to zero at the boundary of superconductor. The schematic plot of $|\psi|^2$ (=n_s) vs. x is shown in Fig. 1.4. In a pure superconductor far below T_c, $\xi_{GL}(T) \approx \xi_o$, the coherence length. Considering ξ_{GL} is temperature-dependent, the two coherence lengths are related but distinct quantities.

The GL parameter is defined as

$$\kappa = \frac{\lambda_L}{\xi_{GL}}. \tag{1.35}$$

From Eqs. (1.10), (1.21), and (1.24) we have

$$\lambda_L^2 = \frac{m^{**}c^2\beta}{4\pi e^{*2}|\alpha|}. \tag{1.36}$$

From Eqs. (1.30), (1.35), and (1.36) we have

$$\kappa = \frac{m^{**}c}{|e^*|\hbar}\left(\frac{\beta}{2\pi}\right)^{\frac{1}{2}}. \tag{1.37}$$

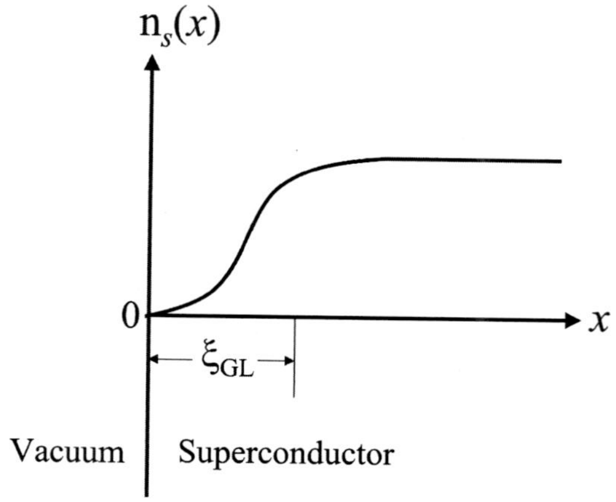

Figure 1.4. The schematic plot of $n_s(x)$ vs. x.

Abrikosov showed in 1957 by the solution of GL equation that for superconductor with $\kappa < 1/\sqrt{2}$ there is discontinuous breakdown of superconductivity in a first order phase transition at thermodynamic critical field H_c, and it is called type-I superconductor [30]. For superconductor with $\kappa > 1/\sqrt{2}$, which is called type-II superconductor, there is a continuous increase in magnetic flux penetration starting at a lower critical field H_{c1} ($< H_c$) and reaching $B = H$ at an upper critical field H_{c2} ($> H_c$), as shown schematically in Fig. 1.5 [30]. $H_{c2} = \sqrt{2}\kappa H_c$, and $H_{c1} = H_c \ln\kappa/(\sqrt{2}\kappa)$ [31]. The London local superconductor ($\lambda_L > \xi_{GL}$) belongs to type-II superconductor. The Pippard nonlocal superconductor ($\lambda_L < \xi_{GL}$) belongs to type-I superconductor. The schematic plot of magnetic induction B vs. applied field H for type-I and type-II superconductors is shown in Fig. 1.5.

Abrikosov further pointed out that in the so-called mixed state, i.e., Schubnikov phase, between H_{c1} and H_{c2}, the flux should penetrate in a regular array of flux tubes, each carrying a quantum of flux [30, 31]

$$\Phi_\circ = \frac{hc}{2|e|} = 2.07 \times 10^{-7}\mathrm{G} - \mathrm{cm}^2. \tag{1.38}$$

Within each unit cell of the array, there is a vortex (flux tube) of supercurrent concentrating the flux toward the vortex center. The triangular array has a slightly lower free energy than the square array.

For the unconventional superconductor at $T < T_c$ the order parameter (effective wavefunction) ψ in GL theory still describes the supercurrent Cooper pairs, which have the long-range phase coherence. Section 3.3 demonstrates that the number density of supercurrent Cooper pairs n_s is a very complex function of the amplitude of Cooper pair wavefunction. However, the order parameter can still be written as $\psi = \sqrt{n_s}\exp(i\phi)$. The n_s is a small quantity at $T \approx T_c$ ($T < T_c$). The GL equation set, Eqs. (1.27) and (1.28), is still correct at $T \approx T_c$ ($T < T_c$) for unconventional superconductor.

It is also a second order phase transition that the Cooper pairs condensed into the chemical potential begin to appear at T^* for unconventional superconductor. The order parameter

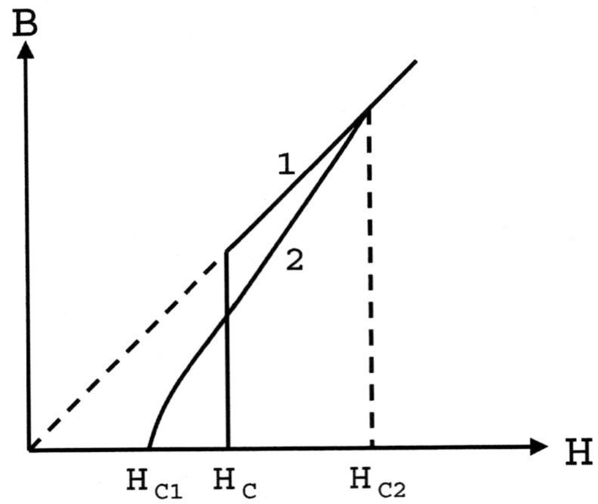

Figure 1.5. The schematic plot of magnetic induction B vs. applied magnetic H for type-I superconductor (curve 1) and type-II superconductor (curve 2).

ψ at $T_c < T < T^*$ is the wavefunction of Cooper pair without the long-range phase coherence. We can write in cases of $T \approx T^*$ $(T < T^*)$ and infinite unconventional superconductor

$$F = F_n(T) + \int d\mathbf{r} \left\{ \alpha_p |\psi|^2 + \frac{1}{2} \beta_p |\psi|^4 \right\}, \tag{1.39}$$

where $\alpha_p = d(T - T^*)$, $d > 0$, and $\beta_p > 0$. Setting $\partial F / \partial |\psi| = 0$ yields

$$|\psi| = 0, \qquad T > T^*, \tag{1.40}$$

$$|\psi| = \left(-\frac{\alpha_p}{\beta_p} \right)^{\frac{1}{2}}, \qquad T \approx T^*(T < T^*). \tag{1.41}$$

In case of existing the boundary effect, the F in Eq. (1.39) should contain kinetic term.

$$F = F_n(T) + \int d\mathbf{r} \left\{ \alpha_p |\psi(\mathbf{r})|^2 + \frac{\beta_p}{2} |\psi(\mathbf{r})|^4 + \frac{1}{2m^{**}} |\hbar \nabla \psi(\mathbf{r})|^2 \right\}, \tag{1.42}$$

where $m^{**} = 2m^*$. Varying $\psi^*(\mathbf{r})$ by $\delta\psi^*(\mathbf{r})$, integrating by parts, and setting $\delta F = 0$, we obtain

$$-\frac{\hbar^2}{2m^{**}} \nabla^2 \psi(\mathbf{r}) + \alpha_p \psi(\mathbf{r}) + \beta_p |\psi(\mathbf{r})|^2 \psi(\mathbf{r}) = 0. \tag{1.43}$$

Imitating the discussions from Eq. (1.29) to Eq. (1.34), we obtain for the half space of unconventional superconductor in the pseudogap state, occupying the region $x > 0$,

$$|\psi(x)| = \left(\frac{|\alpha_p|}{\beta_p} \right)^{\frac{1}{2}} \tanh \left(\frac{x}{\sqrt{2}\xi_{GLp}} \right), \tag{1.44}$$

$$\xi_{GLp}^2 = \frac{\hbar^2}{2m^{**} |\alpha_p|}. \tag{1.45}$$

From Eq. (1.44) we see that ξ_{GLp} is approximately a measure of distance over which the order parameter is from a constant in the interior of unconventional superconductors in the pseudogap state to zero at the boundary of the unconventional superconductor.

1.2.4. Surface Energy

This section will use the GL theory to demonstrate that superconductors with $\kappa > 1/\sqrt{2}$ and $\kappa < 1/\sqrt{2}$ belong to type-II and type-I superconductors, respectively. If the external field H is fixed, then the relevant thermodynamic potential describing the equilibrium state is the Gibbs energy, $G(T,N,H)$.

$$G(T,N,H) = F(T,N,B) - \frac{1}{4\pi}\int d\mathbf{r}\mathbf{H}\cdot\mathbf{B}(\mathbf{r}). \tag{1.46}$$

Let us assume that normal and superconducting phases are separated by an infinite plane boundary (area $S=\infty$) at $x=0$. Superconducting and normal phases are at $+\infty > x > 0$ and $-\infty < x < 0$, respectively. The relevant quantities are only functions of x. The boundary at $x=0$ is not sharp because magnetic field at z direction penetrates into the region of $x>0$, and the order parameter varies from a non-zero constant in the interior of sample to zero at $x=0$. To describe the effect of the complicated boundary we introduce the surface energy, σ_s, defined as the difference between the true Gibbs energy in external field $H=H_c$ (thermodynamic critical field) and the Gibbs energy of a homogeneous superconducting sample without external field per unit area in $y-z$ surface, i.e.,

$$\sigma_s = \frac{1}{S}\left[G(T,N,H) - \int d\mathbf{r} g_s(T,N,0)\right], \tag{1.47}$$

where $G(T,N,H)$ is given by Eq. (1.46), and $g_s(T,N,0)$ is the density of Gibbs energy of the homogeneous superconducting sample without external field. From this definition we can understand the physical meanings of $\sigma_s<0$ and $\sigma_s>0$. If $\sigma_s>0$, then the appearance of normal-superconducting boundary in the interior of superconducting phase under external field is not beneficial in energy. When the external field reaches thermodynamic critical field H_c, the superconducting phase becomes the normal phase. Therefore, the superconductor with $\sigma_s>0$ belongs to type-I superconductor. If $\sigma_s<0$, then the appearance of normal-superconducting boundary in the interior of superconducting phase under external field $H_{c1}<H<H_{c2}$ is beneficial in energy. Therefore, the Abrikosov vortex with a core in normal phase should penetrate into the interior of superconducting phase under $H_{c1}<H<H_{c2}$. The superconductors with $\sigma_s<0$ belong to type-II superconductor.

The magnetic vector potential is $\mathbf{A}=[0,A_y(x),0]$, when the field is at z direction. Substituting Eq. (1.25) into Eqs. (1.46) and (1.47) and considering the density of condensation energy is $f_n(T,N,0)-f_s(T,N,0)=g_n(T,N,0)-g_s(T,N,0)=H_c^2/(8\pi)$, we have

$$\sigma_s = \frac{1}{8\pi}\int dx[B(x)-H_c]^2 + \int_{-\infty}^{+\infty} dx\left[\alpha|\psi(x)|^2 + \frac{\beta}{2}|\psi(x)|^4 + \frac{\hbar^2}{2m^{**}}\left|\frac{d\psi(x)}{dx}\right|^2 + \frac{e^{*2}A_y^2(x)}{2m^{**}c^2}|\psi(x)|^2\right]. \tag{1.48}$$

Using an appropriate gauge transformation, the order parameter ψ can be real in any simply connected superconductor. To eliminate various parameters in the subsequent calculations we rewrite all equations in terms of dimensionless variables:

$$\widetilde{x}=\frac{x}{\lambda_L}, f=\left(\frac{\beta}{|\alpha|}\right)^{1/2}\psi, a=\frac{A_y}{\sqrt{2}H_c\lambda_L}, h=\frac{B}{\sqrt{2}H_c}. \tag{1.49}$$

In terms of these variables the first and second GL equations become

$$f'' + \kappa^2[(1-a^2)f - f^3] = 0 \tag{1.50}$$

and

$$a'' = af^2, \tag{1.51}$$

where the double prime means the second derivative with respect to $\widetilde{x}$. The solutions of Eqs. (1.50) and (1.51) for arbitrary κ must be obtained numerically. The boundary conditions are $f=0$ and $a'=h=1/\sqrt{2}$ for $x=-\infty$ (in the normal phase), and $f=1$ and $a'=0$ for $x=+\infty$ (in the superconducting phase).

To obtain a first integral of Eqs. (1.50) and (1.51) we multiply Eq. (1.50) by f', which leads immediately to

$$\frac{d}{d\widetilde{x}}\left[\frac{f'^2}{\kappa^2}+(1-a^2)f^2-\frac{f^4}{2}\right]+f^2\frac{da^2}{d\widetilde{x}}=0, \tag{1.52}$$

and multiply Eq. (1.51) by a', which yields

$$\frac{da'^2}{d\widetilde{x}}=f^2\frac{da^2}{d\widetilde{x}}. \tag{1.53}$$

Substituting Eq. (1.53) into Eq. (1.52) gives the first integral

$$\frac{f'^2}{\kappa^2}+(1-a^2)f^2-\frac{f^4}{2}+a'^2=\text{constant}=\frac{1}{2}. \tag{1.54}$$

The value of the constant was fixed by the boundary conditions at either $+\infty$ or $-\infty$. From Eq. (1.22) we have the density of condensation energy

$$f_n(T,N,0)-f_s(T,N,0)=|\alpha||\psi|^2-\beta|\psi|^4/2. \tag{1.55}$$

Substituting Eq. (1.24) into Eq. (1.55) yields

$$f_n(T,N,0)-f_s(T,N,0)=|\alpha|^2/(2\beta). \tag{1.56}$$

From $f_n(T,N,0)-f_s(T,N,0)=H_c^2/(8\pi)$ and Eq. (1.56) we obtain

$$\frac{H_c^2}{8\pi}=\frac{\alpha^2}{2\beta}. \tag{1.57}$$

Using Eqs. (1.49) and (1.57), and $\alpha^2/\beta=H_c^2/(8\pi)$, we may rewrite Eq. (1.48) as

$$\sigma_s=\frac{\lambda_L H_c^2}{4\pi}\int_{-\infty}^{+\infty}d\widetilde{x}\left[\left(a'-\frac{1}{\sqrt{2}}\right)^2+(a^2-1)f^2+\frac{f^4}{2}+\frac{f'^2}{\kappa^2}\right]. \tag{1.58}$$

The exact borderline between type-I and type-II superconductors is defined by the condition $\sigma_s = 0$. Integrating the last term f'^2/κ^2 in Eq. (1.58) by parts and using f'' in Eq. (1.50), we obtain

$$\sigma_s = \frac{\lambda_L H_c^2}{4\pi} \int_{-\infty}^{+\infty} d\tilde{x} \left[\left(a' - \frac{1}{\sqrt{2}} \right)^2 - \frac{f^4}{2} \right]. \tag{1.59}$$

If $\sigma_s = 0$, then

$$a' = \frac{1}{\sqrt{2}}(1 - f^2). \tag{1.60}$$

From Eq. (1.60) we obtain

$$a'' = -\sqrt{2} f f'. \tag{1.61}$$

Combining Eqs. (1.61) and (1.51), we have

$$f' = -\frac{af}{\sqrt{2}}. \tag{1.62}$$

Substituting Eqs. (1.62) and (1.60) into Eq. (1.54), we obtain

$$a^2 f^2 \left(1 - \frac{1}{2\kappa^2} \right) = 0. \tag{1.63}$$

Hence $\sigma_s = 0$ at $\kappa = 1/\sqrt{2}$. If $\kappa > 1/\sqrt{2}$ and $\kappa < 1/\sqrt{2}$, then $\sigma_s < 0$ and $\sigma_s > 0$ from Eq. (1.58), respectively. These conclusions are correct for both conventional and unconventional superconductors.

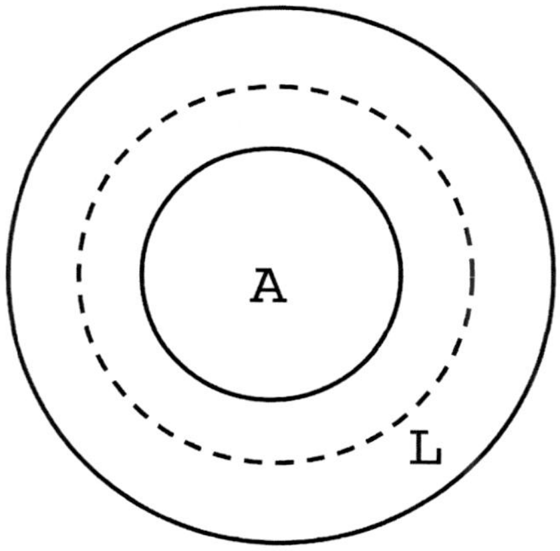

Figure 1.6. The schematic plot of superconducting ring. There is the trapped magnetic flux in the region A.

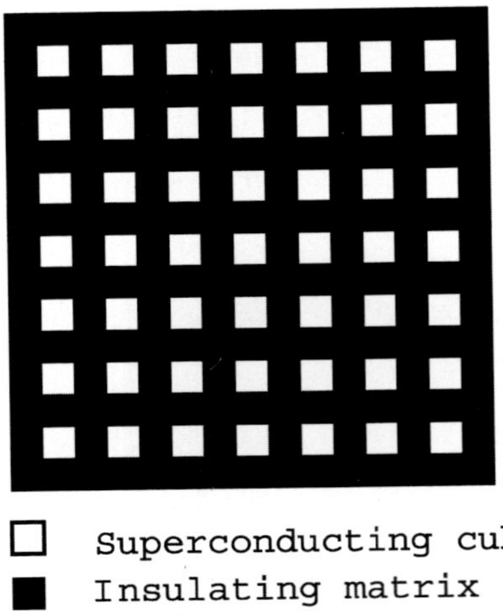

Figure 1.7. The schematic plot of granular superconductor consisting of a lattice of superconducting cubes embedded in an insulating matrix.

1.2.5. Flux Quantization

Let us consider a conventional superconducting ring with the trapped magnetic flux Φ_A, which is schematically shown in Fig. 1.6. The integration contour is shown by L. Since the current sustaining the trapped flux in the ring flows only in a layer of thickness λ_L on the inner surface of the ring, if the ring is thick enough compared to λ_L, then the integration contour can be taken deeper inside the wall of the ring, where $\mathbf{J}_s = 0$ and $\mathbf{B} = 0$.

$$\Phi_A = \int d\mathbf{s} \cdot \mathbf{B}, \tag{1.64}$$

where the surface integral is taken over the region enclosed by the contour L. Therefore, Φ_A is equal to the total magnetic flux through the ring together with the region surrounding the ring where the field penetrates the superconductor. Considering $\mathbf{B}(\mathbf{r}) = \nabla \times \mathbf{A}(\mathbf{r})$ and Stokes theorem yields

$$\Phi_A = \oint_L d\mathbf{l} \cdot \mathbf{A}(\mathbf{r}). \tag{1.65}$$

The $\mathbf{J}_s$ is zero on the integration contour L. The effective wavefunction is $\psi(\mathbf{r}) = |\psi(\mathbf{r})| \exp[i\phi(\mathbf{r})]$. Therefore, from the second GL equation, Eq. (1.28), we obtain

$$\mathbf{A}(\mathbf{r}) = \frac{c\hbar}{e^*} \nabla\phi(\mathbf{r}). \tag{1.66}$$

Substituting Eq. (1.66) into Eq. (1.65) yields

$$\Phi_A = \frac{c\hbar}{e^*}(\phi_1 - \phi_2), \tag{1.67}$$

where $(\phi_1 - \phi_2)$ is a change of the phase in the round trip along the contour L. That the effective wavefunction $\psi(\mathbf{r})$ should be a single-value function requires that $(\phi_1 - \phi_2) = 2\pi n$

and $n = \cdots, -2, -1, 0, 1, 2, \cdots$. Hence, the trapped flux is quantized.

$$\Phi_A = n\Phi_\circ, \tag{1.68}$$

$$\Phi_\circ = \frac{hc}{|e^*|} = 2.07 \times 10^{-7}\mathrm{G} - \mathrm{cm}^2. \tag{1.69}$$

The flux quantization was observed in 1961 [32,33].

Let us consider an unconventional superconducting ring, consisting of the granular superconductor in superconducting state, with the trapped magnetic flux Φ_A at $T < T_c$. For unconventional superconductor, although the amplitude of effective wavefunction $\psi(\mathbf{r})$ in GL theory is a complex function of the gap or the amplitude of Cooper pair wavefunction (See section 3.3.), yet we can write $\psi(\mathbf{r}) = \sqrt{n_s(\mathbf{r})}\exp[i\phi(\mathbf{r})]$ on the zero current contour L. Note that the phase difference between any two nearest-neighbor superconducting grains in the granular superconductor is zero on the zero current contour L (See section 1.1.5.). Using the assumption that the effective wavefunction is a single-value function, and repeating the derivations from Eq. (1.64) to Eq. (1.67), we can also obtain Eqs. (1.68) and (1.69), i.e., the flux quantization. The granular superconductor from which we make the unconventional superconducting ring is schematically shown in Fig. 1.7. The small superconducting grain is a cube with bonder length, for example, 10^{-6} cm. The cube grain is better than the sphere grain because the former has stronger Josephson coupling with the nearest-neighbor cube grains.

Chapter 2

BCS Pairing Theory

2.1. Single Cooper Pair

Cooper showed that if two electrons outside the Fermi surface have a net attractive potential ($-V < 0$), then the two electrons form a bound state so that their total energy is less than $2E_F$ [34]. E_F is Fermi energy. Note that the chemical potential and Fermi energy nearly equal each other. We give the Cooper's mathematical derivation in the following.

At first we make the following four assumptions for convenience. First, $T = 0$. Second, there are only two electrons outside the Fermi surface. Third, the system has translational invariability, so that the mass center momentum of the two electrons, $\mathbf{q}$, is constant. Fourth, the interaction between the two electrons is spin-independent, so that the total spin of the two electrons, S, is constant. Let the two electron wavefunction be $\psi(\mathbf{r}_1, \mathbf{r}_2) = \phi_{\mathbf{q}}(\mathbf{r})\exp(\mathbf{iq} \cdot \mathbf{R})$ where $\mathbf{r} = \mathbf{r}_2 - \mathbf{r}_1$, and $\mathbf{R} = (\mathbf{r}_1 + \mathbf{r}_2)/2$. From Pauli principle, if $\phi_{\mathbf{q}}(\mathbf{r})$ is symmetric, which means when the two electrons exchange positions, the wavefunction does not change sign, then $S = 0$, i.e., spin singlet state of the two electrons; if $\phi_{\mathbf{q}}(\mathbf{r})$ is antisymmetric, then $S = 1$, i.e., spin triplet state of the two electrons.

The form of $\psi(\mathbf{r}_1, \mathbf{r}_2)$ in case of $\mathbf{q} = 0$ is

$$\psi(\mathbf{r}_1, \mathbf{r}_2) = \phi(\mathbf{r}_2 - \mathbf{r}_1). \tag{2.1}$$

Considering that the two electrons are outside the Fermi surface, the expansion of $\phi(\mathbf{r})$ in $\mathbf{k}$ space is

$$\phi(\mathbf{r}_2 - \mathbf{r}_1) = \sum_{\mathbf{k}(|\mathbf{k}| > k_F)} a_{\mathbf{k}} e^{i\mathbf{k}\cdot\mathbf{r}_2 - i\mathbf{k}\cdot\mathbf{r}_1}. \tag{2.2}$$

Since the factors $\exp(i\mathbf{k} \cdot \mathbf{r}_1)$ and $\exp(-i\mathbf{k} \cdot \mathbf{r}_2)$ can be thought of as single-particle states of momentum $\mathbf{k}$ and $-\mathbf{k}$, respectively, we see that the pair wavefunction is a superposition of configurations in each of which a definite pair state ($\mathbf{k}$, $-\mathbf{k}$) is occupied [35]. If $S = 0$, then $\phi(\mathbf{r}_2 - \mathbf{r}_1) = \phi(\mathbf{r}_1 - \mathbf{r}_2)$. From Eq. (2.2) we obtain $a_{\mathbf{k}} = a_{-\mathbf{k}}$. If $S = 1$, then $\phi(\mathbf{r}_2 - \mathbf{r}_1) = -\phi(\mathbf{r}_1 - \mathbf{r}_2)$. From Eq. (2.2) we obtain $a_{\mathbf{k}} = -a_{-\mathbf{k}}$.

To find the eigenvalue of $\psi(\mathbf{r}_1, \mathbf{r}_2)$ we write down the Schrödinger equation.

$$(H_\circ + \widehat{V})\psi = W\psi, \tag{2.3}$$

where H_o is the kinetic energy of the two electron system, relative to the Fermi surface, and $\widehat{V}$ the interaction energy between the two electrons. We need to obtain a Schrödinger equation for $a_{\mathbf{k}}$. We do this as follows.

$$\int \frac{d\mathbf{r}_1}{V_1} \int \frac{d\mathbf{r}_2}{V_2} e^{-i\mathbf{k}\cdot(\mathbf{r}_2-\mathbf{r}_1)} \left\{ \left[-\frac{\hbar^2}{2m^*}\nabla^2_{\mathbf{r}_1} - \frac{\hbar^2}{2m^*}\nabla^2_{\mathbf{r}_2} - 2E_F + V(\mathbf{r}_1,\mathbf{r}_2) \right] \right.$$

$$\left. \sum_{\mathbf{k}'(|\mathbf{k}'|>k_F)} a_{\mathbf{k}'} e^{i\mathbf{k}'\cdot(\mathbf{r}_2-\mathbf{r}_1)} = W \sum_{\mathbf{k}'(|\mathbf{k}'|>k_F)} a_{\mathbf{k}'} e^{i\mathbf{k}'\cdot(\mathbf{r}_2-\mathbf{r}_1)} \right\}, \tag{2.4}$$

where V_1 and V_2 are the system volumes. After integration for Eq. (2.4) we obtain

$$[W - 2\xi(\mathbf{k})]a_{\mathbf{k}} = \sum_{\mathbf{k}'(|\mathbf{k}'|>k_F)} V_{\mathbf{k}\mathbf{k}'} a_{\mathbf{k}'}, \tag{2.5}$$

where $\xi(\mathbf{k}) = \hbar^2 k^2/(2m^*) - E_F$, $V_{\mathbf{k}\mathbf{k}'} =< \mathbf{k}, -\mathbf{k}|V(\mathbf{r}_1,\mathbf{r}_2)|\mathbf{k}', -\mathbf{k}' >$, and E_F is the Fermi energy.

Assume

$$V_{\mathbf{k}\mathbf{k}'} = \begin{cases} -V(<0), & \text{for} \quad 0 < \xi(\mathbf{k}), \xi(\mathbf{k}') < \hbar\omega_c; \\ 0, & \text{otherwise}, \end{cases} \tag{2.6}$$

where ω_c is the upper boundary of interaction. To solve Eq. (2.5) we use a non-perturbation method, i.e., we look for a non-perturbation solution. From Eq. (2.5) we have

$$a_{\mathbf{k}} = V \sum_{\mathbf{k}'(|\mathbf{k}'|>k_F)} \frac{a_{\mathbf{k}'}}{-W + 2\xi(\mathbf{k})}. \tag{2.7}$$

After doing the restrictive summation over $\mathbf{k}$ on both sides of Eq. (2.7) we obtain

$$\frac{1}{V} = \sum_{\mathbf{k}(|\mathbf{k}|>k_F)} \frac{1}{2\xi(\mathbf{k}) - W}. \tag{2.8}$$

Transforming the summation in Eq. (2.8) into integration yields

$$\frac{1}{V} = \int_0^{\hbar\omega_c} d\xi N(\xi) \frac{1}{2\xi - W}. \tag{2.9}$$

Assuming that the density of states $N(\xi)$ is a slow variation function and performing the integration in Eq. (2.9) yield

$$W = \frac{-2\hbar\omega_c}{e^{\frac{2}{N(E_F)V}} - 1}, \tag{2.10}$$

where $N(E_F)$ is the density of states at Fermi surface. Assuming $N(E_F)V < 1/4$, i.e., week coupling, Eq. (2.10) becomes

$$W = -2\hbar\omega_c e^{\frac{-2}{N(E_F)V}} < 0. \tag{2.11}$$

$W < 0$ means that the two electron eigenstate forms a bound state so that their total energy is less than $2E_F$. Note that it is a non-perturbation solution because in the above derivations we do not use any perturbation approximation. If one uses perturbation approach, one definitely can not obtain this solution because $V = 0$ is an inherent singular point of the solution.

The two electrons forming a bound state are called a Cooper pair. If we start from a free electron gas, and turn on the interaction $-V$, then the electrons will group themselves in Cooper pairs giving up energy to the external world. The normal state of the electrons is unstable. The largest number of allowed scattering processes, yielding the maximum lowering of energy, is obtained by pairing electrons with equal and opposite momentum [36]. If the total spin of the two electrons $S = 0$, then the Cooper pair consists of a superposition of $\mathbf{k}\uparrow$ and $-\mathbf{k}\downarrow$. In this case we say that the Cooper pair is in time-reversed state. Until now the Cooper pair of a great part of superconductors is in $S = 0$ state. Note that there are no any restriction for the attractive potential $-V < 0$. The $-V < 0$ may come from electron-phonon interaction, electron-magnon interaction, electron-plasmon interaction, electron-local trap interaction, electron-electron gauge force, and so on. The key point to form the bound state of two electrons is that there is net attractive interaction between two electrons. The carriers can be holes instead of electrons. From the above mentioned arbitrary of $-V < 0$, it is not surprised why more than two hundred different mechanisms were proposed until 1990 to interpret the high-T_c superconductivity in cuprates after its discovery in 1986.

2.2. The Size of a Cooper Pair

Now we estimate the spatial distance between the two electrons forming a Cooper pair. The average square distance, ρ^2, is

$$\rho^2 = \frac{\int d(\mathbf{r}_1 - \mathbf{r}_2)|\psi(\mathbf{r}_1 - \mathbf{r}_2)|^2|\mathbf{r}_1 - \mathbf{r}_2|^2}{\int d(\mathbf{r}_1 - \mathbf{r}_2)|\psi(\mathbf{r}_1 - \mathbf{r}_2)|^2}. \tag{2.12}$$

Inserting Eqs. (2.1) and (2.2) into Eq. (2.12) yields

$$\rho^2 = \frac{\int d(\mathbf{r}_1 - \mathbf{r}_2)|\sum_{\mathbf{k}} a_{\mathbf{k}} \nabla_{\mathbf{k}} e^{i\mathbf{k}\cdot(\mathbf{r}_1 - \mathbf{r}_2)}|^2}{\int d(\mathbf{r}_1 - \mathbf{r}_2)|\sum_{\mathbf{k}} a_{\mathbf{k}} e^{i\mathbf{k}\cdot(\mathbf{r}_1 - \mathbf{r}_2)}|^2}. \tag{2.13}$$

Eq. (2.13) is

$$\rho^2 = \frac{\int d(\mathbf{r}_1 - \mathbf{r}_2)\sum_{\mathbf{k}}\sum_{\mathbf{k}'} a_{\mathbf{k}} a_{\mathbf{k}'} \nabla_{\mathbf{k}} e^{i\mathbf{k}\cdot(\mathbf{r}_1 - \mathbf{r}_2)} \nabla_{\mathbf{k}'} e^{-i\mathbf{k}'\cdot(\mathbf{r}_1 - \mathbf{r}_2)}}{\int d(\mathbf{r}_1 - \mathbf{r}_2)\sum_{\mathbf{k}}\sum_{\mathbf{k}'} a_{\mathbf{k}} a_{\mathbf{k}'} e^{i(\mathbf{k}-\mathbf{k}')\cdot(\mathbf{r}_1 - \mathbf{r}_2)}}. \tag{2.14}$$

Note that the summations over **k** and **k**' in Eqs. (2.13) and (2.14) are constrained in $|\mathbf{k}| > k_F$ and $|\mathbf{k}'| > k_F$. In Eq. (2.14) we transform the summations over **k** and **k**' into $\int d\mathbf{k} V_1/(2\pi)^3$ and $\int d\mathbf{k}' V_2/(2\pi)^3$, respectively, perform integrations by parts for **k** and **k**', and use

$$\frac{V_1}{(2\pi)^3}\int d(\mathbf{r}_1 - \mathbf{r}_2) e^{i(\mathbf{k}-\mathbf{k}')\cdot(\mathbf{r}_1 - \mathbf{r}_2)} = \delta(\mathbf{k} - \mathbf{k}'), \tag{2.15}$$

then we have

$$\rho^2 = \frac{\int d\mathbf{k}|\nabla_{\mathbf{k}} a_{\mathbf{k}}|^2}{\int d\mathbf{k}|a_{\mathbf{k}}|^2}. \tag{2.16}$$

Making variable transformation from $\mathbf{k}$ to ξ [$=\hbar^2k^2/(2m^*) - E_F$], Eq. (2.16) becomes

$$\rho^2 \cong \frac{N(E_F)\left(\frac{\partial\xi}{\partial k}\right)_{\xi=0} \int_0^\infty d\xi \left(\frac{\partial a}{\partial \xi}\right)^2}{N(E_F)\int\limits_0^\infty d\xi a^2}. \tag{2.17}$$

We calculate the quantities appearing in Eq. (2.17) as follows.

$$\left(\frac{\partial\xi}{\partial k}\right)_{\xi=0} = \frac{\partial}{\partial k}\left(\frac{\hbar^2k^2}{2m^*} - E_F\right)_{\xi=0(k=k_F)}$$

$$= \frac{\hbar^2 k_F}{m^*} = \hbar v_F. \tag{2.18}$$

Using Eq. (2.7), we have

$$\int\limits_0^\infty \left(\frac{\partial a}{\partial \xi}\right)^2 = \int\limits_0^\infty d\xi \left[\frac{\partial}{\partial \xi}\frac{C}{(W-2\xi)}\right]^2$$

$$= -\frac{2C^2}{3W^3}, \tag{2.19}$$

where C is a constant.

$$\int\limits_0^\infty d\xi a^2 = \int\limits_0^\infty d\xi \frac{C^2}{(W-2\xi)^2}$$

$$= -\frac{C^2}{2W}. \tag{2.20}$$

So,

$$\rho = \frac{2}{\sqrt{3}}\frac{\hbar v_F}{W} \cong \hbar\frac{\hbar k_F}{m^*}\frac{1}{W}. \tag{2.21}$$

For conventional low-T_c superconductors we take $m^* = m_\circ$ (free electron mass), $k_F = 1/\AA$, and the bound energy $W = 10$ K, then $\rho \cong 10^{-4}$ cm. The value of ρ is equal to the value of coherence length $\xi_\circ$ in Eq. (1.18) of section 1.2.2.

There is a problem. The average Coulomb repulsion is about 1 eV per electron, and W is only about 10^{-3} eV. So, even the former's fluctuation can break so weak bound pair. Especially, the Cooper pairs in the conventional superconductor overlap each other strongly. This long coherence length belongs to strongly overlapping pair case. Why can the Cooper pair still be formed in this case? The Landau Fermi liquid theory can answer this question [35]. In this theory the electron states of a real normal metal are placed in one-to-one correspondence with those of a free electron gas. The electrons in an interaction system, such as a real normal metal, are quasiparticles which are stable excitations in immediate vicinity of the Fermi surface. The coupling between quasiparticles, mediated by phonon in conventional superconductors, for example, is the residual attractive interaction between two quasiparticles, neglected within the Fermi liquid approximation. This residual interaction gives rise to formation of Cooper pair. Exactly speaking, Cooper pair is quasi-electron pairing. On the other hand, for the high-T_c cuprates, such as $YBa_2Cu_3O_{7-\delta}$, $\rho = \xi_o \approx 20-30$ Å. This short coherence length belongs to separating pair case. The reason to form Cooper pairs in this case can be interpreted by Landau Fermi liquid theory as well. The formation of Cooper pair in the high-T_c cuprates is caused by the residual attractive interaction between two quasiparticles, i.e., the two local spin-mediated interaction (See section 3.1.), neglected within the Fermi liquid approximation.

2.3. Pairing Approximation

In sections 2.3, 2.4, and 2.5 we expound BCS pairing theory [2]. Note that there is no, exactly speaking, equation determining T_c in BCS pairing theory (The T_c equation will be given in section 3.3.). In section 2.1 we mentioned that the important scattering processes occur between Cooper pairs with equal and opposite momentum. Assume that the total spin of a Cooper pair is $S = 0$. We consider approximately only the interaction between electrons with $\mathbf{k}\uparrow$ and $-\mathbf{k}\downarrow$, which is called pairing approximation. Then, we can write a so-called reduced Hamiltonian H_{red}.

$$H_{red} = \sum_{\mathbf{k}s} \varepsilon_{\mathbf{k}} n_{\mathbf{k}s} + \sum_{\mathbf{k}\mathbf{k}'} V_{\mathbf{k}'\mathbf{k}} b^+_{\mathbf{k}'} b_{\mathbf{k}}, \tag{2.22}$$

where

$$V_{\mathbf{k}'\mathbf{k}} = <\mathbf{k}', -\mathbf{k}'|\hat{V}|\mathbf{k}, -\mathbf{k}>, \tag{2.23}$$

$$b^+_{\mathbf{k}'} = c^+_{\mathbf{k}'\uparrow} c^+_{-\mathbf{k}'\downarrow}, \tag{2.24}$$

$$b_{\mathbf{k}} = c_{-\mathbf{k}\downarrow} c_{\mathbf{k}\uparrow}, \tag{2.25}$$

where $b^+_{\mathbf{k}'}$ and $b_{\mathbf{k}}$ are the pair creation and destruction operator, respectively.

At $T = 0$, the system is in ground state, and all electrons are paired. We can rewrite the first term in Eq. (2.22) as

$$\sum_{\mathbf{k}s} \varepsilon_{\mathbf{k}} n_{\mathbf{k}s} = \sum_{\mathbf{k}} 2\varepsilon_{\mathbf{k}} b^+_{\mathbf{k}} b_{\mathbf{k}}. \tag{2.26}$$

We prove Eq. (2.26) as follows.

$$\begin{aligned} 2b^+_{\mathbf{k}} b_{\mathbf{k}}|G> &= 2c^+_{\mathbf{k}\uparrow} c^+_{-\mathbf{k}\downarrow} c_{-\mathbf{k}\downarrow} c_{\mathbf{k}\uparrow}|G> \\ &= 2c^+_{\mathbf{k}\uparrow} c_{\mathbf{k}\uparrow} c^+_{-\mathbf{k}\downarrow} c_{-\mathbf{k}\downarrow}|G> \\ &= 2n_{\mathbf{k}\uparrow} n_{-\mathbf{k}\downarrow}|G> \\ &= (n_{\mathbf{k}\uparrow} + n_{-\mathbf{k}\downarrow})|G>, \end{aligned}$$

where $|G>$ means $|$ground state$>$. At $T = 0$, the reduced Hamiltonian , Eq. (2.22), becomes

$$H^{(o)}_{red} = \sum_{\mathbf{k}} 2\varepsilon_{\mathbf{k}} b^+_{\mathbf{k}} b_{\mathbf{k}} + \sum_{\mathbf{k}'\mathbf{k}} V_{\mathbf{k}'\mathbf{k}} b^+_{\mathbf{k}'} b_{\mathbf{k}}. \tag{2.27}$$

Note that the commutation relations between b^+ and b are neither Fermi type nor Bose type. For example,

$$[b_{\mathbf{k}}, b^+_{\mathbf{k}'}]_- = 0, \quad \text{when} \quad \mathbf{k} \neq \mathbf{k}', \tag{2.28}$$

$$[b_{\mathbf{k}}, b^+_{\mathbf{k}}]_- = 1 - (n_{\mathbf{k}\uparrow} + n_{-\mathbf{k}\downarrow}). \tag{2.29}$$

Assume $S = 0$. We look for the BCS ground state wavefunction. Let N denote the number of electrons in a system. Assume that the very large number N is an even number.

$$\psi_{\circ N}(\mathbf{r}_1,\cdots,\mathbf{r}_N) = \widehat{A}\phi(\mathbf{r}_1-\mathbf{r}_2)\cdots\phi(\mathbf{r}_{N-1}-\mathbf{r}_N)(\uparrow 1)(\downarrow 2)\cdots(\uparrow N-1)(\downarrow N), \tag{2.30}$$

where $\widehat{A}$ is an antisymmetric operator. The function $\phi(\mathbf{r}_i - \mathbf{r}_{i+1})$ is a relative coordinate wavefunction of a Cooper pair. All Cooper pairs have the same wavefunction, i.e., all Cooper pairs are in the same state in the ground state wavefunction of system. Let $\phi(\mathbf{r}_i - \mathbf{r}_j) = \sum_{\mathbf{k}} g_{\mathbf{k}} \exp[i\mathbf{k}\cdot(\mathbf{r}_i - \mathbf{r}_j)]$. Eq. (2.30) becomes

$$\begin{aligned}\psi_{\circ N}(\mathbf{r}_1,\cdots,\mathbf{r}_N) &= \sum_{\mathbf{k}_1}\cdots\sum_{\mathbf{k}_{N/2}} g_{\mathbf{k}_1}\cdots g_{\mathbf{k}_{N/2}}\widehat{A}e^{i\mathbf{k}_1\cdot(\mathbf{r}_1-\mathbf{r}_2)}\cdots e^{i\mathbf{k}_{N/2}\cdot(\mathbf{r}_{N-1}-\mathbf{r}_N)} \\ &\quad (\uparrow 1)(\downarrow 2)\cdots(\uparrow N-1)(\downarrow N) \\ &= \sum_{\mathbf{k}_1}\cdots\sum_{\mathbf{k}_{N/2}} g_{\mathbf{k}_1}\cdots g_{\mathbf{k}_{\mathbf{k}_{N/2}}}(SD),\end{aligned} \tag{2.31}$$

where (SD) represents Slater determinant. In second quantization approach Eq. (2.31) becomes

$$|\psi_{\circ N}> = \sum_{\mathbf{k}_1}\cdots\sum_{\mathbf{k}_{N/2}} g_{\mathbf{k}_1}\cdots g_{\mathbf{k}_{N/2}} c^+_{\mathbf{k}_1\uparrow}c^+_{-\mathbf{k}_1\downarrow}\cdots c^+_{\mathbf{k}_{N/2}\uparrow}c^+_{-\mathbf{k}_{N/2}\downarrow}|0>. \tag{2.32}$$

Note that due to the Pauli principle, there is no same $\mathbf{k}$ in Eq. (2.32).

For convenience, instead of $|\psi_{\circ N}>$, which has a fixed electron number, BCS takes

$$|\psi_\circ> = C\prod_{\mathbf{k}}(1 + g_{\mathbf{k}}c^+_{\mathbf{k}\uparrow}c^+_{-\mathbf{k}\downarrow})|0>, \tag{2.33}$$

where $\prod_{\mathbf{k}}$ represents a product of many factors [37]. Set

$$<\psi_\circ|\widehat{N}|\psi_\circ> = N, \tag{2.34}$$

where N is the electron number of a real system. The C in Eq. (2.33) can be determined by the normalization condition, i.e., $<\psi_\circ|\psi_\circ> = 1$.

$$C = \frac{1}{\prod_{\mathbf{k}}(1+|g_{\mathbf{k}}|^2)^{1/2}}. \tag{2.35}$$

Substituting Eq. (2.35) into Eq. (2.33) yields

$$\begin{aligned}|\psi_\circ> &= \prod_{\mathbf{k}}\frac{1+g_{\mathbf{k}}b^+_{\mathbf{k}}}{(1+|g_{\mathbf{k}}|^2)^{1/2}}|0> \\ &= \prod_{\mathbf{k}}(u_{\mathbf{k}} + v_{\mathbf{k}}b^+_{\mathbf{k}})|0>,\end{aligned} \tag{2.36}$$

where $u_{\mathbf{k}}$ is a real number.

To find the ground state energy, let us look for the variational minimum of the average value of $H_{red}^{(\circ)} - \mu\widehat{N}$ over $|\psi_\circ>$, i.e.,

$$\delta W = \delta < \psi_\circ | H_{red}^{(\circ)} - \mu\widehat{N} | \psi_\circ >= 0. \tag{2.37}$$

Let us at first look for the variational parameter, and then make the variational operation.

$$W =< \psi_\circ | H_{red}^{(\circ)} - \mu\widehat{N} | \psi_\circ > . \tag{2.38}$$

Substituting Eqs. (2.33) and (2.27) into Eq. (2.38) yields

$$W = \sum_{\mathbf{k}} 2(\varepsilon_{\mathbf{k}} - \mu)|v_{\mathbf{k}}|^2 + \sum_{\mathbf{k}\mathbf{k}'} V_{\mathbf{k}'\mathbf{k}} u_{\mathbf{k}} v_{\mathbf{k}} u_{\mathbf{k}'} v_{\mathbf{k}'}^*. \tag{2.39}$$

The $u_{\mathbf{k}}$ and $v_{\mathbf{k}}$ in Eq. (2.36) become

$$u_{\mathbf{k}} = \sin(\vartheta_{\mathbf{k}}), \tag{2.40}$$

$$v_{\mathbf{k}} = \cos(\vartheta_{\mathbf{k}})\exp(i\phi), \tag{2.41}$$

where ϕ is the phase of wavefunction of a Cooper pair. So, the variational parameter is $\vartheta_{\mathbf{k}}$. Substituting Eqs. (2.40) and (2.41) into Eqs. (2.39) and (2.37), we have

$$\frac{\partial W}{\partial \vartheta_{\mathbf{k}}} = 0, \tag{2.42}$$

i.e.,

$$-2(\varepsilon_{\mathbf{k}} - \mu)\sin(2\vartheta_{\mathbf{k}}) + \sum_{\mathbf{k}'} V_{\mathbf{k}'\mathbf{k}} \cos(2\vartheta_{\mathbf{k}})\sin(2\vartheta_{\mathbf{k}'}) = 0, \tag{2.43}$$

$$-2(\varepsilon_{\mathbf{k}} - \mu)\tan(2\vartheta_{\mathbf{k}}) = -\sum_{\mathbf{k}} V_{\mathbf{k}'\mathbf{k}} \sin(2\vartheta_{\mathbf{k}'}). \tag{2.44}$$

The extra factor of 2 enters in the second sum of Eq. (2.43) because both $\mathbf{k}$ and $\mathbf{k}'$ in Eq. (2.39) run over any given value $\mathbf{k}''$. Define

$$\Delta_{\mathbf{k}} = -\sum_{\mathbf{k}'} V_{\mathbf{k}'\mathbf{k}} u_{\mathbf{k}'} v_{\mathbf{k}'}$$

$$= -\sum_{\mathbf{k}'} V_{\mathbf{k}'\mathbf{k}} \frac{1}{2} \sin(2\vartheta_{\mathbf{k}'}) e^{i\phi}. \tag{2.45}$$

From the following derivations we know that $\Delta_{\mathbf{k}}$ represents gap function. Thus from Eq. (2.45) we see that the gap function contains a phase factor of Cooper pair wavefunction. Define

$$E_{\mathbf{k}} = \sqrt{(\varepsilon_{\mathbf{k}} - \mu)^2 + |\Delta_{\mathbf{k}}|^2}. \tag{2.46}$$

As we shall see in section 2.4.1, $E_{\mathbf{k}}$ turns out to be the energy required to create a quasiparticle of momentum $\mathbf{k}$ or $-\mathbf{k}$ in the superconducting state for conventional or unconventional superconductor and in the pseudogap state for unconventional superconductor, and $\Delta_{\mathbf{k}}$ is the energy gap for quasiparticle excitation. From Eqs. (2.44) and (2.45) we have

$$\tan(2\vartheta_{\mathbf{k}}) = \frac{|\Delta_{\mathbf{k}}|}{-(\varepsilon_{\mathbf{k}} - \mu)}. \tag{2.47}$$

From Eqs. (2.40), (2.41), (2.46), and (2.47) we have

$$2u_{\mathbf{k}}v_{\mathbf{k}} = \sin(2\vartheta_{\mathbf{k}})e^{i\phi} = \frac{\Delta_{\mathbf{k}}}{E_{\mathbf{k}}}, \tag{2.48}$$

$$-u_{\mathbf{k}}^2 + |v_{\mathbf{k}}|^2 = \cos(2\vartheta_{\mathbf{k}}) = \frac{-(\varepsilon_{\mathbf{k}} - \mu)}{E_{\mathbf{k}}}. \tag{2.49}$$

From Eq. (2.36) we have

$$u_{\mathbf{k}}^2 + |v_{\mathbf{k}}|^2 = 1. \tag{2.50}$$

From Eqs. (2.49) and (2.50) we have

$$u_{\mathbf{k}}^2 = \frac{1}{2}\left(1 + \frac{\varepsilon_{\mathbf{k}} - \mu}{E_{\mathbf{k}}}\right), \tag{2.51}$$

$$|v_{\mathbf{k}}|^2 = \frac{1}{2}\left(1 - \frac{\varepsilon_{\mathbf{k}} - \mu}{E_{\mathbf{k}}}\right). \tag{2.52}$$

From Eqs. (2.48) and (2.45) we have

$$\Delta_{\mathbf{k}} = -\sum_{\mathbf{k}'} V_{\mathbf{k}'\mathbf{k}} \frac{\Delta_{\mathbf{k}'}}{2E_{\mathbf{k}'}}. \tag{2.53}$$

Eq. (2.53) is the famous BCS gap equation at 0 K. For convenience, in the following discussions we take $\Delta_{\mathbf{k}}$ and $v_{\mathbf{k}}$ as real numbers. The trivial solution of Eq. (2.53) is

$$\Delta_{\mathbf{k}} = 0, \tag{2.54}$$

which corresponds to the normal state.

Let us look for the non-zero energy gap at $T = 0$ K, $\Delta(0)$. Assume

$$V_{\mathbf{k}'\mathbf{k}} = \begin{cases} -V < 0, & \text{if} \quad |\varepsilon_{\mathbf{k}} - \mu|, |\varepsilon_{\mathbf{k}'} - \mu| < \hbar\omega_c; \\ 0, & \text{if} \quad \text{otherwise}. \end{cases} \tag{2.55}$$

This approximation for $V_{\mathbf{k}'\mathbf{k}}$ is called BCS cut. If $V_{\mathbf{k}'\mathbf{k}}$ comes from phonon-mediated interaction, then ω_c is the average frequency of phonon, ω_D. Using BCS cut, we have

$$\Delta_{\mathbf{k}} = \begin{cases} \Delta(0), & \text{if} \quad |\varepsilon_{\mathbf{k}} - \mu| < \hbar\omega_c; \\ 0, & \text{if} \quad \text{otherwise}. \end{cases} \tag{2.56}$$

Substituting Eq. (2.56) into Eq. (2.53) gives

$$\begin{aligned} \Delta(0) &= V\sum_{\mathbf{k}'} \frac{\Delta(0)}{2\sqrt{(\varepsilon_{\mathbf{k}'} - \mu)^2 + \Delta^2(0)}} \\ &= N(E_F)V \int_{-\hbar\omega_c}^{\hbar\omega_c} d\xi \frac{\Delta(0)}{2\sqrt{\xi^2 + \Delta^2(0)}} \\ &= N(E_F)V\Delta(0)\sinh^{-1}\frac{\hbar\omega_c}{\Delta(0)}, \end{aligned} \tag{2.57}$$

where $\xi = (\varepsilon_{\mathbf{k}} - \mu)$. From Eq. (2.57) we have

$$\Delta(0) = \frac{\hbar\omega_c}{\sinh\frac{1}{N(E_F)V}}. \tag{2.58}$$

Using weak coupling approximation, $N(E_F)V < 1/4$, Eq. (2.58) becomes

$$\Delta(0) = 2\hbar\omega_c e^{-\frac{1}{N(E_F)V}}. \tag{2.59}$$

Let us look for the condensation energy W_{cond}. Its definition is

$$W_{cond} = W_N - W_S, \tag{2.60}$$

where W_N represents the ground state energy in normal state ($\Delta(0) = 0$), and W_S represents the energy of ground state with $\Delta(0) > 0$. We discuss a unit volume. Substituting Eq. (2.39) into Eq. (2.60) gives

$$W_{cond} = \left[\sum_{\mathbf{k}} 2(\varepsilon_{\mathbf{k}} - \mu)v_{\mathbf{k}}^2 + \sum_{\mathbf{k}'\mathbf{k}} V_{\mathbf{k}'\mathbf{k}} u_{\mathbf{k}} v_{\mathbf{k}} u_{\mathbf{k}'} v_{\mathbf{k}'}\right]\Bigg|_{\Delta(0)=0}$$
$$- \left[\sum_{\mathbf{k}} 2(\varepsilon_{\mathbf{k}} - \mu)v_{\mathbf{k}}^2 + \sum_{\mathbf{k}'\mathbf{k}} V_{\mathbf{k}'\mathbf{k}} u_{\mathbf{k}} v_{\mathbf{k}} u_{\mathbf{k}'} v_{\mathbf{k}'}\right]\Bigg|_{\Delta(0)\neq 0}. \tag{2.61}$$

From Eqs. (2.51) and (2.52) we know that when $\Delta(0) = 0$,

$$v_{\mathbf{k}}^2 = \begin{cases} 1, & \text{if} \quad \varepsilon_{\mathbf{k}} - \mu < 0; \\ 0, & \text{if} \quad \varepsilon_{\mathbf{k}} - \mu > 0. \end{cases}$$

$$u_{\mathbf{k}}^2 = \begin{cases} 1, & \text{if} \quad \varepsilon_{\mathbf{k}} - \mu > 0; \\ 0, & \text{if} \quad \varepsilon_{\mathbf{k}} - \mu < 0. \end{cases} \tag{2.62}$$

Considering Eq. (2.62) and unit volume, the W_N becomes

$$W_N = \sum_{|\mathbf{k}|<k_F} 2(\varepsilon_{\mathbf{k}} - \mu) = N(E_F)\int_{-\hbar\omega_c}^{0} d\xi 2\xi. \tag{2.63}$$

Substituting Eqs. (2.51) and (2.52) into W_S yields

$$W_S = -2N(E_F)\int_0^{\hbar\omega_c} d\xi \frac{\xi^2}{\sqrt{\xi^2 + \Delta^2(0)}} - \Delta^2(0)N(E_F)\int_0^{\hbar\omega_c} d\xi \frac{1}{2\sqrt{\xi^2 + \Delta^2(0)}}. \tag{2.64}$$

Performing the integrations in Eqs. (2.63) and (2.64) yields

$$W_{cond} = -N(E_F)(\hbar\omega_c)^2 + 2N(E_F)\left[\frac{\hbar\omega_c\sqrt{(\hbar\omega_c)^2 + \Delta^2(0)}}{2}\right]. \tag{2.65}$$

If we assume $\hbar\omega_c \gg \Delta(0)$, then Eq. (2.65) becomes

$$W_{cond} = \frac{N(E_F)\Delta^2(0)}{2}. \tag{2.66}$$

Substituting Eq. (2.59) into Eq. (2.66) gives

$$W_{cond} = 2N(E_F)(\hbar\omega_c)^2 e^{-\frac{2}{N(E_F)V}}. \tag{2.67}$$

From $W_{cond} > 0$, we know that the ground state with $\Delta(0) > 0$ is more stable than the normal state. Here, we have to note that $\Delta(0) > 0$ does not mean superconducting state. The $\Delta(0) > 0$ only means that there are Cooper pairs in the system. If the Cooper pairs have the long-range phase coherence, then the system will be in superconducting state. For the greater part of low-T_c superconductors, which belong to the conventional superconductors, the amplitude and the long-range phase coherence of Cooper pairs appear nearly at the same temperature (See section 3.3 for detail.). For these conventional superconductors, the $\Delta(0) > 0$ means nearly superconducting state. However, for unconventional superconductor, even at $T = 0$ K, $\Delta(0) > 0$ does not definitely correspond to superconducting state. For example, for $YBa_2Cu_3O_{7-\delta}$ with $x = 0.05$, $T_c = 0$ K, $T^* = 340$ K, and $\Delta(0) = 34$ meV (See chapter 4.). The x is the $O_{p\sigma}$ hole number of unit cell of Cu lattice in the CuO_2 plane. The Cooper pair consists of two $O_{p\sigma}$ holes in the high-T_c cuprates. At $T = 0$ K, all carriers become Cooper pairs and condense into chemical potential, but there are no the long-range phase coherence for $YBa_2Cu_3O_{7-\delta}$ with $x = 0.05$ (See chapter 4.).

For the type-I conventional superconductors, thermodynamics gives

$$W_N - W_S = \frac{H_c^2}{8\pi}, \tag{2.68}$$

where H_c is the thermodynamic critical magnetic field for destroying superconducting state at 0 K. From Eqs. (2.68) and (2.66) we find

$$H_c = 2\left[\pi N(E_F)\right]^{1/2}\Delta(0). \tag{2.69}$$

This relation is in agreement with experiment.

2.4. Quasiparticle Energy Spectrum

2.4.1. Excitation Energy

From Eq. (2.39), the energy of a pair $(\mathbf{p}\uparrow, -\mathbf{p}\downarrow)$, W_1, is

$$W_1 = 2(\varepsilon_{\mathbf{p}} - \mu)v_{\mathbf{p}}^2 + 2\sum_{\mathbf{k}'} V_{\mathbf{pk}'}u_{\mathbf{k}'}v_{\mathbf{k}'}u_{\mathbf{p}}v_{\mathbf{p}}, \tag{2.70}$$

where the second factor "2" comes from double summation. Substituting Eqs. (2.48) and (2.53) into Eq. (2.70) yields

$$W_1 = 2(\varepsilon_{\mathbf{p}} - \mu)v_{\mathbf{p}}^2 - 2\Delta_{\mathbf{p}}u_{\mathbf{p}}v_{\mathbf{p}}. \tag{2.71}$$

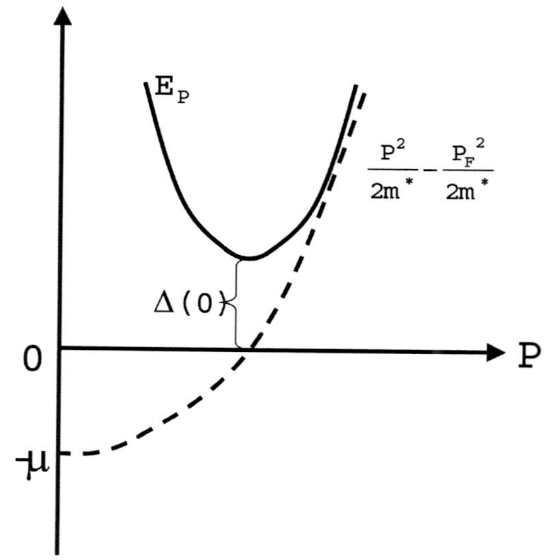

Figure 2.1. The solid line is the schematic plot of $E_{\mathbf{p}}$ vs. p ($=|\mathbf{p}|$). The dashed line is the schematic plot of $[p^2/(2m^*) - p_F^2/(2m^*)]$ vs. p. The minimum energy $\Delta(0)$ is required to create each quasiparticle ($\mathbf{p}\uparrow$) or ($-\mathbf{p}\downarrow$) above the chemical potential.

The single quasi-electron ($\mathbf{p}\uparrow$) excitation energy $W_{\mathbf{p}}$ is defined to be the total excitation energy of the system when the extra electron ($\mathbf{p}\uparrow$) is added to the system. The only effect of adding the extra electron is to block the pair state ($\mathbf{p}\uparrow, -\mathbf{p}\downarrow$) from participating in the pairing interaction (due to the Pauli principle). Thus, W_p is equal to the energy change of the system, putting one quasi-electron ($\mathbf{p}\uparrow$) and taking out a pair ($\mathbf{p}\uparrow, -\mathbf{p}\downarrow$).

$$W_{\mathbf{p}} = (\varepsilon_{\mathbf{p}} - \mu) - \left[2(\varepsilon_{\mathbf{p}} - \mu)v_{\mathbf{p}}^2 - 2\Delta_{\mathbf{p}}u_{\mathbf{p}}v_{\mathbf{p}}\right]. \tag{2.72}$$

Substituting Eqs. (2.52) and (2.48) into Eq. (2.72) yields

$$W_{\mathbf{p}} = E_{\mathbf{p}}. \tag{2.73}$$

We can simply say that $E_{\mathbf{p}}$ is the energy required to create a quasi-electron in state ($\mathbf{p}\uparrow$) or ($-\mathbf{p}\downarrow$). $E_{\mathbf{p}}$ differs from the ($\varepsilon_{\mathbf{p}} - \mu$) clearly in the vicinity of Fermi surface, where $\varepsilon_{\mathbf{p}} = p^2/(2m^*)$. A plot of $E_{\mathbf{p}}$ vs. $|\mathbf{p}|$ is given in Fig. 2.1. Note that when a Cooper pair is destroyed, we always get two quasi-electrons ($\mathbf{p}\uparrow$) and ($-\mathbf{p}\downarrow$) above the chemical potential. If there is no quasi-electron excitation above the chemical potential, then all Cooper pairs condense into the chemical potential, which is called BCS model and is shown in Fig. 2.2. BCS model means Bose-Einstein condensation of Cooper pairs.

2.4.2. Momentum Distribution in BCS Ground State

Let us study the average occupation number condensed into the Bloch state $\mathbf{k}\uparrow$, $< n_{\mathbf{k}\uparrow} >$.

$$< n_{\mathbf{k}\uparrow} > = < \psi_\circ | c_{\mathbf{k}\uparrow}^+ c_{\mathbf{k}\uparrow} | \psi_\circ > . \tag{2.74}$$

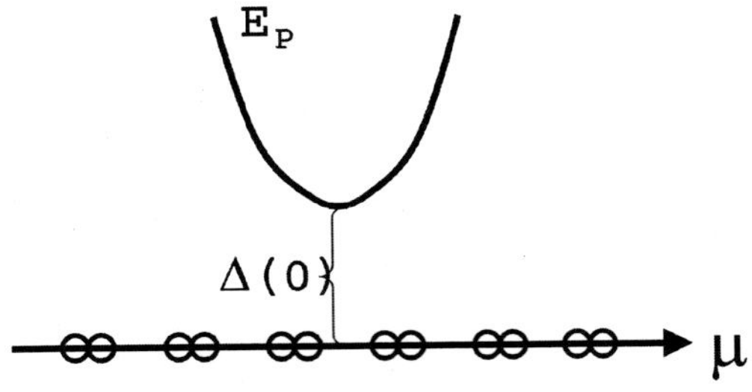

Figure 2.2. BCS model. At BCS ground state all Cooper pairs reside in the chemical potential. The excitation spectrum is shown as a curve above the chemical potential.

Substituting Eq. (2.36) into Eq. (2.74) yields

$$< n_{\mathbf{k}\uparrow} >= |v_{\mathbf{k}}|^2. \tag{2.75}$$

$|v_{\mathbf{k}}|^2$ has been given in Eq. (2.52). The average occupation number condensed into the Bloch state $\mathbf{k}\uparrow$ is equal to the occupation probability of the pair state $(\mathbf{k}\uparrow, -\mathbf{k}\downarrow)$ because $< n_{\mathbf{k}\uparrow} >=< n_{-\mathbf{k}\downarrow} >$ for BCS ground state. The schematic plot of $< n_{\mathbf{k}\uparrow} >$ vs. k $(=|\mathbf{k}|)$ is shown in Fig. 2.3. Note that, in BCS ground state, even at absolute zero there are vacancies with $k < k_F$ and occupied states with $k > k_F$. According to the Cooper's conclusion in section 2.1, one might at first think that there is no limit to the number of electrons which may be raised from $k < k_F$ to form Cooper pairs with a resultant lowering of the total energy, so that we should end up with all the electrons having $k > k_F$. This, however, is clearly absurd, and the reason is not hard to find. In order that a Cooper pair may be scattered from $(\mathbf{k}\uparrow, -\mathbf{k}\downarrow)$ to $(\mathbf{k}'\uparrow, -\mathbf{k}'\downarrow)$ the state $(\mathbf{k}\uparrow, -\mathbf{k}\downarrow)$ must be first occupied and the state $(\mathbf{k}'\uparrow, -\mathbf{k}'\downarrow)$ must empty. As more and more electrons form Cooper pairs with $k > k_F$, the chance of finding the states $(\mathbf{k}'\uparrow, -\mathbf{k}'\downarrow)$ empty becomes progressively smaller and smaller, so the number of scattering processes which may take place is reduced, with a consequent decrease in the magnitude of the negative potential energy. We know that the kinetic energy increases when two electrons go from $k < k_F$ to $k > k_F$, and the potential energy of the two electrons reduces when the two electrons form a Cooper pair. Eventually a condition is reached in which the lowering of the potential energy is insufficient to outweigh the increase in kinetic energy, and it is no longer possible to lower the total energy of electrons by forming Cooper pairs. In mathematics, the equilibrium occupation probability of the pair state is determined by the variational minimum condition of total energy, i.e., Eq. (2.42). The transition region, where $|v_{\mathbf{k}}|^2$ goes about from 1 to 0, has an energy width $\Delta(0)$ and a momentum width δk approximately.

$$\frac{\hbar^2}{2m^*}\left[(k_F+\delta k)^2 - k_F^2\right] = \Delta(0). \tag{2.76}$$

$$\delta k = \frac{\Delta(0)}{\hbar v_F}. \tag{2.77}$$

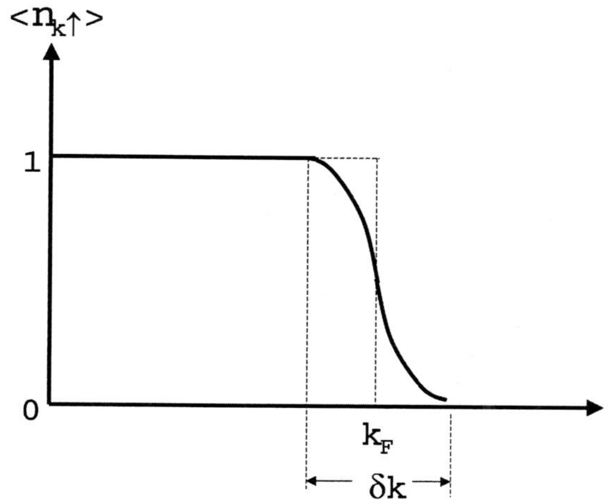

Figure 2.3. The schematic plot of $< n_{\mathbf{k}\uparrow} >$ vs. k. $\delta k = \Delta(0)/(\hbar v_F) \cong 1/\xi_\circ$. The occupation number in the normal state is 1 for $k < k_F$ and 0 for $k > k_F$. The Fermi surface is smeared by the pairing correlation.

Let us add two electrons to the states $\mathbf{k}\uparrow$ and $-\mathbf{k}\downarrow$ in the condensed state $|\psi_{\circ N}>$. What is the probability amplitude for obtaining the state $|\psi_{\circ N+2}>$? This is the quantity

$$F_{\mathbf{k}} = |<\psi_{\circ N+2}|c^+_{\mathbf{k}\uparrow}c^+_{-\mathbf{k}\downarrow}|\psi_{\circ N}>| = |<\psi_\circ|c^+_{\mathbf{k}\uparrow}c^+_{-\mathbf{k}\downarrow}|\psi_\circ>|. \quad (2.78)$$

We call $F_{\mathbf{k}}$ the condensation amplitude in state $\mathbf{k}$. Substituting Eq. (2.36) into Eq. (2.78) yields

$$F_{\mathbf{k}} = u_{\mathbf{k}}|v_{\mathbf{k}}|. \quad (2.79)$$

$u_{\mathbf{k}}$ and —$v_{\mathbf{k}}$— have been given in Eqs. (2.51) and (2.52). The schematic plot of $F_{\mathbf{k}}$ vs. k is shown in Fig. 2.4. From Fig. 2.4 we see that $F_{\mathbf{k}}$ does not vanish only in the transition zone $\delta k \cong 1/\xi_\circ$ and has maximum at $k = k_F$. From Figs. 2.3 and 2.4 we see that the Fermi surface of normal state is "smeared" by the pairing correlation.

2.5. Quasiparticle Excitation

2.5.1. Bogoliubov-Valatin Transformation

This transformation is convenient to describe elementary excitation of superconductors in superconducting or pseudogap state. Define [38] [39]

$$\gamma^+_{\mathbf{p}\uparrow} = u_{\mathbf{p}}c^+_{\mathbf{p}\uparrow} - v_{\mathbf{p}}c_{-\mathbf{p}\downarrow}. \quad (2.80)$$

$$\gamma_{-\mathbf{p}\downarrow} = u_{\mathbf{p}}c_{-\mathbf{p}\downarrow} + v_{\mathbf{p}}c^+_{\mathbf{p}\uparrow}. \quad (2.81)$$

$\gamma^+_{\mathbf{p}\uparrow}$ and $\gamma_{-\mathbf{p}\downarrow}$ are quasiparticle (elementary excitation), which is not in pairing, creation and annihilation operator, respectively, in both superconducting state or pseudogap state. Eqs.

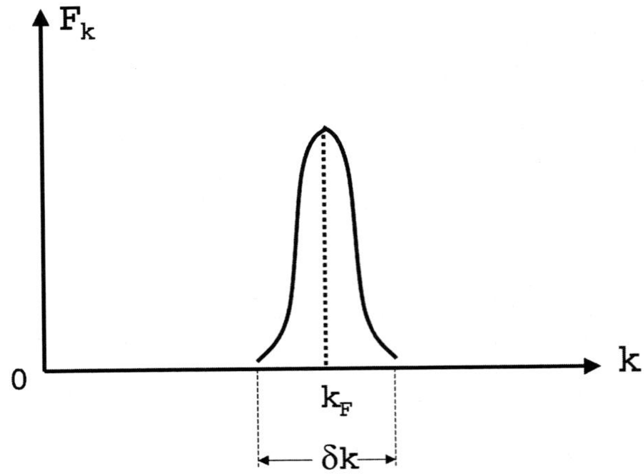

Figure 2.4. The schematic plot of $F_{\mathbf{k}}$ vs. k.

(2.80) and (2.81) are the Bogoliubov-Valatin transformation. The corresponding Hermitian conjugate operators are

$$\gamma_{\mathbf{p}\uparrow} = u_{\mathbf{p}} c_{\mathbf{p}\uparrow} - v_{\mathbf{p}} c^{+}_{-\mathbf{p}\downarrow}. \tag{2.82}$$

$$\gamma^{+}_{-\mathbf{p}\downarrow} = u_{\mathbf{p}} c^{+}_{-\mathbf{p}\downarrow} + v_{\mathbf{p}} c_{\mathbf{p}\uparrow}. \tag{2.83}$$

We can show that γ and γ^{+} are the destruction and creation operators of quasiparticles on the BCS ground state. For example,

$$\begin{aligned} \gamma^{+}_{\mathbf{p}\uparrow}|\psi_{circ}> &= (u_{\mathbf{p}} c^{+}_{\mathbf{p}\uparrow} - v_{\mathbf{p}} c_{-\mathbf{p}\downarrow})(u_{\mathbf{p}} + v_{\mathbf{p}} c^{+}_{\mathbf{p}\uparrow} c^{+}_{-\mathbf{p}\downarrow}) \prod_{\mathbf{k}(\neq\mathbf{p})} (u_{\mathbf{k}} + v_{\mathbf{k}} c^{+}_{\mathbf{k}\uparrow} c^{+}_{-\mathbf{k}\downarrow})|0> \\ &= (u^2_{\mathbf{p}} c^{+}_{\mathbf{p}\uparrow} + v^2_{\mathbf{p}} c^{+}_{\mathbf{p}\uparrow}) \prod_{\mathbf{k}(\neq\mathbf{p})} (u_{\mathbf{k}} + v_{\mathbf{k}} c^{+}_{\mathbf{k}\uparrow} c^{+}_{-\mathbf{k}\downarrow})|0> \\ &= c^{+}_{\mathbf{p}\uparrow} \prod_{\mathbf{k}(\neq\mathbf{p})} (u_{\mathbf{k}} + v_{\mathbf{k}} c^{+}_{\mathbf{k}\uparrow} c^{+}_{-\mathbf{k}\downarrow})|0> \\ &= |\psi_{\mathbf{p}\uparrow} >. \end{aligned} \tag{2.84}$$

$|\psi_{\mathbf{p}\uparrow} >$ represents a state in which there are the pairs without $(\mathbf{p}\uparrow, -\mathbf{p}\downarrow)$ and a single quasiparticle $(\mathbf{p}\uparrow)$. The single quasiparticle created by $\gamma^{+}_{\mathbf{p}\uparrow}$ is called a Bogolon. The quasiparticle operators are of Fermion type because it is easy to show that they satisfy the following anticommutation relation.

$$\left[\gamma_{\mathbf{p}s}, \gamma^{+}_{\mathbf{p}'s'}\right]_{+} = \delta_{\mathbf{p}\mathbf{p}'}\delta_{ss'}. \tag{2.85}$$

$$\left[\gamma_{\mathbf{p}s}, \gamma_{\mathbf{p}'s'}\right]_{+} = \left[\gamma^{+}_{\mathbf{p}s}, \gamma^{+}_{\mathbf{p}'s'}\right]_{+} = 0. \tag{2.86}$$

2.5.2. Partition Function $Z_{\mathbf{red}}$

The definition of partition function Z_{red} is as follows.

$$Z_{red} \equiv Tr\left[e^{-\beta(H_{red}-\mu\widehat{N})}\right], \tag{2.87}$$

where Tr means to sum the matrix elements in diagonal line of $\exp[-\beta(H_{red}-\mu\widehat{N})]$, and β=1/($k_B T$). Using Eq. (2.22),

$$H_{red}-\mu\widehat{N} = \sum_{\mathbf{k}s}\xi_{\mathbf{k}} n_{\mathbf{k}s} + \sum_{\mathbf{k}\mathbf{k}'} V_{\mathbf{k}'\mathbf{k}} b^{+}_{\mathbf{k}'} b_{\mathbf{k}}, \tag{2.88}$$

where

$$\xi_{\mathbf{k}} = \varepsilon_{\mathbf{k}} - \mu. \tag{2.89}$$

Making the Bogoliubov-Valatin transformation, Eq. (2.88) becomes

$$H_{red}-\mu\widehat{N} = H_M + H_I + 2\sum_{\mathbf{k}}\xi_{\mathbf{k}} v^2_{\mathbf{k}} + \sum_{\mathbf{k}\mathbf{k}'} V_{\mathbf{k}'\mathbf{k}} u_{\mathbf{k}} v_{\mathbf{k}} u_{\mathbf{k}'} v_{\mathbf{k}'}, \tag{2.90}$$

$$H_M = \sum_{\mathbf{k}} E_{\mathbf{k}}(\gamma^{+}_{\mathbf{k}\uparrow}\gamma_{\mathbf{k}\uparrow} + \gamma^{+}_{\mathbf{k}\downarrow}\gamma_{\mathbf{k}\downarrow}). \tag{2.91}$$

H_M is quasiparticle Hamiltonian. $E_{\mathbf{k}}$ is only an introduced parameter at this step.

$$H_I = \sum_{\mathbf{k}}\left[\xi_{\mathbf{k}}(u^2_{\mathbf{k}} - v^2_{\mathbf{k}}) - 2u_{\mathbf{k}} v_{\mathbf{k}}\sum_{\mathbf{k}'} V_{\mathbf{k}'\mathbf{k}} u_{\mathbf{k}'} v_{\mathbf{k}'} - E_{\mathbf{k}}\right](\gamma^{+}_{\mathbf{k}\uparrow}\gamma_{\mathbf{k}\uparrow} + \gamma^{+}_{\mathbf{k}\downarrow}\gamma_{\mathbf{k}\downarrow})$$

$$+\sum_{\mathbf{k}\mathbf{k}'} V_{\mathbf{k}'\mathbf{k}} u_{\mathbf{k}} v_{\mathbf{k}} u_{\mathbf{k}'} v_{\mathbf{k}'}(\gamma^{+}_{\mathbf{k}'\uparrow}\gamma_{\mathbf{k}'\uparrow} + \gamma^{+}_{\mathbf{k}'\downarrow}\gamma_{\mathbf{k}'\downarrow})(\gamma^{+}_{\mathbf{k}\uparrow}\gamma_{\mathbf{k}\uparrow} + \gamma^{+}_{\mathbf{k}\downarrow}\gamma_{\mathbf{k}\downarrow}). \tag{2.92}$$

To derive Eq. (2.90), we express $c_{\mathbf{p}s}$ and $c^{+}_{\mathbf{p}s}$ by $\gamma_{\mathbf{p}s}$ and $\gamma^{+}_{\mathbf{p}s}$.

$$c^{+}_{\mathbf{p}\uparrow} = u_{\mathbf{p}}\gamma^{+}_{\mathbf{p}\uparrow} + v_{\mathbf{p}}\gamma_{-\mathbf{p}\downarrow}. \tag{2.93}$$

$$c_{\mathbf{p}\uparrow} = u_{\mathbf{p}}\gamma_{\mathbf{p}\uparrow} + v_{\mathbf{p}}\gamma^{+}_{-\mathbf{p}\downarrow}. \tag{2.94}$$

$$c_{-\mathbf{p}\downarrow} = u_{\mathbf{p}}\gamma_{-\mathbf{p}\downarrow} - v_{\mathbf{p}}\gamma^{+}_{\mathbf{p}\uparrow}. \tag{2.95}$$

$$c^{+}_{-\mathbf{p}\downarrow} = u_{\mathbf{p}}\gamma^{+}_{-\mathbf{p}\downarrow} - v_{\mathbf{p}}\gamma_{\mathbf{p}\uparrow}. \tag{2.96}$$

Substituting Eqs. (2.93) - (2.96) into Eq. (2.88), assuming $\xi_{\mathbf{k}} = \xi_{-\mathbf{k}}$, keeping only the quasiparticle number conservation terms because we will use the $(H_{red}-\mu\widehat{N})$ to get the trace in Eq. (2.87), and using the anticommutation relations, we obtain

$$\sum_{\mathbf{k}s}\xi_{\mathbf{k}} n_{\mathbf{k}s} = \sum_{\mathbf{k}}\xi_{\mathbf{k}}(u^2_{\mathbf{k}} - v^2_{\mathbf{k}})(\gamma^{+}_{\mathbf{k}\uparrow}\gamma_{\mathbf{k}\uparrow} + \gamma^{+}_{-\mathbf{k}\downarrow}\gamma_{-\mathbf{k}\downarrow}) + 2\sum_{\mathbf{k}}\xi_{\mathbf{k}} v^2_{\mathbf{k}}. \tag{2.97}$$

$$\sum_{\mathbf{k}\mathbf{k}'} V_{\mathbf{k}'\mathbf{k}} c^{+}_{\mathbf{k}'\uparrow} c^{+}_{-\mathbf{k}'\downarrow} c_{-\mathbf{k}\downarrow} c_{\mathbf{k}\uparrow} = -2\sum_{\mathbf{k}\mathbf{k}'} V_{\mathbf{k}'\mathbf{k}} u_{\mathbf{k}'} v_{\mathbf{k}'} u_{\mathbf{k}} v_{\mathbf{k}}(\gamma^{+}_{\mathbf{k}\uparrow}\gamma_{\mathbf{k}\uparrow} + \gamma^{+}_{\mathbf{k}\downarrow}\gamma_{\mathbf{k}\downarrow})$$

$$+\sum_{\mathbf{k}\mathbf{k}'} V_{\mathbf{k}'\mathbf{k}} u_{\mathbf{k}'} v_{\mathbf{k}'} u_{\mathbf{k}} v_{\mathbf{k}}$$

$$+\sum_{\mathbf{kk}'} V_{\mathbf{k}'\mathbf{k}} u_{\mathbf{k}'} v_{\mathbf{k}'} u_{\mathbf{k}} v_{\mathbf{k}} (\gamma^{+}_{\mathbf{k}'\uparrow}\gamma_{\mathbf{k}'\uparrow} + \gamma^{+}_{\mathbf{k}'\downarrow}\gamma_{\mathbf{k}'\downarrow})(\gamma^{+}_{\mathbf{k}\uparrow}\gamma_{\mathbf{k}\uparrow} + \gamma^{+}_{\mathbf{k}\downarrow}\gamma_{\mathbf{k}\downarrow}). \tag{2.98}$$

Substituting Eqs. (2.97) and (2.98) into Eq. (2.88) yields Eq. (2.90).

H_I is still very complicate, so it is very difficult to get the exact Z_{red}. In the following we look for the average of H_I in terms of the density matrix operator of quasiparticles, $\exp(-\beta H_M)$. Denote the $H_{red} - \mu\widehat{N}$ after averaging of H_I still by $H_{red} - \mu\widehat{N}$, i.e.,

$$H_{red} - \mu\widehat{N} = H_M + C. \tag{2.99}$$

$$\begin{aligned} C = & \frac{\mathrm{Tr}\left[e^{-\beta H_M} H_I\right]}{\mathrm{Tr}\left[e^{-\beta H_M}\right]} + 2\sum_{\mathbf{k}} \xi_{\mathbf{k}} v_{\mathbf{k}}^2 + \sum_{\mathbf{kk}'} V_{\mathbf{k}'\mathbf{k}} u_{\mathbf{k}} v_{\mathbf{k}} u_{\mathbf{k}'} v_{\mathbf{k}'} \\ = & \, 2\sum_{\mathbf{k}} \xi_{\mathbf{k}} v_{\mathbf{k}}^2 + \sum_{\mathbf{k}} \left[\xi_{\mathbf{k}}(u_{\mathbf{k}}^2 - v_{\mathbf{k}}^2) - E_{\mathbf{k}}\right] 2f(E_{\mathbf{k}}) \\ & + \sum_{\mathbf{kk}'} V_{\mathbf{k}'\mathbf{k}} u_{\mathbf{k}'} v_{\mathbf{k}'} u_{\mathbf{k}} v_{\mathbf{k}} \left[1 - 2f(E_{\mathbf{k}})\right]\left[1 - 2f(E_{\mathbf{k}'})\right], \end{aligned} \tag{2.100}$$

where $f(E_{\mathbf{k}})$ is the Fermi distribution of the quasiparticles, $f(E_{\mathbf{k}}) = 1/[\exp(\beta E_{\mathbf{k}}) + 1]$.

The free energy function, F, is

$$F = -k_B T \ln Z_{red}. \tag{2.101}$$

Substituting Eq. (2.99) into (2.87) yields

$$\begin{aligned} F = & -k_B T \ln \mathrm{Tr}\left[e^{-\beta(H_M + C)}\right] = C - k_B T \ln \mathrm{Tr}\left(e^{-\beta \sum_{\mathbf{k}\alpha} E_{\mathbf{k}} \gamma^{+}_{\mathbf{k}\alpha}\gamma_{\mathbf{k}\alpha}}\right) \\ = & \, C - k_B T \ln \prod_{\mathbf{k}\alpha} \mathrm{Tr}\left(e^{-\beta E_{\mathbf{k}} \gamma^{+}_{\mathbf{k}\alpha}\gamma_{\mathbf{k}\alpha}}\right) = C - k_B T \ln \prod_{\mathbf{k}} \left(1 + e^{-\beta E_{\mathbf{k}}}\right)^2 \\ = & \, C + 2k_B T \sum_{\mathbf{k}} \ln\left[1 - f(E_{\mathbf{k}})\right]. \end{aligned} \tag{2.102}$$

To obtain the real free energy function, we should get the variational minimum value of F. The variational variables are $E_{\mathbf{k}}$ and $v_{\mathbf{k}}$. From the two minimum conditions and $u_{\mathbf{k}}^2 + v_{\mathbf{k}}^2 = 1$ we can obtain the equations for $u_{\mathbf{k}}$, $v_{\mathbf{k}}$, and energy gap. Substituting Eqs. (2.102) and (2.100) into

$$\frac{\partial F}{\partial E_{\mathbf{k}}} = 0, \tag{2.103}$$

we obtain

$$\begin{aligned} & 2k_B T \frac{-\frac{\partial f(E_{\mathbf{k}})}{\partial E_{\mathbf{k}}}}{1 - f(E_{\mathbf{k}})} - 2f(E_{\mathbf{k}}) + 2\left[\xi_{\mathbf{k}}(u_{\mathbf{k}}^2 - v_{\mathbf{k}}^2) - E_{\mathbf{k}}\right] \frac{\partial f(E_{\mathbf{k}})}{\partial E_{\mathbf{k}}} \\ & + 2u_{\mathbf{k}} v_{\mathbf{k}} \sum_{\mathbf{k}'} V_{\mathbf{k}'\mathbf{k}} u_{\mathbf{k}'} v_{\mathbf{k}'} \left[1 - 2f(E_{\mathbf{k}'})\right]\left[-2\frac{\partial f(E_{\mathbf{k}})}{\partial E_{\mathbf{k}}}\right] = 0. \end{aligned} \tag{2.104}$$

Considering $f(E_{\mathbf{k}}) = 1/[\exp(\beta E_{\mathbf{k}}) + 1]$, we have

$$2k_B T \frac{-\frac{\partial f(E_{\mathbf{k}})}{\partial E_{\mathbf{k}}}}{1 - f(E_{\mathbf{k}})} - 2f(E_{\mathbf{k}}) = 0. \tag{2.105}$$

So,

$$E_{\mathbf{k}} = \xi_{\mathbf{k}}(u_{\mathbf{k}}^2 - v_{\mathbf{k}}^2) - 2u_{\mathbf{k}}v_{\mathbf{k}}\sum_{\mathbf{k}'} V_{\mathbf{k}'\mathbf{k}}u_{\mathbf{k}'}v_{\mathbf{k}'}\left[1 - f(E_{\mathbf{k}'})\right]. \tag{2.106}$$

Considering $u_{\mathbf{k}} = \sqrt{1 - v_{\mathbf{k}}^2}$ and

$$\frac{\partial F}{\partial v_{\mathbf{k}}} = 0, \tag{2.107}$$

we obtain

$$2\xi_{\mathbf{k}}u_{\mathbf{k}}v_{\mathbf{k}} = -(u_{\mathbf{k}}^2 - v_{\mathbf{k}}^2)\sum_{\mathbf{k}'} V_{\mathbf{k}'\mathbf{k}}u_{\mathbf{k}'}v_{\mathbf{k}'}\left[1 - 2f(E_{\mathbf{k}'})\right]. \tag{2.108}$$

Let

$$-\sum_{\mathbf{k}'} V_{\mathbf{k}'\mathbf{k}}u_{\mathbf{k}'}v_{\mathbf{k}'}\left[1 - 2f(E_{\mathbf{k}'})\right] = \Delta_{\mathbf{k}}, \tag{2.109}$$

then Eq. (2.106) becomes

$$E_{\mathbf{k}} = \xi_{\mathbf{k}}(u_{\mathbf{k}}^2 - v_{\mathbf{k}}^2) + 2u_{\mathbf{k}}v_{\mathbf{k}}\Delta_{\mathbf{k}}, \tag{2.110}$$

and Eq. (2.108) becomes

$$2\xi_{\mathbf{k}}u_{\mathbf{k}}v_{\mathbf{k}} = (u_{\mathbf{k}}^2 - v_{\mathbf{k}}^2)\Delta_{\mathbf{k}}. \tag{2.111}$$

Substituting Eq. (2.111) into Eq. (2.110) gives

$$E_{\mathbf{k}} = (u_{\mathbf{k}}^2 - v_{\mathbf{k}}^2)\frac{\xi_{\mathbf{k}}^2 + \Delta_{\mathbf{k}}^2}{\xi_{\mathbf{k}}}. \tag{2.112}$$

Let

$$E_{\mathbf{k}} = \sqrt{\xi_{\mathbf{k}}^2 + \Delta_{\mathbf{k}}^2}, \tag{2.113}$$

then Eq. (2.112) becomes

$$u_{\mathbf{k}}^2 - v_{\mathbf{k}}^2 = \frac{\xi_{\mathbf{k}}}{E_{\mathbf{k}}}. \tag{2.114}$$

We have already known

$$u_{\mathbf{k}}^2 + v_{\mathbf{k}}^2 = 1. \tag{2.115}$$

The solutions of Eqs. (2.114) and (2.115) are

$$u_{\mathbf{k}}^2 = \frac{1}{2}\left(1 + \frac{\xi_{\mathbf{k}}}{E_{\mathbf{k}}}\right), \tag{2.116}$$

and

$$v_{\mathbf{k}}^2 = \frac{1}{2}\left(1 - \frac{\xi_{\mathbf{k}}}{E_{\mathbf{k}}}\right). \tag{2.117}$$

Substituting Eqs. (2.116) and (2.117) into Eq. (2.109) yields

$$\Delta_{\mathbf{k}} = -\sum_{\mathbf{k}'} V_{\mathbf{k}'\mathbf{k}}\frac{\Delta_{\mathbf{k}'}}{2E_{\mathbf{k}'}}\tanh\frac{\beta E_{\mathbf{k}'}}{2}. \tag{2.118}$$

Eq. (2.118) is the famous BCS gap ($\Delta_{\mathbf{k}}$) equation at finite temperature [2]. Here $\Delta_{\mathbf{k}}$ is a real number. However, we can use a similar method as to obtain Eq. (2.53) to show that $\Delta_{\mathbf{k}}$ in Eq. (2.118) is proportional to $\exp(i\phi)$.

2.5.3. Discussion on the Gap Equation

If $V_{\mathbf{k}'\mathbf{k}}$ is approximated by Eq. (2.55), then the $\Delta(T)$ in Eq. (2.118) has a simple non-trivial solution as follows. For convenience, we take $\hbar = k_B = 1$.

$$1 = \frac{N(E_F)V}{2} \int_{-\omega_c}^{\omega_c} d\xi \frac{1}{\sqrt{\xi^2 + \Delta^2(T)}} \tanh \frac{\sqrt{\xi^2 + \Delta^2(T)}}{2T}. \tag{2.119}$$

The T^* equation for unconventional superconductors can be obtained from Eq. (2.119) by using zero-gap condition, i.e., $\Delta(T^*) = 0$. The T_c equation for conventional superconductors can be obtained from Eq. (2.119) by using zero-gap condition, i.e., $\Delta(T_c) = 0$, approximately (The reason for "approximately" can be found in section 3.3.). For example, if $\Delta(T^*) = 0$ in case of unconventional superconductors, then Eq. (2.119) is transformed into

$$1 = N(E_F)V \int_0^{\omega_c} d\xi \frac{1}{\xi} \tanh \frac{\xi}{2T^*}$$
$$= N(E_F)V \int_0^{\omega_c/(2T^*)} \tanh x d\ln x. \tag{2.120}$$

Assume $\omega_c/(2T^*) \gg 1$, then $\tanh[\omega_c/(2T^*)] \cong 1$. Eq. (2.120) becomes

$$1 \cong N(E_F)V \left[\ln \frac{\omega_c}{2T^*} - \int_0^{\infty} dx \frac{\ln x}{(\mathrm{ch} x)^2} \right]$$
$$= N(E_F)V \left[\ln \frac{\omega_c}{2T^*} + \ln \frac{4\gamma}{\pi} \right], \tag{2.121}$$

where $\gamma = 1.78$. From Eq. (2.121) we have

$$T^* = 1.13 \omega_c e^{-\frac{1}{N(E_F)V}}. \tag{2.122}$$

Reconsidering $\hbar \neq k_B \neq 1$, then

$$T^* = 1.13 \frac{\hbar \omega_c}{k_B} e^{-\frac{1}{N(E_F)V}}. \tag{2.123}$$

Because we use $\omega_c \gg T^*$, which means $N(E_F)V < 1/4$, we call Eq. (2.123) the formula of T^* in weak coupling approximation.

From Eqs. (2.59) and (2.123) we have

$$\frac{2\Delta(0)}{T^*} = 3.53, \tag{2.124}$$

where $\Delta(0)$ is the gap at 0 K. For weak coupling unconventional superconductors the "3.53" is a universal constant. However, for the unconventional superconductors the ratio $2\Delta(0)/T_c$

is meaningless because the value of T_c does not come from the gap equation (See section 3.3.).

For the conventional superconductors the gap, which is proportional to amplitude of wavefunction of a Cooper pair (See section 2.6.1.), and the long-range phase coherence between Cooper pairs appear nearly at the same temperature, i.e., $T^* \approx T_c$. The reasons are as follows. The coherence length ξ_o is the average size of a Cooper pair, i.e., the average distance between the two electrons within a Cooper pair. In the conventional superconductors, such as Sn, Hg, and Al, a Cooper pair has an average size, ξ_o, about 10^4 Å. Assume that $\xi_o = 10^4$ Å and the electron number per cubic centimeter is $n = 10^{22}/\mathrm{cm}^3$. We know that at $T = T^*$ the gap , and thus the number of Cooper pairs, are zero. It is imaginable that at $T = T_c$, which is only a little less than T^*, there are 10^{13} electrons in one cm^3 to become Cooper pairs. In this case, at $T = T_c$ within the space of single Cooper pair there exist 10^1 other electrons, each being correlated in Cooper pairs. Because of the strong overlap of Cooper pairs there is a strong coupling between Cooper pairs, and thus the Cooper pairs in static superconducting state have the same unified phase. The meaning of the long-range phase coherence for the conventional superconductors in static superconducting state is that there is the same unified phase for all Cooper pairs. Therefore, the appearance of the Cooper pairs and the long-range phase coherence among Cooper pairs are nearly at the same temperature. Because the gap is proportional to the amplitude of Cooper pair, the $\Delta(T) > 0$ means nearly the appearance of Cooper pairs with the long-range phase coherence, i.e., the appearance of superconductivity, for the conventional superconductor. However, strictly speaking, the BCS pairing theory in this chapter is T_c-independent in mathematics. For the conventional superconductors, repeating the same derivations as to obtain T^* for the unconventional superconductors, we have obviously

$$T_c \cong T^* = 1.13\omega_c e^{-\frac{1}{N(E_F)V}}, \tag{2.125}$$

and

$$\frac{2\Delta(0)}{T_c} \cong \frac{2\Delta(0)}{T^*} = 3.53, \tag{2.126}$$

where $T_c \cong T^*$ for conventional superconductor comes from the purely physical analysis.

If $V_{\mathbf{k}'\mathbf{k}}$ is approximated by Eq. (2.55), then the $\Delta(T)$ in Eq. (2.118) has a simple non-trivial solution as follows.

$$\frac{1}{N(E_F)V} = \int_0^{\omega_c} d\xi \frac{1}{\sqrt{\xi^2+\Delta^2(T)}} \tanh \frac{\sqrt{\xi^2+\Delta^2(T)}}{2T}$$

$$= \int_0^{\omega_c} d\xi \frac{1}{\sqrt{\xi^2+\Delta^2(T)}} + \int_0^{\omega_c} d\xi \left[\frac{1}{\sqrt{\xi^2+\Delta^2(T)}} \tanh \frac{\sqrt{\xi^2+\Delta^2(T)}}{2T} - \frac{1}{\sqrt{\xi^2+\Delta^2(T)}} \right]. \tag{2.127}$$

For the second term of Eq. (2.127) we make the reduction of fractions to a common denominator. We then obtain

$$\frac{1}{N(E_F)V} = \int_0^{\omega_c} d\xi \frac{1}{\sqrt{\xi^2+\Delta^2(T)}} - \int_0^{\omega_c} d\xi \frac{1}{\sqrt{\xi^2+\Delta^2(T)}} \frac{2}{e^{\sqrt{\xi^2+\Delta^2(T)}/T}+1}. \tag{2.128}$$

If we assume $\omega_c \gg \Delta(0)$, then Eq. (2.128) at 0 K is

$$\frac{1}{N(E_F)V} = \int_0^{\omega_c} d\xi \frac{1}{\sqrt{\xi^2+\Delta^2(0)}}$$

$$\cong \ln\frac{2\omega_c}{\Delta(0)}. \tag{2.129}$$

Substituting Eq. (2.129) into Eq. (2.128) yields

$$\ln\frac{\Delta(0)}{\Delta(T)} \cong 2\int_0^{\infty} d\xi \frac{1}{\sqrt{\xi^2+\Delta^2(T)}} \frac{1}{e^{\sqrt{\xi^2+\Delta^2(T)}/T}+1}. \tag{2.130}$$

Making variable transformation from ξ to E $[=\sqrt{\xi^2+\Delta^2(T)}]$ yields

$$\ln\frac{\Delta(0)}{\Delta(T)} = 2\int_{\Delta}^{\infty} dE \frac{1}{\sqrt{E^2-\Delta^2}} \frac{1}{e^{E/T}+1}. \tag{2.131}$$

Making variable transformation from E to ϕ $(E/\Delta = \mathrm{ch}\phi)$ yields

$$\ln\frac{\Delta(0)}{\Delta(T)} = 2\int_0^{\infty} d\phi \frac{1}{e^{\Delta \mathrm{ch}\phi/T}+1}$$

$$= 2\int_0^{\infty} d\phi e^{-\frac{\Delta \mathrm{ch}\phi}{T}} \frac{1}{1+e^{-\frac{\Delta \mathrm{ch}\phi}{T}}}$$

$$= 2\int_0^{\infty} d\phi e^{-\frac{\Delta \mathrm{ch}\phi}{T}} \sum_{m=0}^{\infty} (-1)^m e^{-m\frac{\Delta \mathrm{ch}\phi}{T}}. \tag{2.132}$$

Using the integral formula in Ref. [40] yields

$$\ln\frac{\Delta(0)}{\Delta(T)} = 2\sum_{m=1}^{\infty} (-1)^{m+1} K_\circ\left(\frac{m\Delta}{T}\right), \tag{2.133}$$

where $K_\circ(x)$ is the zero order Bessel function. The Δ in Eqs. (2.131), (2.132), and (2.133) represents $\Delta(T)$.

At low temperature limit $T \ll T^*$ for unconventional superconductor, $\Delta(T)/T \gg 1$. The asymptotic expansion of $K_\circ(x)$ $(x \to \infty)$ is

$$K_\circ \sim \sqrt{\frac{\pi}{2x}} e^{-x} \left(1 - \frac{1}{8x} + \frac{9}{128x^3}\right). \tag{2.134}$$

$$\ln\frac{\Delta(0)}{\Delta(T)} \cong \sqrt{\frac{2\pi T}{\Delta(T)}} e^{-\frac{\Delta(T)}{T}} \left[1 - \frac{T}{8\Delta(T)}\right]. \tag{2.135}$$

The left hand of Eq. (2.135) is

$$\ln\frac{\Delta(0)}{\Delta(T)} = -\ln\left[1+\frac{\Delta(T)-\Delta(0)}{\Delta(0)}\right]$$
$$\cong -\frac{\Delta(T)-\Delta(0)}{\Delta(0)}. \tag{2.136}$$

Combining Eqs. (2.135) and (2.136) yields

$$\Delta(T) = \Delta(0) - \Delta(0)\sqrt{\frac{2\pi T}{\Delta(T)}}\left[1-\frac{T}{8\Delta(T)}\right]e^{-\frac{\Delta(T)}{T}}. \tag{2.137}$$

Considering low temperature limit, Eq. (2.137) becomes

$$\Delta(T) = \Delta(0) - \sqrt{2\pi T\Delta(0)}\left[1-\frac{T}{8\Delta(0)}\right]e^{-\frac{\Delta(0)}{T}}. \tag{2.138}$$

At high temperature limit $T \approx T^*$ ($T < T^*$), we can also obtain an analytical expression of $\Delta(T)$ for unconventional superconductors. In this case $\Delta(T)/T \ll 1$, we can not use the expansion in Eq. (2.134). Instead we use

$$\frac{\tanh x}{x} = \sum_{n=-\infty}^{\infty}\frac{1}{x^2+[\pi(n+1/2)]^2}. \tag{2.139}$$

Using Eq. (2.139) and setting $\omega_n = (2n+1)\pi T$, Eq. (2.127) becomes

$$\frac{1}{N(E_F)V} = T\sum_{n=-\infty}^{\infty}\int_{-\omega_c}^{\omega_c} d\xi\frac{1}{\omega_n^2+\xi^2+\Delta^2(T)}$$
$$\cong T\sum_{n=-\infty}^{\infty}\int_{-\omega_c}^{\omega_c} d\xi\frac{1}{\omega_n^2+\xi^2}\left[1-\frac{\Delta^2(T)}{\omega_n^2+\xi^2}\right]. \tag{2.140}$$

The summation over n in the first term of Eq. (2.140) can be calculated by the residue theorem, and it is

$$2T\sum_{-\infty}^{\infty}\int_0^{\omega_c} d\xi\frac{1}{\omega_n^2+\xi^2} = \int_0^{\omega_c} d\xi\left(\tanh\frac{\xi}{2T}\right)\frac{1}{\xi}. \tag{2.141}$$

The second term can be calculated as follows.

$$T\sum_n\int_{-\omega_c}^{\omega_c} d\xi\frac{-\Delta^2(T)}{(\omega_n^2+\xi^2)^2} \cong -2T\Delta^2(T)\sum_{n=0}^{\infty}\int_{-\infty}^{\infty} d\xi\frac{1}{(\omega_n^2+\xi^2)^2}$$
$$= -\frac{\Delta^2(T)}{\pi^2T^2}\sum_{n=0}^{\infty}\frac{1}{(2n+1)^3}$$
$$= -\frac{\Delta^2(T)}{\pi^2T^2}\frac{2^3-1}{2^3}\zeta(3), \tag{2.142}$$

where $\zeta(3)$ is Rimann's zeta function. Substituting Eqs. (2.141) and (2.142) into Eq. (2.140), we have

$$\frac{1}{N(E_F)V}=\int_0^{\omega_c}d\xi\frac{1}{\xi}\tanh\frac{\xi}{2T}-\frac{\Delta^2(T)}{\pi^2T^2}\frac{7}{8}\zeta(3). \tag{2.143}$$

Using the same method as to obtain T^* from Eq. (2.120), we obtain from Eq. (2.143)

$$\ln\frac{T}{T^*}=-\frac{7\zeta(3)}{8}\frac{\Delta^2(T)}{(\pi T)^2}. \tag{2.144}$$

Considering

$$\ln\frac{T}{T^*}\cong\frac{T-T^*}{T^*} \tag{2.145}$$

yields

$$\Delta(T)=\pi T^*\sqrt{\frac{8}{7\zeta(3)}}\sqrt{1-\frac{T}{T^*}}\cong 3.06T^*\sqrt{1-\frac{T}{T^*}}. \tag{2.146}$$

At low temperature limit $T \ll T_c$, we have the same equation as Eq. (2.138) for conventional superconductors. At high temperature limit $T \approx T_c$ $(T < T_c)$, we have for conventional superconductors

$$\Delta(T)\cong 3.06T_c\sqrt{1-\frac{T}{T_c}} \tag{2.147}$$

because T_c is nearly equal to T^* for conventional superconductors.

2.5.4. Specific Heat Jump at T*

The specific heat jumps occurred at T^* for unconventional and conventional superconductors are the same. Let us study the unconventional superconductors.

From Eq. (2.113) we know that in the unconventional superconductors there is a quasiparticle excitation with energy $E_{\mathbf{k}}=\sqrt{\xi_{\mathbf{k}}^2+\Delta_{\mathbf{k}}^2(T)}$. For simplicity, in the following we assume that $\Delta_{\mathbf{k}}(T)$ is independent of $\mathbf{k}$. The quasiparticles satisfy the free Fermion distribution $f(E_{\mathbf{k}})=1/[\exp(\beta E_{\mathbf{k}})+1]$. The entropy of this ideal Fermion gas in superconducting or pseudogap state is

$$S_{es}=-2k_B\sum_{\mathbf{k}}\{[1-f(E_{\mathbf{k}})]\ln[1-f(E_{\mathbf{k}})]+f(E_{\mathbf{k}})\ln f(E_{\mathbf{k}})\}. \tag{2.148}$$

The specific heat can be written as

$$C_{es}=T\frac{dS_{es}}{dT}=-\beta\frac{dS_{es}}{d\beta}. \tag{2.149}$$

Substituting Eq. (2.148) into Eq. (2.149) yields

$$C_{es}=2\beta k_B\sum_{\mathbf{k}}-\frac{\partial f(E_{\mathbf{k}})}{\partial E_{\mathbf{k}}}\left(E_{\mathbf{k}}^2+\frac{1}{2}\beta\frac{d\Delta^2}{d\beta}\right). \tag{2.150}$$

At $T = T^*$, the $\Delta(T) \to 0$, and thus the first term of Eq. (2.150) reduces to the electronic specific heat of usual normal state

$$C_{en} = \frac{2\pi^2}{3} N(E_F) k_B^2 T. \tag{2.151}$$

The second term is finite just below T^*, where $d\Delta^2/dT$ is large, but it is zero above T^*, giving rise to a discontinuity of the quasiparticle specific heat at T^*. The size of the discontinuity is

$$(C_{es} - C_{en})|_{T=T^*} = N(E_F) \left(-\frac{d\Delta^2(T)}{dT} \right) \bigg|_{T=T^*}. \tag{2.152}$$

Substituting Eq. (2.146) into Eq. (2.152) yields

$$(C_{es} - C_{en})|_{T=T^*} = (3.06 k_B)^2 N(E_F) T^*. \tag{2.153}$$

Combining Eqs. (2.151) and (2.153), we have

$$\frac{(C_{es} - C_{en})|_{T=T^*}}{C_{en}|_{T=T^*}} = \frac{(3.06)^2}{2\pi^2/3} = 1.43. \tag{2.154}$$

Most people think that the specific heat jump given by Eq. (2.154) takes place just at T_c, and that there is no specific heat jump at T^*, related to the Cooper-bound-pair formation. Even when people discovered the specific heat jump at $T > T_c$ for the high-T_c cuprates, which belong to one kind of the unconventional superconductors, they always attributed it to some cause other than that related to the Cooper-bound-pair formation, or just simply neglected it. Ref. [41] discussed the possible experimental verifications for the specific heat jump at T^* in the high-T_c cuprates. Inderhees *et al.* said that they found a specific heat jump at 89 K for their overdoped $T_c = 89$ K $YBa_2Cu_3O_{7-\delta}$ sample [42]. However, we notice that there is clearly another specific heat jump at 93 K in their data for specific heat versus temperature [42]. Considering the errors in experiment of Ref. [42], the anisotropy of the gap in the high-T_c cuprates, and the approximate property of theoretical value of T^*, it is suggested that the jump observed at 93 K in Ref. [42] is the jump at T^*. Dunlap *et al.* found a specific heat jump at 31 K for their underdoped $T_c = 31$ K $La_{2-x}Sr_xCuO_4$ sample [43]. Dunlap *et al.* also noticed that there is a phase transition at $T \approx 80$ K, which is not transformed in heating cycle. Fig. 2 in Ref. [44] and Fig. 4.23 in this book indicate $T^* = 90$ K for $La_{2-x}Sr_xCuO_4$ with $T_c = 31$ K. Considering the three undetermined factors in experiment and theory, it is suggested that the observed jump at $T \approx 80$ K in Ref. [43] is the jump at T^*. Nagata *et al.* found that in their $La_{2-x}Sr_xCuO_4$ sample there are two specific heat jumps at $T_c = 30$ K and $T^* = 95$ K [45]. If we assume that the maximum value of T_c of the sample in Ref. [45] is 35 K, then from the empirical formula of T_c we obtain the number of $O_{p\sigma}$ hole in one unit cell of Cu^{++} lattice in the CuO_2 plane is $x = 0.118$. From Fig. 4.23 of this book we know $T^* = 95$ K. Therefore, the specific heat jump at T^* in Ref. [45] comes from formation of Cooper pairs without the long-range phase coherence.

It is very difficult to measure the electronic specific heat at high temperature, where T^* usually appears, since the specific heats of the high-T_c cuprates are dominated by phonons [18]. Ref. [44] pointed out that for the n-type cuprate $Nd_{2-x}Ce_xCuO_4$ with $x = 0.12$, $T^* =$ 43 K and $T_c = 0$ K. We think $Nd_{2-x}Ce_xCuO_4$ is the best candidate for verifying the specific heat jump at T^*. From Fig. 4.8 we know that the overdoped high-T_c cuprates (Bi2212 and Y123) are also one of the best candidates for verifying the specific heat jump at T^*.

For the high-T_c cuprates at T_c the long-range phase difference locking among Cooper pairs begins to appear, i.e., the supercurrent Cooper pairs begin to appear (See section 3.3.). Therefore, there is another specific heat jump at T_c for the high-T_c cuprates.

From chapter 1 we know that the conventional superconductors have three characteristic temperatures: T^*, $T_{c,1}$, and T_c. T^* is a temperature at which the Cooper pairs without the long-range phase coherence begin to appear. $T_{c,1}$ is a temperature at which the Cooper pairs with the long-range phase difference locking (one type of the long-range phase coherence) begin to appear. T_c is a temperature at which the Cooper pairs with the unified phase (another type of the long-range phase coherence) begin to appear. Actually, Eq. (2.154) does not depend on the unconventional superconductors. Eq. (2.154) is also correct at T^* for the conventional superconductors. Because the supercurrent Cooper pairs begin to appear at $T_{c,1}$, there is a second order phase transition at $T_{c,1}$, and thus there is a specific heat jump at $T_{c,1}$. According to our theory, there is no second order phase transition at T_c, and thus there is no specific heat jump at T_c. Because the three characteristic temperatures are in a very narrow temperature interval (10^{-3} K), people think that the specific heat jump is at T_c for the conventional superconductors, and do not notice the difference among T^*, $T_{c,1}$, and T_c.

2.6. Application of Quantum Field Theory

2.6.1. Gorkov Equation Set and Gap

The references of section 2.6 are Refs. [46,47]. The effective two-body interaction Hamiltonian H_I is

$$H_I = -\frac{V}{2}\int d\mathbf{r}\psi_\alpha^+(\mathbf{r})\psi_\beta^+(\mathbf{r})\psi_\beta(\mathbf{r})\psi_\alpha(\mathbf{r}), \tag{2.155}$$

where $-V$ is the effective attractive interaction constant, $V > 0$, the double spin subscripts mean spin summation, and $\psi_\alpha(\mathbf{r})$ is the field annihilation operator. The free Hamiltonian is ($\hbar = c = k_B = 1$)

$$H_\circ = \int d\mathbf{r}(-1)\psi_\alpha^+(\mathbf{r})\frac{\nabla^2}{2m^*}\psi_\alpha(\mathbf{r}). \tag{2.156}$$

The total Hamiltonian is

$$H = H_\circ + H_I. \tag{2.157}$$

In the field theory at $T > 0$ K, the field operators in Heisenberg representation are

$$\widetilde{\psi}(x) = e^{\tau(H-\mu N)}\psi(\mathbf{r})e^{-\tau(H-\mu N)}, \tag{2.158}$$

$$\widetilde{\overline{\psi}}(x) = e^{\tau(H-\mu N)}\psi^+(\mathbf{r})e^{-\tau(H-\mu N)}, \tag{2.159}$$

where $x = (\mathbf{r},\tau)$, and $0 < \tau < 1/T = \beta$.

The Green functions $G(x,x')$ and $G(x',x)$, and Gorkov functions $F^+(x,x')$ and $F(x,x')$ are defined as follows.

$$G(x,x') = -< T_\tau\left(\widetilde{\psi}_\uparrow(x)\widetilde{\overline{\psi}}_\uparrow(x')\right)>, \tag{2.160}$$

$$G(x',x) = - < T_\tau \left(\widetilde{\psi}_\downarrow (x') \widetilde{\overline{\psi}}_\downarrow (x) \right) >, \tag{2.161}$$

$$F^+(x,x') = - < T_\tau \left(\widetilde{\overline{\psi}}_\downarrow (x) \widetilde{\overline{\psi}}_\uparrow (x') \right) >, \tag{2.162}$$

$$F(x,x') = - < T_\tau \left(\widetilde{\psi}_\uparrow (x) \widetilde{\psi}_\downarrow (x') \right) >, \tag{2.163}$$

where $< \cdots >$ represents statistical average in grand canonical ensemble, and T_τ is the imaginary-time-ordering operator. If a system is of translational invariant property, then the above four functions depend on $\mathbf{r} - \mathbf{r}'$. From the definitions in Eqs. (2.158) - (2.163) we know that the above four functions depend on $\tau - \tau'$. We introduce a new function $\Delta(\mathbf{r})$, which is defined as

$$\Delta(\mathbf{r}) = VF(\mathbf{r}, \tau + 0^+, \mathbf{r}, \tau). \tag{2.164}$$

We can show that

$$\Delta(\mathbf{r}) = -V < \psi_\uparrow(\mathbf{r}) \psi_\downarrow(\mathbf{r}) > . \tag{2.165}$$

The demonstration for Eq. (2.165) is as follows. Substituting Eqs. (2.163) and (2.158) into Eq. (2.164) yields

$$\begin{aligned} \Delta(\mathbf{r}) &= -V < T_\tau \left(\widetilde{\psi}_\uparrow (\mathbf{r}\tau + 0^+) \widetilde{\psi}_\downarrow (\mathbf{r}\tau) \right) > \\ &= -V < \widetilde{\psi}_\uparrow (\mathbf{r}\tau) \widetilde{\psi}_\downarrow (\mathbf{r}\tau) > \\ &= -V < \psi_\uparrow(\mathbf{r}) \psi_\downarrow(\mathbf{r}) > . \end{aligned} \tag{2.166}$$

From Eq. (2.165) we know that $\Delta(\mathbf{r})$ is proportional to the wavefunction describing the motion of mass-center of a Cooper pair.

Next, we demonstrate that the four equations satisfied by $G(x,x')$, $F^+(x,x')$, $F(x,x')$, and $G(x',x)$ can be symbolically written in the following matrix form.

$$\begin{pmatrix} \left\{ -\frac{\partial}{\partial \tau} + \frac{[\nabla - ie\mathbf{A}(\mathbf{r})]^2}{2m^*} + \mu \right\} & \Delta(\mathbf{r}) \\ \Delta^*(\mathbf{r}) & \left\{ -\frac{\partial}{\partial \tau} - \frac{[\nabla + ie\mathbf{A}(\mathbf{r})]^2}{2m^*} - \mu \right\} \end{pmatrix} \begin{pmatrix} G(x,x') & F(x,x') \\ F^+(x,x') & -G(x',x) \end{pmatrix}$$

$$= \begin{pmatrix} 1 & 0 \\ 0 & 1 \end{pmatrix} \delta(x - x'). \tag{2.167}$$

Eq. (2.167) is the Gorkov equation set at $T > 0$ K. We show, for example,

$$\left\{ \frac{\partial}{\partial \tau} + \frac{[\nabla + ie\mathbf{A}(\mathbf{r})]^2}{2m^*} + \mu \right\} F^+(x,x') - \Delta^*(\mathbf{r}) G(x,x') = 0. \tag{2.168}$$

At first, from the definition of the field operator in Heisenberg representation we have the following equation.

$$\frac{\partial}{\partial \tau} \widetilde{\overline{\psi}}_\downarrow (\mathbf{r}\tau) = \left[\frac{1}{2m^*} (i\nabla - e\mathbf{A})^2 - \mu \right] \widetilde{\overline{\psi}}_\downarrow (\mathbf{r}\tau) - V \widetilde{\overline{\psi}}_\downarrow (\mathbf{r}\tau) \widetilde{\overline{\psi}}_\delta (\mathbf{r}\tau) \widetilde{\psi}_\delta (\mathbf{r}\tau). \tag{2.169}$$

According to the definition in Eq. (2.162)

$$\frac{\partial F^{+}(x,x')}{\partial \tau} = \frac{\partial}{\partial \tau}(-1) < T_{\tau}\left(\widetilde{\overline{\psi}}_{\downarrow}(\mathbf{r}\tau)\,\widetilde{\overline{\psi}}_{\uparrow}(\mathbf{r}'\tau')\right) >$$

$$= -\frac{\partial}{\partial \tau} < \vartheta(\tau-\tau')\,\widetilde{\overline{\psi}}_{\downarrow}(\mathbf{r}\tau)\,\widetilde{\overline{\psi}}_{\uparrow}(\mathbf{r}'\tau') - \vartheta(\tau'-\tau)\,\widetilde{\overline{\psi}}_{\uparrow}(\mathbf{r}'\tau')\,\widetilde{\overline{\psi}}_{\downarrow}(\mathbf{r}\tau) >$$

$$= - < T_{\tau}\left(\frac{\partial\,\widetilde{\overline{\psi}}_{\downarrow}(\mathbf{r}\tau)}{\partial \tau}\,\widetilde{\overline{\psi}}_{\uparrow}(\mathbf{r}'\tau')\right) >, \tag{2.170}$$

where $\vartheta(\tau-\tau')$ is a step function. $\vartheta(\tau-\tau') = 1$ if $(\tau-\tau') > 0$. $\vartheta(\tau-\tau') = 0$ if $(\tau-\tau') < 0$. Substituting Eq. (2.169) into Eq. (2.170) yields

$$\frac{\partial F^{+}(x,x')}{\partial \tau} =$$

$$- < T_{\tau}\left\{\left[\left[\frac{1}{2m^{*}}(i\nabla - e\mathbf{A})^{2} - \mu\right]\widetilde{\overline{\psi}}_{\downarrow}(x) - V\,\widetilde{\overline{\psi}}_{\downarrow}(x)\,\widetilde{\overline{\psi}}_{\delta}(x)\,\widetilde{\psi}_{\delta}(x)\right]\widetilde{\overline{\psi}}_{\uparrow}(x')\right\} >. \tag{2.171}$$

Keeping only the term which has contribution to the pairing state, Eq. (2.171) becomes

$$\left[\frac{\partial}{\partial \tau} - \frac{1}{2m^{*}}(i\nabla - e\mathbf{A})^{2} + \mu\right] F^{+}(x,x')$$

$$-V < T_{\tau}\left(\widetilde{\overline{\psi}}_{\downarrow}(x)\,\widetilde{\overline{\psi}}_{\uparrow}(x)\right) >< T_{\tau}\left(\widetilde{\psi}_{\uparrow}(x)\,\widetilde{\overline{\psi}}_{\uparrow}(x')\right) >= 0, \tag{2.172}$$

where

$$< T_{\tau}\left(\widetilde{\overline{\psi}}_{\downarrow}(x)\,\widetilde{\overline{\psi}}_{\uparrow}(x)\right) >=<\widetilde{\overline{\psi}}_{\downarrow}(x)\,\widetilde{\overline{\psi}}_{\uparrow}(x) > . \tag{2.173}$$

Eq. (2.173) can be rewritten as

$$< T_{\tau}\left(\widetilde{\overline{\psi}}_{\downarrow}(x)\,\widetilde{\overline{\psi}}_{\uparrow}(x)\right) >=< \psi_{\downarrow}^{+}(\mathbf{r})\psi_{\uparrow}^{+}(\mathbf{r}) >=$$

$$< \psi_{\uparrow}(\mathbf{r})\psi_{\downarrow}(\mathbf{r}) >^{*}= \frac{\Delta^{*}(\mathbf{r})}{-V}. \tag{2.174}$$

For the last equality we have used Eq. (2.166). Substituting Eqs. (2.174) and (2.160) into Eq. (2.172) gives the Eq. (2.168).

To show that the $\Delta(\mathbf{r})$ in energy-momentum representation is the gap, we take the following two equations from Eq. (2.167), and neglect the magnetic field. Assume that the system is of translational invariant property. In this case $\Delta(\mathbf{r})$ is independent of $\mathbf{r}$.

$$\left(-\frac{\partial}{\partial \tau} + \frac{\nabla^{2}}{2m^{*}} + \mu\right) G(x-x') + \Delta F^{+}(x-x') = \delta(x-x'). \tag{2.175}$$

$$\left(\frac{\partial}{\partial \tau} + \frac{\nabla^{2}}{2m^{*}} + \mu\right) F^{+}(x-x') - \Delta^{*} G(x-x') = 0. \tag{2.176}$$

Using the following three equations, we can transform Eqs. (2.175) and (2.176) into energy-momentum representation. Assume that the volume of the system is a unit.

$$F^+(x-x') = \frac{1}{(2\pi)^3} T \sum_n e^{-i\omega_n(\tau-\tau')} \int d\mathbf{p} e^{i\mathbf{p}\cdot(\mathbf{r}-\mathbf{r}')} F^+_{\omega_n}(\mathbf{p}), \tag{2.177}$$

$$G(x-x') = \frac{1}{(2\pi)^3} T \sum_n e^{-i\omega_n(\tau-\tau')} \int d\mathbf{p} e^{i\mathbf{p}\cdot(\mathbf{r}-\mathbf{r}')} G_{\omega_n}(\mathbf{p}), \tag{2.178}$$

$$\delta(x-x') = \frac{1}{(2\pi)^3} T \sum_n e^{-i\omega_n(\tau-\tau')} \int d\mathbf{p} e^{i\mathbf{p}\cdot(\mathbf{r}-\mathbf{r}')}. \tag{2.179}$$

The Eqs. (2.175) and (2.176) in the energy-momentum representation are

$$[i\omega_n - \xi(\mathbf{p})]\, G_{\omega_n}(\mathbf{p}) + \Delta F^+_{\omega_n}(\mathbf{p}) = 1, \tag{2.180}$$

$$[i\omega_n + \xi(\mathbf{p})]\, F^+_{\omega_n}(\mathbf{p}) + \Delta^* G_{\omega_n}(\mathbf{p}) = 0. \tag{2.181}$$

The solutions of Eqs. (2.180) and (2.181) are

$$G_{\omega_n}(\mathbf{p}) = -\frac{i\omega_n + \xi(\mathbf{p})}{\omega_n^2 + \xi^2(\mathbf{p}) + |\Delta|^2}, \tag{2.182}$$

$$F^+_{\omega_n}(\mathbf{p}) = \frac{\Delta^*}{\omega_n^2 + \xi^2(\mathbf{p}) + |\Delta|^2}. \tag{2.183}$$

Using Eq. (2.183), we can obtain the expression of Δ^*, and thus know the physical meaning of Δ^*.

$$\Delta^* = VF^+(0^+) = V\frac{T}{(2\pi)^3} \sum_n \int d\mathbf{p} e^{i\mathbf{0}\cdot\mathbf{p}} e^{-i\omega_n 0^+} F^+_{\omega_n}(\mathbf{p}), \tag{2.184}$$

where $\exp(i\mathbf{0}\cdot\mathbf{p}) = \exp(-i\omega_n 0^+) = 1$. Substituting Eq. (2.183) into Eq. (2.184), we have

$$\begin{aligned}\Delta^* &= V\frac{T}{(2\pi)^3} \sum_n \int d\mathbf{p} \frac{\Delta^*}{\omega_n^2 + \xi^2(\mathbf{p}) + |\Delta|^2} \\ &= V\frac{T\Delta^*}{(2\pi)^3} \int d\mathbf{p} \frac{1}{2\pi i} \oint d\omega \frac{1}{\omega^2 + \xi^2(\mathbf{p}) + |\Delta|^2} \frac{1}{e^{\frac{i\omega}{T}} + 1} \frac{-i}{T},\end{aligned} \tag{2.185}$$

where the integration contour is shown in Fig. 2.5. The contour surrounds the first order poles at $\omega = (2n+1)\pi T$ $(n = -\infty, \cdots, -1, 0, +1, \cdots, +\infty)$ on the real axis. To calculate Eq. (2.185), we transform the integration contour in Fig. 2.5 into the integration contour shown in Fig. 2.6. The contours in Fig. 2.6 surround $\omega = i\sqrt{\xi^2(\mathbf{p}) + |\Delta|^2}$ and $\omega = -i\sqrt{\xi^2(\mathbf{p}) + |\Delta|^2}$, respectively. Eq. (2.185) can be rewritten as

$$\Delta^* = V\frac{T}{(2\pi)^3} \int d\mathbf{p} \frac{1}{2\pi i} \frac{i}{T} \oint d\omega \frac{\Delta^*}{\omega - i\sqrt{\xi^2(\mathbf{p}) + |\Delta|^2}} \frac{1}{\omega + i\sqrt{\xi^2(\mathbf{p}) + |\Delta|^2}} \frac{1}{e^{\frac{i\omega}{T}} + 1}. \tag{2.186}$$

Carrying out the contour integration in Eq. (2.186), we obtain

$$\Delta^* = V\frac{T}{(2\pi)^3} \int d\mathbf{p} \frac{\Delta^*}{2T\sqrt{\xi^2(\mathbf{p}) + |\Delta|^2}} \left(\frac{1}{e^{-\frac{\sqrt{\xi^2(\mathbf{p})+|\Delta|^2}}{T}} + 1} - \frac{1}{e^{\frac{\sqrt{\xi^2(\mathbf{p})+|\Delta|^2}}{T}} + 1} \right)$$

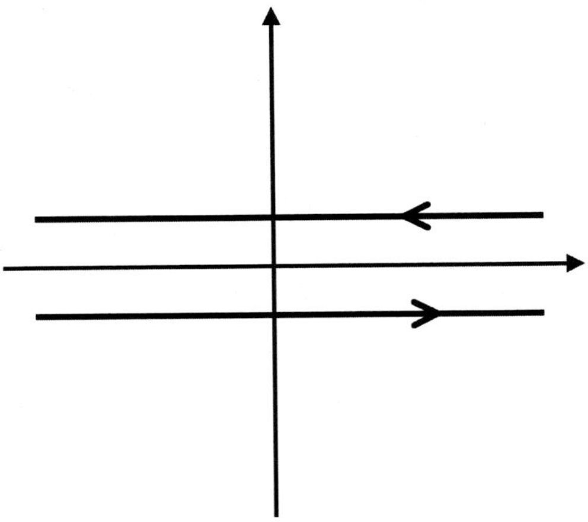

Figure 2.5. The integration contour of Eq. (2.185) in the complex-ω plane.

$$= \frac{V}{(2\pi)^3} \int d\mathbf{p} \frac{\Delta^*}{2\sqrt{\xi^2(\mathbf{p}) + |\Delta|^2}} \tanh \frac{\sqrt{\xi^2(\mathbf{p}) + |\Delta|^2}}{2T}. \tag{2.187}$$

Eq. (2.187) is just the famous BCS gap equation, Eq. (2.118). Therefore, we realize that when $\Delta(\mathbf{r})$ of Eq. (2.164) is transformed into energy-momentum representation, the Δ represents the BCS gap of quasiparticle excitation. Considering that $\Delta(\mathbf{r})$ is proportional to the wavefunction describing the motion of mass-center of a Cooper pair, the amplitude of wavefunction of a Cooper pair and the gap are often used synonymously.

Here, we have to mention the gapless superconductivity. The energy gap in the quasiparticle excitation spectrum of a superconducting state can be reduced by a number of factors; for example, incorporation of magnetic impurities. It is found that magnetic impurities lower the superconducting transition temperature as well as reduce the energy gap, but the superconducting transition temperature can still be above zero at some impurity concentration while the energy gap is reduced to zero. This case is called gapless superconductivity. The gapless superconductivity indicates that the superconducting state does not manifest itself by the existence of a gap in the excitation spectrum, but rather by the presence of condensed Cooper pairs with the long-range phase coherence. For the gapless superconductivity, the gap is not proportional to the amplitude of a Cooper pair. We do not concern the gapless superconductivity in all other paragraphs of this book.

2.6.2. Microscopic Derivation of Ginzburg-Landau Equations

Let us at first discuss the conventional superconductor at $T \approx T_c$ ($T < T_c$). Assume that the field is weak. Note that at $T \approx T_c$ the gap is small for conventional superconductor (See section 2.5.3.). (However, the gap at T_c is generally not small for unconventional superconductor.) We will show that for conventional superconductor the order parameter in GL theory is just proportional to $\Delta(\mathbf{r})$, which is proportional to the mass-center wavefunction

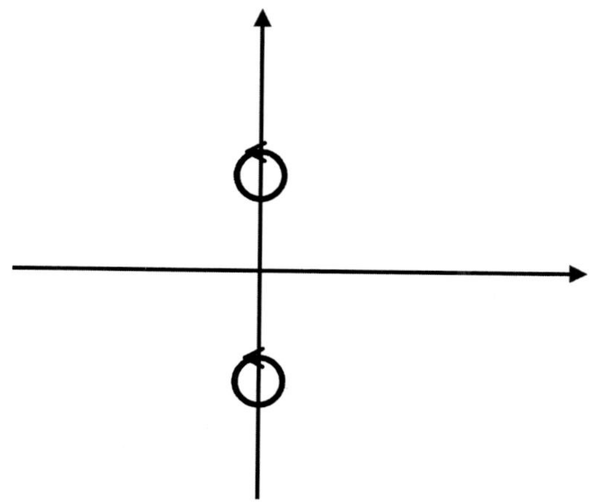

Figure 2.6. The integration contour of Eq. (2.186) in the complex-ω plane.

of a Cooper pair. The Fourier transformation of Eq. (2.167) with respect to $\tau-\tau'$ is

$$\begin{pmatrix} \left\{i\omega_n - \frac{[i\nabla+e\mathbf{A}(\mathbf{r})]^2}{2m^*}+\mu\right\} & \Delta(\mathbf{r}) \\ \Delta^*(\mathbf{r}) & \left\{i\omega_n + \frac{[i\nabla-e\mathbf{A}(\mathbf{r})]^2}{2m^*}-\mu\right\} \end{pmatrix} \begin{pmatrix} G_{\omega_n}(\mathbf{r},\mathbf{r}') & F_{\omega_n}(\mathbf{r},\mathbf{r}') \\ F^+_{\omega_n}(\mathbf{r},\mathbf{r}') & -G_{-\omega_n}(\mathbf{r}',\mathbf{r}) \end{pmatrix}$$

$$= \begin{pmatrix} 1 & 0 \\ 0 & 1 \end{pmatrix} \delta(\mathbf{r}-\mathbf{r}'), \tag{2.188}$$

where

$$\Delta^*(\mathbf{r}) = VT\sum_n F^+_{\omega_n}(\mathbf{r},\mathbf{r}), \tag{2.189}$$

$$\Delta(\mathbf{r}) = VT\sum_n F_{\omega_n}(\mathbf{r},\mathbf{r}). \tag{2.190}$$

Assuming in the same magnetic field as in the superconducting state, from Eq. (2.188) we know that a normal metal has the following equations.

$$\left\{i\omega_n + \frac{1}{2m^*}\left[\nabla_\mathbf{r} - ie\mathbf{A}(\mathbf{r})\right]^2+\mu\right\} \overset{\sim(\circ)}{G}_{\omega_n}(\mathbf{r},\mathbf{r}') = \delta(\mathbf{r}-\mathbf{r}'), \tag{2.191}$$

$$\left\{i\omega_n - \frac{1}{2m^*}\left[\nabla_\mathbf{r} + ie\mathbf{A}(\mathbf{r})\right]^2-\mu\right\} (-1)\overset{\sim(\circ)}{G}_{-\omega_n}(\mathbf{r}',\mathbf{r}) = \delta(\mathbf{r}-\mathbf{r}'). \tag{2.192}$$

The equation satisfied by $G_{\omega_n}(\mathbf{r},\mathbf{r}')$ is as follows.

$$G_{\omega_n}(\mathbf{r},\mathbf{r}') = \frac{1}{i\omega_n + \frac{1}{2m^*}\left[\nabla_\mathbf{r} - ie\mathbf{A}(\mathbf{r})\right]^2+\mu}\left[\delta(\mathbf{r}-\mathbf{r}') - \Delta(\mathbf{r})F^+_{\omega_n}(\mathbf{r},\mathbf{r}')\right]. \tag{2.193}$$

Substituting Eq. (2.191) into Eq. (2.193) two times, we have

$$G_{\omega_n}(\mathbf{r},\mathbf{r}') = \overset{\sim(\circ)}{G}_{\omega_n}(\mathbf{r},\mathbf{r}') - \frac{\int \delta(\mathbf{r}-\mathbf{l})\Delta(\mathbf{l})F^+_{\omega_n}(\mathbf{l},\mathbf{r}')d\mathbf{l}}{i\omega_n + \frac{1}{2m^*}\left[\nabla_\mathbf{r} - ie\mathbf{A}(\mathbf{r})\right]^2+\mu}$$

$$=\overset{\sim(\circ)}{G_{\omega_n}}(\mathbf{r},\mathbf{r}') - \int \overset{\sim(\circ)}{G_{\omega_n}}(\mathbf{r},\mathbf{l})\Delta(\mathbf{l})F^{+}_{\omega_n}(\mathbf{l},\mathbf{r}')d\mathbf{l}. \tag{2.194}$$

By the same technique as to obtain Eq. (2.194) we have

$$F^{+}_{\omega_n}(\mathbf{r},\mathbf{r}') = \int \overset{\sim(\circ)}{G_{-\omega_n}}(\mathbf{l},\mathbf{r})\Delta^*(\mathbf{l})G_{\omega_n}(\mathbf{l},\mathbf{r}')d\mathbf{l}. \tag{2.195}$$

We can show that the Green function of normal metal without field, $G^{(\circ)}_{\omega_n}(\mathbf{r}-\mathbf{r}')$, is

$$G^{(\circ)}_{\omega_n}(\mathbf{r}-\mathbf{r}') = \begin{cases} -\frac{m^*}{2\pi R}e^{ip_F R-\frac{|\omega_n|}{v_F}R}, & \omega_n > 0, \\ -\frac{m^*}{2\pi R}e^{-ip_F R-\frac{|\omega_n|}{v_F}R}, & \omega_n < 0, \end{cases} \tag{2.196}$$

where $R = |\mathbf{R}| = |\mathbf{r}-\mathbf{r}'|$. The demonstration for Eq. (2.196) is as follows.

$$G^{(\circ)}_{\omega_n}(\mathbf{r}-\mathbf{r}') = \frac{1}{(2\pi)^3}\int d\mathbf{p}e^{i\mathbf{p}\cdot(\mathbf{r}-\mathbf{r}')}G^{(\circ)}_{\omega_n}(\mathbf{p})$$

$$= \frac{1}{(2\pi)^3}\int d\mathbf{p}e^{i\mathbf{p}\cdot(\mathbf{r}-\mathbf{r}')}\frac{1}{i\omega_n-\xi(\mathbf{p})}. \tag{2.197}$$

After integration over θ and ϕ we have

$$G^{(\circ)}_{\omega_n}(\mathbf{r}-\mathbf{r}') = \frac{1}{(2\pi)^2}\frac{1}{iR}\int \frac{1}{i\omega_n-\xi(\mathbf{p})}\left(e^{ipR}-e^{-ipR}\right)pdp. \tag{2.198}$$

Considering $\xi(\mathbf{p}) = p^2/(2m^*) - p_F^2/(2m^*) \cong v_F(p-p_F)$, Eq. (2.198) becomes

$$G^{(\circ)}_{\omega_n}(\mathbf{r}-\mathbf{r}') = \frac{m^*}{(2\pi)^2 iR}\int_{-\infty}^{\infty} d\xi \frac{e^{iR\left(\frac{\xi}{v_F}+p_F\right)} - e^{-iR\left(\frac{\xi}{v_F}+p_F\right)}}{i\omega_n-\xi}. \tag{2.199}$$

Using contour integration method, Eq. (2.199) becomes Eq. (2.196).

For $\overset{\sim(\circ)}{G_{\omega_n}}(\mathbf{r},\mathbf{r}')$ we try the following solution.

$$\overset{\sim(\circ)}{G_{\omega_n}}(\mathbf{r},\mathbf{r}') = e^{i\phi(\mathbf{r},\mathbf{r}')}G^{(\circ)}_{\omega_n}(\mathbf{r}-\mathbf{r}'). \tag{2.200}$$

Substituting Eq. (2.200) into Eq. (2.191), using the equation satisfied by $G^{(\circ)}_{\omega_n}$, and considering that the field is weak and varies slowly in space, and that the electron circular radius is much larger than the penetration depth, we obtain an approximate equation

$$\left[\nabla_{\mathbf{r}}\phi(\mathbf{r},\mathbf{r}') - e\mathbf{A}(\mathbf{r})\right]\cdot(\mathbf{r}-\mathbf{r}') = 0. \tag{2.201}$$

Because $(\mathbf{r}-\mathbf{r}')$ is arbitrary, so

$$\nabla_{\mathbf{r}}\phi(\mathbf{r},\mathbf{r}') - e\mathbf{A}(\mathbf{r}) = 0. \tag{2.202}$$

It is easy to verify that the solution of Eq. (2.202) is

$$\phi(\mathbf{r},\mathbf{r}') = e\mathbf{A}(\mathbf{r})\cdot(\mathbf{r}-\mathbf{r}'). \tag{2.203}$$

Considering that the gap Δ^* is small at $T < T_c$ and $T \approx T_c$, we obtain from Eq. (2.195)

$$F^{+}_{\omega_n}(\mathbf{l},\mathbf{r}') = \int \overset{\sim(\circ)}{G}_{-\omega_n}(\mathbf{m},\mathbf{l})\Delta^*(\mathbf{m})\overset{\sim(\circ)}{G}_{\omega_n}(\mathbf{m},\mathbf{r}')d\mathbf{m}. \tag{2.204}$$

Substituting Eq. (2.204) into Eq. (2.194) yields

$$G_{\omega_n}(\mathbf{r},\mathbf{r}') = \overset{\sim(\circ)}{G}_{\omega_n}(\mathbf{r},\mathbf{r}') -$$

$$\int\int \overset{\sim(\circ)}{G}_{\omega_n}(\mathbf{r},\mathbf{l})\Delta(\mathbf{l})\overset{\sim(\circ)}{G}_{-\omega_n}(\mathbf{m},\mathbf{l})\Delta^*(\mathbf{m})\overset{\sim(\circ)}{G}_{\omega_n}(\mathbf{m},\mathbf{r}')d\mathbf{m}d\mathbf{l}. \tag{2.205}$$

Substituting Eq. (2.205) into Eqs. (2.195) and (2.189), we have

$$\Delta^*(\mathbf{r}) = VT\sum_n \int \overset{\sim(\circ)}{G}_{\omega_n}(\mathbf{l},\mathbf{r})\Delta^*(\mathbf{l})\overset{\sim(\circ)}{G}_{-\omega_n}(\mathbf{l},\mathbf{r})d\mathbf{l}$$

$$-VT\sum_n \int\int\int \overset{\sim(\circ)}{G}_{\omega_n}(\mathbf{l},\mathbf{m})\Delta(\mathbf{m})\overset{\sim(\circ)}{G}_{\omega_n}(\mathbf{s},\mathbf{r})\Delta^*(\mathbf{s})\overset{\sim(\circ)}{G}_{-\omega_n}(\mathbf{s},\mathbf{m})$$

$$\Delta^*(\mathbf{l})\overset{\sim(\circ)}{G}_{-\omega_n}(\mathbf{l},\mathbf{r})d\mathbf{l}d\mathbf{m}d\mathbf{s}. \tag{2.206}$$

At first we make the summation over n in the first term of Eq. (2.206). Substituting Eqs. (2.196), (2.200), and (2.203) into the first term, we have

$$VT\sum_n \overset{\sim(\circ)}{G}_{-\omega_n}(\mathbf{l},\mathbf{r})\overset{\sim(\circ)}{G}_{\omega_n}(\mathbf{l},\mathbf{r}) = e^{2ie\mathbf{A}(\mathbf{r})\cdot(\mathbf{l}-\mathbf{r})}K_\circ(\mathbf{l}-\mathbf{r}). \tag{2.207}$$

$$K_\circ(\mathbf{l}-\mathbf{r}) = VT\sum_n \left(\frac{m^*}{2\pi R}\right)^2 e^{-\frac{2|\omega_n|R}{v_F}}$$
$$= \frac{VTm^{*2}}{(2\pi R)^2}\frac{1}{\sinh\left(\frac{2\pi TR}{v_F}\right)}, \tag{2.208}$$

where $R = |\mathbf{R}| = |\mathbf{l}-\mathbf{r}|$. Substituting Eqs. (2.207) and (2.208) into the first term, I, of Eq. (2.206), setting $\mathbf{R} = \mathbf{l}-\mathbf{r}$, and making expansions of weak field and small Δ, we obtain

$$I = \int K_\circ(\mathbf{R})e^{2ie\mathbf{A}(\mathbf{r})\cdot\mathbf{R}}\Delta^*(\mathbf{R}+\mathbf{r})d\mathbf{R}$$

$$= \int K_\circ(\mathbf{R})\left\{1 + 2ie\mathbf{A}(\mathbf{r})\cdot\mathbf{R} - 2e^2[\mathbf{A}(\mathbf{r})\cdot\mathbf{R}]^2\right\}$$

$$\left\{1 + \mathbf{R}\cdot\nabla + \frac{1}{2}(\mathbf{R}\cdot\nabla)^2\right\}\Delta^*(\mathbf{r})d\mathbf{R}. \tag{2.209}$$

Considering that the integration of $\mathbf{R}$ over $\mathbf{R}$ is zero, and $\mathbf{A}(\mathbf{r})$ and $\nabla\Delta^*(\mathbf{r})$ are small, we obtain

$$I = \Delta^*(\mathbf{r})\int K_\circ(R)d\mathbf{R} + \frac{1}{6}(\nabla + 2ie\mathbf{A})^2\Delta^*(\mathbf{r})\int K_\circ(R)R^2 d\mathbf{R}. \tag{2.210}$$

Let us calculate the integrals containing $K_\circ(R)$ in Eq. (2.210).

$$\int K_\circ(R)d\mathbf{R} = V\int d\mathbf{R}T\sum_n G^{(\circ)}_{\omega_n}(\mathbf{R})G^{(\circ)}_{-\omega_n}(\mathbf{R})$$

$$= V\int d\mathbf{R}T\sum_n \int\int \frac{d\mathbf{p}d\mathbf{p}'}{(2\pi)^6}\frac{e^{i(\mathbf{p}\cdot\mathbf{R}+\mathbf{p}'\cdot\mathbf{R})}}{[i\omega_n - \xi(\mathbf{p})]\,[-i\omega_n - \xi(\mathbf{p}')]}$$

$$= VT\sum_n \int \frac{d\mathbf{p}}{(2\pi)^3}\frac{1}{\omega_n^2+\xi^2(\mathbf{p})}$$

$$= \frac{V}{2(2\pi)^3}\int d\mathbf{p}\frac{1}{\xi(\mathbf{p})}\tanh\left[\frac{\xi(\mathbf{p})}{2T}\right]$$

$$= \frac{m^* p_F}{2\pi^2}V\int_0^{\omega_c} d\xi\frac{1}{\xi}\tanh\frac{\xi}{2T}$$

$$= VN(E_F)\ln\frac{2\gamma\omega_c}{\pi T}, \tag{2.211}$$

where $\gamma = 1.78$, and the density of states at Fermi surface $N(E_F) = m^* p_F/(2\pi^2)$. We cut the divergent integration over ξ at $|\xi| = \omega_c$ as usual. Considering

$$T_c = \frac{2\omega_c\gamma}{\pi}e^{-\frac{1}{N(E_F)V}}, \tag{2.212}$$

and $T \approx T_c$, Eq. (2.211) can be approximately equal to

$$\int K_\circ(R)d\mathbf{R} = 1 + N(E_F)V\ln\left(1+\frac{T_c - T}{T}\right)$$

$$\cong 1 + N(E_F)V\frac{T_c - T}{T_c} \tag{2.213}$$

Substituting Eq. (2.208) into the second integral of Eq. (2.210), we have

$$\int K_\circ R^2 d\mathbf{R} = \frac{Vm^{*2}T}{(2\pi)^2}\int d\mathbf{R}\frac{1}{\sinh\left(\frac{2\pi TR}{v_F}\right)}$$

$$= \frac{Vm^{*2}T}{\pi}\int_0^\infty dR\frac{R^2}{\sinh\left(\frac{2\pi TR}{v_F}\right)}$$

$$= \frac{2Vm^{*2}T}{\pi}\int_0^\infty dRR^2\sum_{n=0}^\infty e^{-\frac{2\pi TR}{v_F}(2n+1)}$$

$$= \frac{4Vm^{*2}T}{\pi}\sum_{n=0}^\infty\left[\frac{v_F}{4\pi T(2n+1)}\right]^3$$

$$= \frac{7\zeta(3)Vv_F^2 N(E_F)}{8(\pi T)^2}, \tag{2.214}$$

where $\zeta(3)$ is Riemann Zeta function. In comparison of the second term with the first term in Eq. (2.210), the second term is smaller because the second term contains the weak field $\mathbf{A}(\mathbf{r})$ and the variation of $\Delta^*(\mathbf{r})$. Therefore, the T^2 in Eq. (2.214) can be taken approximately as T_c^2. Substituting Eqs. (2.213) and (2.214) into Eq. (2.210) yields

$$I = \Delta^*(\mathbf{r}) + \Delta^*(\mathbf{r})N(E_F)V\frac{T_c - T}{T_c}$$

$$+\frac{1}{6}(\nabla + 2ie\mathbf{A})^2\Delta^*(\mathbf{r})\frac{7\zeta(3)v_F^2 N(E_F)V}{8(\pi T_c)^2}. \tag{2.215}$$

The second term of Eq. (2.206) has $\Delta(\mathbf{m})\Delta^*(\mathbf{s})\Delta^*(\mathbf{l})$ which is very small for conventional superconductor when $T \approx T_c$. Therefore, the Δ and $\tilde{G}^{(o)}_{\omega_n}$ may take zero-order approximation. The second term II becomes

$$II = -|\Delta(\mathbf{r})|^2\Delta^*(\mathbf{r})VT\sum_n \int\int\int d\mathbf{l}d\mathbf{m}d\mathbf{s}$$

$$G^{(o)}_{\omega_n}(\mathbf{l}-\mathbf{m})G^{(o)}_{\omega_n}(\mathbf{s}-\mathbf{r})G^{(o)}_{-\omega_s}(\mathbf{s}-\mathbf{m})G^{(o)}_{-\omega_n}(\mathbf{l}-\mathbf{r}). \tag{2.216}$$

Expressing $G^{(o)}_{\omega_n}$ by Fourier expansion and carrying out the space integration, we have

$$II = -VT|\Delta(\mathbf{r})|^2\Delta^*(\mathbf{r})\sum_n \int \frac{d\mathbf{k}}{(2\pi)^3}\frac{1}{[\omega_n^2 + \xi^2(\mathbf{k})]^2}$$

$$= -N(E_F)V\frac{7\zeta(3)}{8}\frac{1}{(\pi T)^2}|\Delta(\mathbf{r})|^2\Delta^*(\mathbf{r}). \tag{2.217}$$

Because $T \approx T_c$, the T in Eq. (2.217) can be taken as T_c.
Substituting Eqs. (2.215) and (2.217) into Eq. (2.206), we have

$$\Delta^*(\mathbf{r})N(E_F)V\frac{T_c - T}{T} + \frac{1}{6}[\nabla + 2ie\mathbf{A}(\mathbf{r})]^2\Delta^*(\mathbf{r})\frac{7\zeta(3)v_F^2}{8(\pi T_c)^2}N(E_F)V$$

$$-N(E_F)V\frac{7\zeta(3)}{8(\pi T_c)^2}|\Delta(\mathbf{r})|^2\Delta^*(\mathbf{r}) = 0. \tag{2.218}$$

Taking out the $N(E_F)V$, noting $E_F = m^*v_F^2/2$, and introducing $\eta = 7\zeta(3)E_F/[6(\pi T_c)^2]$, Eq. (2.218) becomes

$$\left\{\frac{1}{4m^*}[\nabla + 2ie\mathbf{A}(\mathbf{r})]^2 + \frac{1}{\eta}\left[\frac{T_c - T}{T_c} - \frac{7\zeta(3)}{8(\pi T_c)^2}|\Delta(\mathbf{r})|^2\right]\right\}\Delta^*(\mathbf{r}) = 0. \tag{2.219}$$

Taking complex conjugation of Eq. (2.219), and introducing

$$\psi(\mathbf{r}) = \sqrt{\frac{7\zeta(3)n}{8(\pi T_c)^2}}\Delta(\mathbf{r}), \tag{2.220}$$

Eq. (2.219) becomes

$$\left\{-\frac{1}{4m^*}[\nabla - 2ie\mathbf{A}(\mathbf{r})]^2 + \frac{1}{\eta}\left[-\frac{T_c - T}{T_c} + \frac{1}{n}|\psi(\mathbf{r})|^2\right]\right\}\psi(\mathbf{r}) = 0, \tag{2.221}$$

where $n = k_F^3/(3\pi^2)$ is the electron density. The first equation of GL equation set in section 1.2.3 is ($\hbar = c = k_B = 1$)

$$-\frac{1}{2m^{**}}\left[\nabla - ie^*\mathbf{A}(\mathbf{r})\right]^2\psi(\mathbf{r}) + \alpha\psi(\mathbf{r}) + \beta|\psi(\mathbf{r})|^2\psi(\mathbf{r}) = 0. \tag{2.222}$$

Making comparison between Eqs. (2.222) and (2.221), we have that m^{**} and e^* are the mass and charge of the Cooper electron pair, $m^{**} = 2m^*$, $e^* = 2e$, and

$$\alpha = -\frac{1}{\eta}\frac{T_c - T}{T_c} = -\frac{6\pi^2 T_c}{7\zeta(3)E_F}(T_c - T), \tag{2.223}$$

$$\beta = \frac{6\pi^2 T_c^2}{7\zeta(3)E_F n}. \tag{2.224}$$

From similar derivation as to obtain the first GL equation, we have the following second GL equation [48].

$$\mathbf{J}(\mathbf{r}) = \frac{e^*}{i2m^{**}}\left[\psi^*(\mathbf{r})\nabla\psi(\mathbf{r}) - \psi(\mathbf{r})\nabla\psi^*(\mathbf{r})\right] - \frac{e^{*2}}{m^{**}}|\psi(\mathbf{r})|^2\mathbf{A}(\mathbf{r}), \tag{2.225}$$

where $m^{**} = 2m^*$ and $e^* = 2e$. The supercurrent carrier looks like a quasiparticle with charge $2e$ ($e < 0$) and mass $2m^*$.

To obtain GL equation set, Ginsburg and Landau assumed that there is an order parameter $\psi(\mathbf{r})$ which is a unified effective wavefunction for all supercurrent Cooper pairs, and thus the phases of all supercurrent Cooper pairs are of long-range phase coherence. What does the long-range phase coherence mean for a supercurrent-carrying state? The phase factor of a supercurrent Cooper pair consists of two parts. For the static superconducting state the phases of all Cooper pairs in conventional superconductor are the same (See section 2.5.3.), and thus every Cooper pair has the same phase factor $\exp(i\phi)$, where ϕ is the phase of a Cooper pair, and is independent of space and time. For a supercurrent-carrying state the centers of mass of all supercurrent Cooper pairs have the same momentum $\hbar\mathbf{q}$ (See section 2.8.), where $\mathbf{q}$ is a wave vector, and move as independent traveling waves. The phase factor of the traveling wave has the form of $\exp(i\mathbf{q}\cdot\mathbf{r})$. Therefore, the total phase factor of a supercurrent Cooper pair is $\exp[i(\phi + \mathbf{q}\cdot\mathbf{r})]$. This phase factor extends over indefinitely large distance, i.e., long-range phase coherence. In fact, the wave travels unscattered through the whole volume of the superconducting material. In case of a persistent current in a ring, therefore, the long-range phase coherence extends over centimeters; in the case of the winding of a superconducting solenoid over miles. Note that the ϕ in the phase factor is also important for the long-range phase coherence in supercurrent-carrying state. If the ϕ is arbitrary for every Cooper pair, then the average value of the wavefunctions of many Cooper pairs at any space point is zero, and thus there is no macroscopic supercurrent in the system.

However, for unconventional superconductors the cases are completely different in the following two aspects. First, at $T \approx T_c$ ($T < T_c$) the gap is, generally, not a small quantity. The gap is a small quantity only at $T \approx T^*$ ($T < T^*$). So, all the small gap approximations in the above microscopic derivation for conventional superconductor can not be made at $T \approx T_c$ ($T < T_c$) for unconventional superconductor. Second, section 3.3 shows that the amplitude of the order parameter in GL theory is a very complex function of the gap at $T \approx T_c$ ($T < T_c$). Therefore, the microscopic derivations of GL equation set at $T \approx T_c$ ($T < T_c$) for unconventional superconductors are still lacking. Note that the phenomenological GL equation, satisfied by supercurrent Cooper pairs, is still correct for the unconventional superconductors at $T \approx T_c$ ($T < T_c$) because the supercurrent Cooper pair, i.e., the order parameter in GL theory, begins to appear at T_c (See section 3.3.), and thus the number of supercurrent Cooper pairs at $T \approx T_c$ and $T < T_c$ is a small quantity.

From the above microscopic derivations for the GL equation set we know that the order parameter $\psi(\mathbf{r})$ of GL theory is proportional to the gap function $\Delta(\mathbf{r})$ [See Eq. (2.220).], and $\Delta(\mathbf{r})$ is proportional to the wavefunction describing the motion of mass-center of a single Cooper pair [See Eqs. (2.166), (2.184), and(2.190).]. Therefore, in principle, the GL equation set is an equation set satisfied by a single Cooper pair. Because there is the long-range phase coherence between Cooper pairs in the conventional superconductors at $T < T_c$, the wavefunction of Cooper pair can be normalized to total number of supercurrent Cooper pairs, and thus $|\psi(\mathbf{r})| = \sqrt{n_s(\mathbf{r})}$ [See Eq. (1.21).], where $n_s(\mathbf{r})$ is the number density of supercurrent Cooper pairs.

From section 3.3 we know that for the high-T_c cuprates, one kind of the unconventional superconductor, every Cooper pair has its own phase at $T_c < T < T^*$. The wavefunction of a single Cooper pair in the pseudogap state can be normalized to 1, and thus $|\psi(\mathbf{r})| = \sqrt{p(\mathbf{r})}$, where $p(\mathbf{r})$ is the probability density of a Cooper pair. $p(\mathbf{r})$ is a small quantity at $T \approx T^*$ ($T < T^*$). We may imitate the above microscopic derivations at $T \approx T_c$ ($T < T_c$) for conventional superconductors to obtain GL-like equation at $T \approx T^*$ ($T < T^*$) for any single Cooper pair in the pseudogap state, which should be

$$-\frac{1}{2m^{**}}\left[\nabla - ie^*\mathbf{A}(\mathbf{r})\right]^2\psi(\mathbf{r}) + \alpha_p\psi(\mathbf{r}) + \beta_p|\psi(\mathbf{r})|^2\psi(\mathbf{r}) = 0, \tag{2.226}$$

where $m^{**} = 2m^*$ and

$$\alpha_p = -\frac{6\pi^2 T^*}{7\zeta(3)E_F}(T^* - T), \tag{2.227}$$

$$\beta_p = \frac{6\pi^2 T^{*2}}{7\zeta(3)E_F n}, \tag{2.228}$$

$$\psi(\mathbf{r}) = \sqrt{\frac{7\zeta(3)n}{8(\pi T^*)^2}}\Delta(\mathbf{r}). \tag{2.229}$$

The equation of electric current density of one Cooper pair in the pseudogap state is the same as Eq. (2.225)

We may imitate the section 1.2.3 to discuss the case involving an inhomogeneous distribution, generated by the presence of boundary, of the Cooper pairs in the pseudogap state. Assume that the pseudogap state occupies the half space $x > 0$. We further assume that $|\psi(0)| = 0$ at the interface. Repeating the derivations in section 1.2.3 we have

$$|\psi(x)| = \left(\frac{|\alpha_p|}{\beta_p}\right)^{1/2}\tanh\left(\frac{x}{\sqrt{2}\xi_{GLp}}\right), \tag{2.230}$$

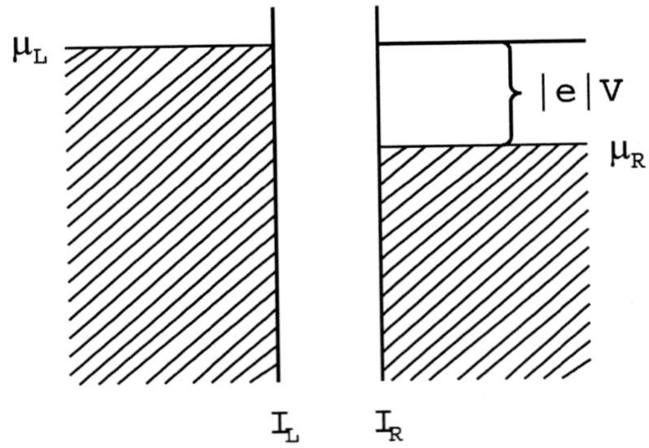

Figure 2.7. Tunneling junction. The two conductors (electrodes) are at the two sides of insulating layer. V represents bias. I_L (I_R) represents the interface between left (right) side conductor and insulating layer. μ_L (μ_R) represents the chemical potential at left (right) side conductor.

$$\xi^2_{GLp} = \frac{\hbar^2}{2m^{**}|\alpha_p|}. \tag{2.231}$$

From Eqs. (2.230) and (2.231) we see that ξ_{GLp} is the distance over which the amplitude of Cooper pair is from a constant in the interior of material to zero at the boundary of the material.

That any single Cooper pair in the pseudogap state satisfies its own GL equation set is called microscopic superconductivity. The word "superconductivity" in all literatures before this book means macroscopic superconductivity. Therefore, the idea "microscopic superconductivity" proposed in this section and sections 2.7.4 and 2.8 has significance in both theory and application (See section 4.2.3.). As we know, the macroscopic superconductivity has a great influence on many fields of physics. The microscopic superconductivity will do so as well.

2.7. Theory of Electron Tunneling

2.7.1. General Theory of Tunneling

Electron tunneling between two conductors is shown in Fig. 2.7. The total Hamiltonian of this tunneling is [49]

$$H = H_R + H_L + H_T, \tag{2.232}$$

where the H_T is the tunneling Hamiltonian, and it is [49]

$$H_T = \sum_{\mathbf{kp}\sigma} \left[T_{\mathbf{kp}} c^+_{\mathbf{k}\sigma} c_{\mathbf{p}\sigma} + h.c. \right]. \tag{2.233}$$

$c^+_{\mathbf{k}\sigma}$ and $c_{\mathbf{k}\sigma}$ belong to the right side of the insulating layer. $c^+_{\mathbf{p}\sigma}$ and $c_{\mathbf{p}\sigma}$ belong to the left side. The operators on two sides are independent. Let N_L denote the total number of electrons on the left side. The total tunneling current $I(t)$ is [50]

$$I(t) = |e| < \frac{dN_L}{dt} >, \tag{2.234}$$

where $< \cdots >$ means statistical average. We define the direction of current from left side to right side to be positive. We use the interaction representation in the field theory, where H_T is treated as interaction, and unperturbed Hamiltonian is $H_\circ = H_R + H_L$. From linear response theory [50, 51], we have

$$I(t) = -|e| i \int_{-\infty}^{t} dt' < \left[\frac{dN_L(t)}{dt}, H_T(t') \right] >_\circ, \tag{2.235}$$

where $< \cdots >_\circ$ denotes the expected value in the unperturbed system.

$$H_T(t') = e^{iH_\circ t'} H_T e^{-iH_\circ t'}. \tag{2.236}$$

If we define

$$K = H_R - \mu_R N_R + H_L - \mu_L N_L, \tag{2.237}$$

then

$$H_T(t') = e^{iKt'} e^{i(\mu_R N_R + \mu_L N_L)t'} H_T e^{-i(\mu_R N_R + \mu_L N_L)t'} e^{-iKt'}. \tag{2.238}$$

Substituting Eq. (2.233) into Eq. (2.238) and using $\exp(i\mu_L N_L t') c_{\mathbf{p}} \exp(-i\mu_L N_L t') = \exp(-i\mu_L t')$ etc., we have

$$H_T(t') = e^{iKt'} \sum_{\mathbf{k}'\mathbf{p}'\sigma'} \left[e^{i(\mu_R - \mu_L)t'} T_{\mathbf{k}'\mathbf{p}'} c^+_{\mathbf{k}'\sigma'} c_{\mathbf{p}'\sigma'} + h.c. \right] e^{-iKt'}. \tag{2.239}$$

Considering $\mu_R - \mu_L = -|e|V$ in Fig. 2.7 and using $c^+_{\mathbf{k}'\sigma'}(t') = \exp[i(H_R - \mu_R N_R)t'] c^+_{\mathbf{k}'\sigma'} \exp[-i(H_R - \mu_R N_R)t']$ etc., we have

$$H_T(t') = \sum_{\mathbf{k}'\mathbf{p}'\sigma'} \left[e^{-i|e|Vt'} T_{\mathbf{k}'\mathbf{p}'} c^+_{\mathbf{k}'\sigma'}(t') c_{\mathbf{p}'\sigma'}(t') + e^{i|e|Vt'} T^*_{\mathbf{k}'\mathbf{p}'} c^+_{\mathbf{p}'\sigma'}(t') c_{\mathbf{k}'\sigma'}(t') \right]. \tag{2.240}$$

$$\begin{aligned} \frac{dN_L(t)}{dt} &= e^{iH_\circ t} \frac{dN_L}{dt} e^{-iH_\circ t} = e^{iH_\circ t} i[H, N_L] e^{-iH_\circ t} \\ &= e^{iH_\circ t} i[H_T, N_L] e^{-iH_\circ t}. \end{aligned} \tag{2.241}$$

Making the same derivations as for Eq. (2.240), we have

$$\frac{dN_L(t)}{dt} = i \sum_{\mathbf{k}\mathbf{p}\sigma} \left[e^{-i|e|Vt} T_{\mathbf{k}\mathbf{p}} c^+_{\mathbf{k}\sigma}(t) c_{\mathbf{p}\sigma}(t) - e^{i|e|Vt} T^*_{\mathbf{k}\mathbf{p}} c^+_{\mathbf{p}\sigma}(t) c_{\mathbf{k}\sigma}(t) \right]. \tag{2.242}$$

Substituting Eqs. (2.240) and (2.242) into Eq. (2.235) gives

$$I(t) = -i^2 |e| \int_{-\infty}^{t} dt' < \left[\sum_{\mathbf{k}\mathbf{p}\sigma} \left[T_{\mathbf{k}\mathbf{p}} e^{-i|e|Vt} c^+_{\mathbf{k}\sigma}(t) c_{\mathbf{p}\sigma}(t) - T^*_{\mathbf{k}\mathbf{p}} e^{i|e|Vt} c^+_{\mathbf{p}\sigma}(t) c_{\mathbf{k}\sigma}(t) \right], \right.$$

$$\sum_{\mathbf{k}'\mathbf{p}'\sigma'}\left[T_{\mathbf{k}'\mathbf{p}'}e^{-i|e|Vt'}c^{+}_{\mathbf{k}'\sigma'}(t')c_{\mathbf{p}'\sigma'}(t')+T^{*}_{\mathbf{k}'\mathbf{p}'}e^{i|e|Vt'}c^{+}_{\mathbf{p}'\sigma'}(t')c_{\mathbf{k}'\sigma'}(t')\right]\Bigg] >_{\circ}. \qquad (2.243)$$

Define

$$A(t)=\sum_{\mathbf{kp}\sigma}T_{\mathbf{kp}}c^{+}_{\mathbf{k}\sigma}(t)c_{\mathbf{p}\sigma}(t). \qquad (2.244)$$

Eq. (2.243) becomes

$$I(t)=|e|\int_{-\infty}^{+\infty}dt'\theta(t-t')\Big\{e^{i|e|V(t'-t)}<[A(t),A^{+}(t')]>_{\circ}$$
$$-e^{-i|e|V(t'-t)}<[A^{+}(t),A(t')]>_{\circ}+e^{-i|e|V(t+t')}<[A(t),A(t')]>_{\circ}$$
$$-e^{i|e|V(t+t')}<[A^{+}(t),A^{+}(t')]>_{\circ}\Big\}, \qquad (2.245)$$

where $\theta(t-t')$ is a step function. The Eq. (2.245) contains two terms I_p and I_{pp}. Define

$$I_p=|e|\int_{-\infty}^{+\infty}dt'\theta(t-t')\Big\{e^{i|e|V(t'-t)}<[A(t),A^{+}(t')]>_{\circ}$$
$$-e^{-i|e|V(t'-t)}<[A^{+}(t),A(t')]>_{\circ}\Big\} \qquad (2.246)$$

and

$$I_{pp}=|e|\int_{-\infty}^{+\infty}dt'\theta(t-t')\Big\{e^{-i|e|V(t'+t)}<[A(t),A(t')]>_{\circ}$$
$$-\,e^{i|e|V(t'+t)}<[A^{+}(t),A^{+}(t')]>_{\circ}\Big\}. \qquad (2.247)$$

I_p represents the single particle tunneling current. I_{pp} represents the tunneling current of pairing particles. Considering that $<[A(t),A^{+}(t')]>_{\circ}$ is only dependent on $(t-t')$, and making variable transformation $t-t'=t''\to t$, I_p becomes

$$I_p=|e|\int_{-\infty}^{+\infty}dt\theta(t)\Big\{e^{-i|e|Vt}<[A(t),A^{+}(0)]>_{\circ}-e^{i|e|Vt}<[A^{+}(t),A(0)]>_{\circ}\Big\}. \qquad (2.248)$$

Define a retarded correlation function

$$\chi_{ret}(t)=-i\theta(t)<[A(t),A^{+}(0)]>_{\circ}. \qquad (2.249)$$

Then I_p can be expressed as

$$I_p=i|e|\left\{\int_{-\infty}^{+\infty}dte^{-i|e|Vt}\chi_{ret}(t)-\int_{-\infty}^{+\infty}dt\left[e^{-i|e|Vt}\chi_{ret}(t)\right]^{*}\right\}$$
$$=-2|e|Im\left[\chi_{ret}(|e|V)\right], \qquad (2.250)$$

where *Im* means taking imaginary part. Define a correlation function

$$\chi(i\omega_n) = -\int_0^\beta d\tau e^{i\omega_n\tau} < T_\tau[A(\tau)A^+(0)] >_\circ, \tag{2.251}$$

where $\beta = 1/T$, and the Matsubara frequency $\omega_n = 2n\pi T$ ($k_B = 1$) since the correlation function has pairs of fermion operators. Making analytical continuation $i\omega_n \to |e|V + i0^+$, we obtain $\chi_{ret}(|e|V)$. Eq. (2.250) becomes

$$I_p = -2|e|Im\left[\chi(|e|V + i0^+)\right]. \tag{2.252}$$

From Eqs. (2.251) and (2.244) we have

$$\chi(i\omega_n) = -\sum_{\mathbf{kp}\sigma\mathbf{k'p'}\sigma'} T_{\mathbf{kp}}T^*_{\mathbf{k'p'}}\int_0^\beta d\tau e^{i\omega_n\tau} < T_\tau c^+_{\mathbf{k}\sigma}(\tau)c_{\mathbf{p}\sigma}(\tau)c^+_{\mathbf{p'}\sigma'}c_{\mathbf{k'}\sigma'} >_\circ. \tag{2.253}$$

Considering that the operators of right and left sides of the insulating layer are independent, we have

$$\chi(i\omega_n) = \sum_{\mathbf{kp}\sigma}|T_{\mathbf{kp}}|^2\int_0^\beta d\tau e^{i\omega_n\tau} < T_\tau c_{\mathbf{k}\sigma}c^+_{\mathbf{k}\sigma}(\tau) >_\circ < T_\tau c_{\mathbf{p}\sigma}(\tau)c^+_{\mathbf{p}\sigma} >_\circ. \tag{2.254}$$

Introducing the Green functions at right and left sides of the junction [50, 51], Eq. (2.254) becomes

$$\begin{aligned}\chi(i\omega_n) &= \sum_{\mathbf{kp}\sigma}|T_{\mathbf{kp}}|^2\int_0^\beta d\tau e^{i\omega_n\tau}G_{R\sigma}(\mathbf{k},-\tau)G_{L\sigma}(\mathbf{p},\tau)\\ &= \sum_{\mathbf{kp}\sigma}|T_{\mathbf{kp}}|^2\frac{1}{\beta}\sum_{\omega'_n}G_{R\sigma}(\mathbf{k},i\omega'_n - i\omega_n)G_{L\sigma}(\mathbf{p},i\omega'_n).\end{aligned} \tag{2.255}$$

2.7.2. N-I-N Tunneling Junction

Here N and I represent normal metal and insulator, respectively. For normal metal the Green functions in Eq. (2.255) are

$$G_{R\sigma}(\mathbf{k},i\omega'_n - i\omega_n) = \frac{1}{i\omega'_n - i\omega_n - \xi(\mathbf{k})}, \tag{2.256}$$

$$G_{L\sigma}(\mathbf{p},i\omega'_n) = \frac{1}{i\omega'_n - \xi(\mathbf{p})}, \tag{2.257}$$

where $\xi(\mathbf{k}) = \hbar^2k^2/(2m^*) - \mu_R$ and $\xi(\mathbf{p}) = \hbar^2p^2/(2m^*) - \mu_L$.

Substituting Eqs. (2.256) and (2.257) into Eq. (2.255) gives

$$\chi(i\omega_n) = 2\sum_{\mathbf{kp}}|T_{\mathbf{kp}}|^2\frac{1}{\beta}\sum_{\omega'_n}\frac{1}{i\omega'_n - i\omega_n - \xi(\mathbf{k})}\frac{1}{i\omega'_n - \xi(\mathbf{p})}$$

$$= 2\sum_{\mathbf{kp}} |T_{\mathbf{kp}}|^2 \frac{f[\xi(\mathbf{k})] - f[\xi(\mathbf{p})]}{i\omega_n + \xi(\mathbf{k}) - \xi(\mathbf{p})}, \tag{2.258}$$

where f represents the Fermi distribution, and the factor "2" comes from summation over spins.

Substituting Eq. (2.258) into Eq. (2.252) gives

$$I_p = 4\pi|e| \sum_{\mathbf{kp}} |T_{\mathbf{kp}}|^2 \delta[|e|V + \xi(\mathbf{k}) - \xi(\mathbf{p})]\{f[\xi(\mathbf{k})] - f[\xi(\mathbf{p})]\}. \tag{2.259}$$

Making the approximate transformation from summation to integration $\sum_{\mathbf{k}} \cdots \to N_R(E_F) \int d\xi(\mathbf{k}) \cdots$ etc. and assuming that $|T_{\mathbf{kp}}|^2$ is a constant $|T_{con}|^2$, Eq. (2.259) becomes

$$I_p = 4\pi|e| N_R(E_F) N_L(E_F) |T_{con}|^2 \int_{-E_F}^{\infty} d\xi(\mathbf{k}) \{f[\xi(\mathbf{k})] - f[\xi(\mathbf{k}) + |e|V]\}. \tag{2.260}$$

The lower limit "$-E_F$" corresponds to $|\mathbf{k}| = 0$. At $T = 0$ K

$$I_p = 4\pi|e| N_R(E_F) N_L(E_F) |T_{con}|^2 \int_{-|e|V}^{0} d\xi(\mathbf{k})$$

$$= 4\pi|e|^2 N_R(E_F) N_L(E_F) |T_{con}|^2 V$$

$$= \sigma_\circ V, \tag{2.261}$$

$$\sigma_\circ = 4\pi|e|^2 N_R(E_F) N_L(E_F) |T_{con}|^2, \tag{2.262}$$

where $\sigma_\circ$ is conductance (inverse resistance). $I_p > 0$ indicates that the direction of electron flow is from the right side of insulator to the left side of insulator, and thus the current direction is from the left to the right.

From Eq. (2.261) we know that the N-I-N junction behaves as a simple resister with a conductance $\sigma_\circ$. Its I-V characteristic is shown in Fig. 2.8.

2.7.3. Josephson Tunneling

Josephson tunneling means tunneling of supercurrent Cooper pairs. Josephson tunneling junction consists of left superconductor in superconducting state (S), insulating layer (I), and right superconductor in superconducting state (S). The Josephson tunneling current can be obtained from Eq. (2.247).

$$I_{pp} = |e| \int_{-\infty}^{+\infty} dt' \theta(t - t') \Big\{ e^{-i|e|V(t'+t)} < [A(t), A(t')] >_\circ$$

$$- e^{i|e|V(t'+t)} < [A^+(t), A^+(t')] >_\circ \Big\}$$

$$= |e| \Bigg\{ e^{-2i|e|Vt} \int_{-\infty}^{+\infty} dt' \theta(t - t') e^{i|e|V(t-t')} < [A(t), A(t')] >_\circ$$

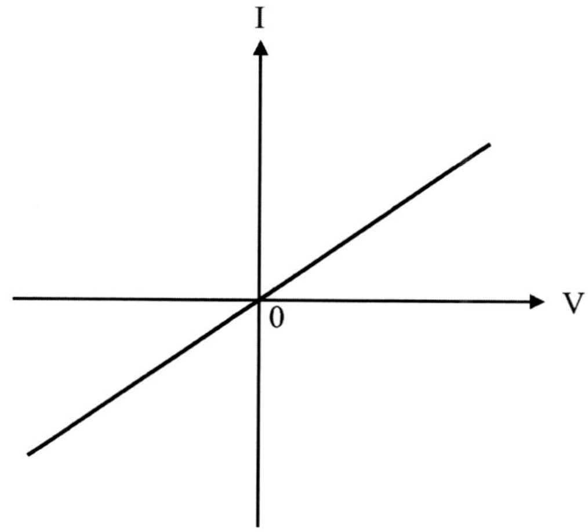

Figure 2.8. The I-V characteristic of N-I-N junction.

$$- e^{2i|e|Vt} \int_{-\infty}^{+\infty} dt' \theta(t-t') e^{-i|e|V(t-t')} < [A^+(t), A^+(t')] >_\circ \Bigg\} . \tag{2.263}$$

To calculate Eq. (2.263) we introduce a retarded correlation function

$$\Phi_{ret}(t-t') = -i\theta(t-t') < [A(t), A(t')] >_\circ . \tag{2.264}$$

Eq. (2.263) becomes

$$I_{pp} = -2|e| Im \left[e^{-2i|e|Vt} \Phi_{ret}(|e|V) \right], \tag{2.265}$$

where

$$\Phi_{ret}(|e|V) = \int_{-\infty}^{+\infty} e^{-i|e|Vt} \Phi_{ret}(t) dt.$$

Let us calculate the retarded correlation function $\Phi_{ret}(|e|V)$. As usual, we start from calculating the corresponding correlation function $\Phi(i\omega_n)$.

$$\Phi(i\omega_n) = -\int_0^{\beta} d\tau e^{i\omega_n \tau} < T_\tau A(\tau) A(0) >_\circ . \tag{2.266}$$

Substituting the definition of A in Eq. (2.244) into Eq. (2.266), and introducing Gorkov F functions, we have

$$\Phi(i\omega_n) = 2 \sum_{\mathbf{kp}} T_{\mathbf{kp}} T_{-\mathbf{k}-\mathbf{p}} \frac{1}{\beta} \sum_{\omega_n'} F^+(\mathbf{k}, i\omega_n') F(\mathbf{p}, i\omega_n' - i\omega_n). \tag{2.267}$$

The factor "2" in the right side of Eq. (2.267) comes from spin summation. $T_{-\mathbf{k}-\mathbf{p}} = T^*_{\mathbf{kp}}$ [52]. $F^+(\mathbf{k}, i\omega'_n)$ is given by Eq. (2.183). The summation over frequency can be obtained by the usual method of contour integral. The result is

$$\Phi(i\omega_n) = \sum_{\mathbf{kp}} \frac{\Delta_L(\mathbf{p})\Delta^*_R(\mathbf{k})}{2} \frac{|T_{\mathbf{kp}}|^2}{E_{\mathbf{k}}E_{\mathbf{p}}} \left\{ [1 - f(E_{\mathbf{p}}) - f(E_{\mathbf{k}})] \left(\frac{1}{i\omega_n + E_{\mathbf{p}} + E_{\mathbf{k}}} \right. \right.$$

$$\left. \left. - \frac{1}{i\omega_n - E_{\mathbf{p}} - E_{\mathbf{k}}} \right) + [f(E_{\mathbf{p}}) - f(E_{\mathbf{k}})] \left(\frac{1}{i\omega_n + E_{\mathbf{p}} - E_{\mathbf{k}}} - \frac{1}{i\omega_n - E_{\mathbf{p}} + E_{\mathbf{k}}} \right) \right\}. \quad (2.268)$$

From Eqs. (2.220) and (2.229), we know that the gap is proportional to the wavefunction of Cooper pair. Therefore, the gap is a complex function as well as the wavefunction.

$$\Delta_L(\mathbf{p}) = |\Delta_L(\mathbf{p})| e^{i\phi_L}. \quad (2.269)$$

$$\Delta^*_R(\mathbf{k}) = |\Delta_R(\mathbf{k})| e^{i\phi_R}. \quad (2.270)$$

The ϕ_L (ϕ_R) represents the unified phase of supercurrent Cooper pairs at left (right) side of the insulating layer for the conventional superconductors. The ϕ_L (ϕ_R) represents the phase of supercurrent Cooper pairs at or near the left (right) interface (See Fig. 2.7.) for the unconventional superconductors. Assume that there is phase difference locking between ϕ_L and ϕ_R.

If we assume $T = 0$, then $f(E_{\mathbf{k}}) = f(E_{\mathbf{p}}) = 0$. Assume that the superconductors on the two sides of the insulating layer are same. Therefore, $|\Delta_L(\mathbf{p})| = |\Delta_R(\mathbf{k})|$.

Considering

$$\Phi_{ret}(|e|V) = \Phi(i\omega_n \to |e|V + i0^+), \quad (2.271)$$

Eq. (2.265) becomes

$$I_{pp} = -2|e| Im \left[e^{-2i|e|Vt} e^{i(\phi_L - \phi_R)} \sum_{\mathbf{kp}} \frac{|\Delta_R(\mathbf{k})||\Delta_L(\mathbf{p})|}{2} \frac{|T_{\mathbf{kp}}|^2}{E_{\mathbf{k}}E_{\mathbf{p}}} \right.$$

$$\left. \left(\frac{1}{|e|V + i0^+ + E_{\mathbf{p}} + E_{\mathbf{k}}} - \frac{1}{|e|V + i0^+ - E_{\mathbf{p}} - E_{\mathbf{k}}} \right) \right]. \quad (2.272)$$

Assume $|\Delta_R(\mathbf{k})| = |\Delta_L(\mathbf{p})| = \Delta$, where Δ is a real number, and $|T_{\mathbf{kp}}|^2 = |T_{con}|^2$. If $|\Delta_R(\mathbf{k})|$ and $|\Delta_L(\mathbf{p})|$ are anisotropic, then we take Δ as the average on Fermi surface, which is a good approximation [53]. Make transformation from summation to integration. We obtain

$$I_{pp} = -2|e| \frac{\Delta^2}{2} |T_{con}|^2 Im \left[e^{-2i|e|Vt} e^{-i(\phi_R - \phi_L)} N_L(E_F) N_R(E_F) \right.$$

$$\left. \int_{-\mu_R}^{\infty} d\xi(\mathbf{k}) \int_{-\mu_L}^{\infty} d\xi(\mathbf{p}) \frac{1}{E_{\mathbf{k}}E_{\mathbf{p}}} \left(\frac{1}{|e|V + E_{\mathbf{p}} + E_{\mathbf{k}} + i0^+} - \frac{1}{|e|V - E_{\mathbf{p}} - E_{\mathbf{k}} + i0^+} \right) \right]. \quad (2.273)$$

Make the following variable transformations

$$\int_{-\mu_R}^{\infty} d\xi(\mathbf{k}) = 2 \int_{0}^{\infty} d\xi(\mathbf{k}) = 2 \int_{\Delta}^{\infty} dE \frac{d\xi(\mathbf{k})}{dE}, \quad (2.274)$$

$$E = \sqrt{\xi^2(\mathbf{k}) + \Delta^2}, \tag{2.275}$$

$$\int_{-\mu_L}^{\infty} d\xi(\mathbf{p}) = 2\int_{0}^{\infty} d\xi(\mathbf{p}) = 2\int_{\Delta}^{\infty} dE' \frac{d\xi(\mathbf{p})}{dE'}, \tag{2.276}$$

$$E' = \sqrt{\xi^2(\mathbf{p}) + \Delta^2}. \tag{2.277}$$

Note that $\delta(|e|V + E_{\mathbf{p}} + E_{\mathbf{k}}) = 0$ because $|e|V > 0$, $E_{\mathbf{p}} \geq \Delta$, and $E_{\mathbf{k}} \geq \Delta$. For simplicity, we assume $|e|V < 2\Delta$. Because $E_{\mathbf{p}} \geq \Delta$ and $E_{\mathbf{p}} \geq \Delta$, we have $\delta(|e|V - E_{\mathbf{p}} - E_{\mathbf{k}}) = 0$. Considering all the above factors, and substituting Eq. (2.262) into Eq. (2.273), we have

$$I_{pp} = \frac{\sigma_\circ \Delta^2}{|e|\pi} \int_{\Delta}^{\infty} \frac{dE}{E} \frac{d\xi(\mathbf{k})}{dE} \int_{\Delta}^{\infty} \frac{dE'}{E'} \frac{d\xi(\mathbf{p})}{dE'} \left(\frac{1}{|e|V + E + E'} - \frac{1}{|e|V - E - E'} \right)$$

$$\sin(\phi_R - \phi_L + 2|e|Vt). \tag{2.278}$$

Note that if we do not assume $|e|V < 2\Delta$, then I_{pp} will contain two expressions. One is given by Eq. (2.278) and another is

$$-2|e| \sum_{\mathbf{kp}} \frac{|\Delta_R(\mathbf{k})||\Delta_L(\mathbf{p})|}{2} \frac{|T_{\mathbf{kp}}|^2}{E_{\mathbf{k}} E_{\mathbf{p}}} \delta(|e|V - E_{\mathbf{k}} - E_{\mathbf{p}}) \cos(\phi_R - \phi_L + 2|e|Vt).$$

Performing the integration in Eq. (2.278) at $|e|V < 2\Delta$ gives

$$I_{pp} = \frac{\sigma_\circ}{|e|} \Delta K \left(\frac{|e|V}{2\Delta} \right) \sin(\phi_R - \phi_L + 2|e|Vt), \tag{2.279}$$

where $K(|e|V/(2\Delta))$ is elliptical integral. If $|e|V > 2\Delta$, then the elliptical integral is an imaginary number. Eq. (2.279) is the Eq. (1.3). Note that we use unit $\hbar = 1$ in Eq. (2.279). At $V = 0$

$$\begin{aligned} I_{pp} &= \frac{\sigma_\circ \pi \Delta}{2|e|} \sin(\phi_R - \phi_L) \\ &= I_c \sin(\phi_R - \phi_L), \end{aligned} \tag{2.280}$$

which is the Eq. (1.2). I_{pp} is the Josephson tunneling supercurrent in S-I-S junction. The I-V characteristic of Josephson tunneling is shown in Fig. 2.9.

The easiest way to evaluate the amplitude of the Josephson tunneling supercurrent at finite temperature is to use a different sequence of steps. In Eq. (2.267) we perform the integration over **k** and **p** before the summation over frequency.

Using

$$\int_{-\infty}^{+\infty} d\xi(\mathbf{k}) F^+(\mathbf{k}, i\omega'_n) = \int_{-\infty}^{+\infty} d\xi(\mathbf{k}) \frac{\Delta e^{-i\phi_R}}{\xi^2(\mathbf{k}) + \Delta^2 + \omega'^2_n} = \frac{\pi \Delta e^{-i\phi_R}}{\sqrt{\Delta^2 + \omega'^2_n}}, \tag{2.281}$$

we have

$$\Phi(i\omega_n) = 2\pi^2 \Delta^2 |T_{con}|^2 e^{-i(\phi_R - \phi_L)} N_L(E_F) N_R(E_F)$$

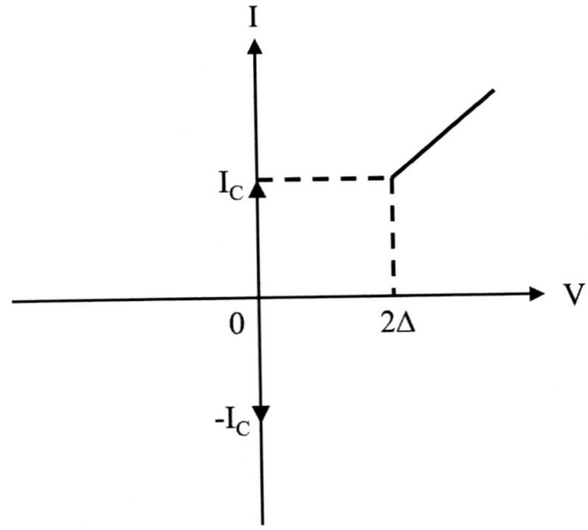

Figure 2.9. The I-V characteristic of Josephson tunneling. The line rising abruptly at $|e|V = 2\Delta$ is the single-particle tunneling curve in Fig. 2.8. The arrows on the vertical axis shows that a supercurrent can flow with no voltage at $I < I_c$. According to our theory, there is an oscillatory current at $V > 0$, which can not be shown in this dc I-V characteristic.

$$\frac{1}{\beta}\sum_{\omega_n'}\frac{1}{\sqrt{\Delta^2+\omega_n'^2}\sqrt{\Delta^2+(\omega_n'-\omega_n)^2}}. \tag{2.282}$$

Assume $V = 0$, i.e., we neglect the influence of V on amplitude of I_{pp}. $i\omega_n \rightarrow |e|V + i0^+ \rightarrow i0^+$. Eq. (2.282) becomes

$$\Phi(\omega_n \rightarrow 0^+) = 2\pi^2\Delta^2|T_{con}|^2 e^{-i(\phi_R-\phi_L)}N_L(E_F)N_R(E_F)\frac{1}{\beta}\sum_{\omega_n'}\frac{1}{\Delta^2+\omega_n'^2}. \tag{2.283}$$

After summation over frequency Eq. (2.283) becomes

$$\Phi(\omega_n \rightarrow 0^+) = 2\pi^2\Delta^2|T_{con}|^2 e^{-i(\phi_R-\phi_L)}N_L(E_F)N_R(E_F)\frac{1}{2\Delta}\tanh\frac{\Delta}{2T}$$

$$= \Phi_{ret}(|e|V = 0). \tag{2.284}$$

Substituting Eq. (2.284) into Eq. (2.265) gives

$$I_{pp} = -2|e|Im\left[e^{-2i|e|Vt}\Phi_{ret}(|e|V = 0)\right]$$

$$= \frac{\sigma_o\pi\Delta}{2|e|}\tanh\frac{\Delta}{2T}\sin(\phi_R - \phi_L + 2|e|Vt). \tag{2.285}$$

Eq. (2.285) is the Eq. (1.3). Eq. (2.285) shows that even at $T > 0$ K the bias-induced oscillatory current still exists. In Eq. (2.285) we have not yet consider the influence of voltage on the amplitude of Josephson tunneling current.

2.7.4. Bias-Induced Oscillatory Current in P-I-P Junction

Section 1.1.5 emphasized that according to our physical mechanism, the ac Josephson effect does not need that the Cooper pairs in the two electrodes of the junction have the long-range phase coherence and that phase difference across the insulating layer is locked. According to this mechanism, section 1.1.6 predicted that a novel ac effect, like the ac Josephson effect, can occur in a series of weakly coupled systems, such as P-I-P junction. Here, P represents a material in the pseudogap state. This section will give a microscopic theory for the bias-induced oscillatory current in P-I-P junction. The theory in this section can be found in [54, 55].

I_{pp} in Eq. (2.247) represents the tunneling current of pairing electrons. Of course, we can use Eq. (2.247) to calculate the tunneling current of Cooper pairs without the long-range phase coherence. Section 2.7.3 has used Eq. (2.247) to evaluate the tunneling current of Cooper pairs with the long-range phase coherence. For the Cooper pairs without the long-range phase coherence we still have Eqs. (2.263) - (2.268). However, we do not have Eqs. (2.269) and (2.270). They becomes in this case

$$\Delta_L(\mathbf{p}) = |\Delta_L(\mathbf{p})|e^{i\phi_{L,arbitrary}}, \tag{2.286}$$

$$\Delta_R(\mathbf{k}) = |\Delta_R(\mathbf{k})|e^{i\phi_{R,arbitrary}}, \tag{2.287}$$

because the phase arbitrariness of the Cooper pairs in the pseudogap state the ϕ_L in Eq. (2.269) and ϕ_R in Eq. (2.270) become $\phi_{L,arbitrary}$ and $\phi_{R,arbitrary}$, respectively. Therefore, Eq. (2.272) becomes

$$I_{pp} = -2|e|Im\left[e^{-2i|e|Vt}e^{i(\phi_{L,arbitrary}-\phi_{R,arbitrary})}\sum_{\mathbf{kp}}\frac{|\Delta_L(\mathbf{p})||\Delta_R(\mathbf{k})|}{2}\right.$$

$$\left.\frac{|T_{\mathbf{kp}}|^2}{E_{\mathbf{k}}E_{\mathbf{p}}}\left(\frac{1}{|e|V+i0^+ +E_{\mathbf{p}}+E_{\mathbf{k}}}-\frac{1}{|e|V+i0^+ -E_{\mathbf{p}}-E_{\mathbf{k}}}\right)\right]. \tag{2.288}$$

Making the same derivations as from Eq. (2.272) to Eq. (2.279), we obtain at $T = 0$ K and $|e|V < 2\Delta$

$$I_{pp} = \frac{\sigma_o}{|e|}\Delta K\left(\frac{|e|V}{2\Delta}\right)\sin\left(\phi_{R,arbitrary} - \phi_{L,arbitrary} + 2|e|Vt\right). \tag{2.289}$$

Eq. (2.289) shows that the bias-induced oscillatory current should occur in the P-I-P junction. Note that if $|e|V > 2\Delta$, then I_{pp} will contain two expressions. The reason is the same as that in section 2.7.3.

Making the same derivations as for Eq. (2.285), we obtain at $T > 0$ K

$$I_{pp} = \frac{\sigma_o\pi\Delta}{2|e|}\tanh\frac{\Delta}{2T}\sin(\phi_{R,arbitrary} - \phi_{L,arbitrary} + 2|e|Vt). \tag{2.290}$$

At $V = 0$ and $T = 0$ we have critical current $I_c = \sigma_o\Delta K(0)/|e|$. At $V = 0$ and $T > 0$ we have critical current $I_c = \sigma_o\pi\Delta\tanh[\Delta/(2T)]/(2|e|)$. Because these critical currents occur at $V = 0$, they are superconducting critical currents. Because $\phi_{R,arbitrary}$ and $\phi_{L,arbitrary}$ are arbitrary, the average value of I_{pp} on macroscopic scale is zero. Therefore, I_{pp} in Eqs.

(2.289) and (2.290) represents microscopic supercurrent or microscopic superconductivity of Cooper pairs without the long-range phase coherence. Note that one can not use Eqs. (2.289) and (2.290) to conclude $\Delta = 0$ if (macroscopic) supercurrent is zero, because the macroscopic average value of $\sin(\phi_{R,arbitrary} - \phi_{L,arbitrary})$ is also zero. Although the bias-induced oscillatory current in P-I-P junction is difficult to be found, the corresponding external radiation with frequency $2|e|V/\hbar$ should be easy to be found. The bias-induced oscillatory current is without loss, and thus it also belongs to microscopic superconductivity.

By the way, Josephson asserted that in equations similar to Eqs. (2.279) and (2.285) $(\phi_R - \phi_L)$ as a whole may be arbitrary [56]. Ref. [57] tried to demonstrate the validity of Josephson's assertion. The current in Ref. [57] is a fourth order effect of the tunneling Hamiltonian, while the current in Ref. [56] is a second order effect. No further discussion on the Josephson's assertion is found in the published literature. In Eqs. (2.289) and (2.290) the current is a second order effect because $\sigma_\circ$ is proportional to $|T_{\mathbf{kq}}|^2$ [See Eq. (2.262).], and ϕ_R and ϕ_L are arbitrary respectively. Therefore, Josephson's assertion is only a special case of Eqs. (2.289) and (2.290).

2.7.5. N-I-S Tunneling Junction

Here N, I, and S represent normal metal, insulator, and superconductor in superconducting or pseudogap state, respectively. We shall take the right side of insulator as superconductor and the left side of insulator as normal metal [50]. The following derivations are dependent on existence of the gap or pseudogap, and are independent of the long-range phase coherence of Cooper pairs. In section 2.7.2 we take the Green functions in Eq. (2.255) as free particle Green functions. In this section we take them as exact Green functions. In this case the summation in Eq. (2.255) becomes

$$S = \frac{1}{\beta}\sum_{\omega_n'} G_{R\sigma}(\mathbf{k}, i\omega_n' - i\omega_n) G_{L\sigma}(\mathbf{p}, i\omega_n')$$

$$= \frac{1}{\beta}\sum_{\omega_n'} \frac{1}{i\omega_n' - i\omega_n - \xi(\mathbf{k}) - \Sigma(\mathbf{k}, i\omega_n' - i\omega_n)} \frac{1}{i\omega_n' - \xi(\mathbf{p}) - \Sigma(\mathbf{p}, i\omega_n')}, \tag{2.291}$$

where Σ represents proper self-energy, $\omega_n' = (2n+1)\pi T$, and $\omega_n = 2n\pi T$. The Eq. (2.291) may be written as a contour integral.

$$S = \frac{-1}{2\pi i}\oint dz \frac{1}{z - \xi(\mathbf{p}) - \Sigma(\mathbf{p}, z)} \frac{1}{z - i\omega_n - \xi(\mathbf{k}) - \Sigma(\mathbf{k}, z - i\omega_n)} f(z), \tag{2.292}$$

The contour is shown in Fig. 2.10, and $f(z)$ is Fermi distribution. The integral contour in Fig. 2.10 can be transformed into Fig. 2.11. Along the integral contour in Fig. 2.11 Eq. (2.292) becomes

$$S = \frac{1}{2\pi i}\oint dz \frac{1}{z - \xi(\mathbf{p}) - \Sigma(\mathbf{p}, z)} \frac{1}{z - i\omega_n - \xi(\mathbf{k}) - \Sigma(\mathbf{k}, z - i\omega_n)} f(z). \tag{2.293}$$

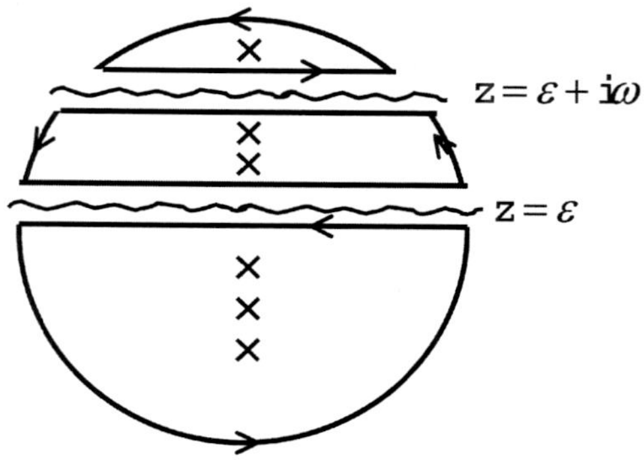

Figure 2.10. The integral contour for Eq. (2.292) at complex z plane. ε is real number. Lines $z = \varepsilon + i\omega_n$ and $z = \varepsilon$ are branch cuts. Branch cut means $G(\mathbf{k}, \varepsilon + i0^+) \neq G(\mathbf{k}, \varepsilon - i0^+)$. At $z = i(2n+1)\pi T$ there are first order poles.

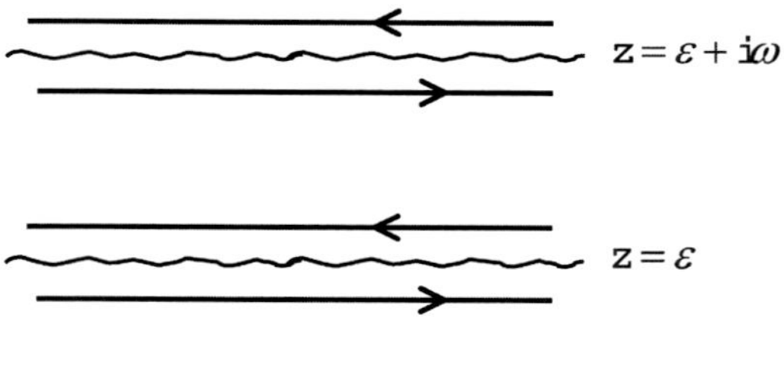

Figure 2.11. The integral contour for Eq. (2.293).

Eq. (2.293) may be rewritten as

$$S = \frac{1}{2\pi i}\left\{\int_{-\infty}^{+\infty} d\varepsilon f(\varepsilon) G_{L\sigma}(\mathbf{p}, \varepsilon - i0^+) G_{R\sigma}(\mathbf{k}, \varepsilon - i\omega_n - i0^+)\right.$$

$$- \int_{-\infty}^{+\infty} d\varepsilon f(\varepsilon) G_{L\sigma}(\mathbf{p}, \varepsilon + i0^+) G_{R\sigma}(\mathbf{k}, \varepsilon - i\omega_n + i0^+)$$

$$+ \int_{-\infty}^{+\infty} d\varepsilon f(\varepsilon + i\omega_n) G_{L\sigma}(\mathbf{p}, \varepsilon + i\omega_n - i0^+) G_{R\sigma}(\mathbf{k}, \varepsilon - i0^+)$$

$$\left. - \int_{-\infty}^{+\infty} d\varepsilon f(\varepsilon + i\omega_n) G_{L\sigma}(\mathbf{p}, \varepsilon + i\omega_n + i0^+) G_{R\sigma}(\mathbf{k}, \varepsilon + i0^+)\right\}. \quad (2.294)$$

Substituting Eqs. (2.294) and (2.255) into Eq. (2.252), we obtain the single electron tunneling current

$$I_p = -2|e| \sum_{\mathbf{kp}\sigma} |T_{\mathbf{kp}}|^2 Im_{i\omega_n \to |e|V + i0^+} \frac{1}{2\pi i}$$

$$\left\{ \int_{-\infty}^{+\infty} d\varepsilon f(\varepsilon) G_{R\sigma}(\mathbf{k},\varepsilon - i\omega_n) \left[G_{L\sigma}(\mathbf{p},\varepsilon - i0^+) - G_{L\sigma}(\mathbf{p},\varepsilon + i0^+)\right] \right.$$

$$\left. + \int_{-\infty}^{+\infty} d\varepsilon f(\varepsilon) G_{L\sigma}(\mathbf{p},\varepsilon + i\omega_n) \left[G_{R\sigma}(\mathbf{k},\varepsilon - i0^+) - G_{R\sigma}(\mathbf{k},\varepsilon + i0^+)\right] \right\}. \tag{2.295}$$

Introducing real spectral functions $A_{R\sigma}(\mathbf{k},x)$ and $A_{L\sigma}(\mathbf{p},x)$,

$$G_{R\sigma}(\mathbf{k},\varepsilon - i0^+) = \int \frac{dx}{2\pi} \frac{A_{R\sigma}(\mathbf{k},x)}{\varepsilon - i0^+ - x} \tag{2.296}$$

and

$$G_{L\sigma}(\mathbf{p},\varepsilon - i0^+) = \int \frac{dx}{2\pi} \frac{A_{L\sigma}(\mathbf{p},x)}{\varepsilon - i0^+ - x}, \tag{2.297}$$

Eq. (2.295) becomes

$$I_p = -2|e| \sum_{\mathbf{kp}\sigma} |T_{\mathbf{kp}}|^2 Im_{i\omega_n \to |e|V + i0^+} \frac{1}{2\pi i} \left\{ \int_{-\infty}^{+\infty} d\varepsilon f(\varepsilon) G_{R\sigma}(\mathbf{k},\varepsilon - i\omega_n) i A_{L\sigma}(\mathbf{p},\varepsilon) \right.$$

$$\left. + \int_{-\infty}^{+\infty} d\varepsilon f(\varepsilon) G_{L\sigma}(\mathbf{p},\varepsilon + i\omega_n) i A_{R\sigma}(\mathbf{k},\varepsilon) \right\}.$$

$$= -\frac{|e|}{\pi} \sum_{\mathbf{kp}\sigma} |T_{\mathbf{kp}}|^2 Im \int_{-\infty}^{+\infty} d\varepsilon f(\varepsilon) \left\{ G_{R\sigma}(\mathbf{k},\varepsilon - |e|V - i0^+) A_{L\sigma}(\mathbf{p},\varepsilon) \right.$$

$$\left. + G_{L\sigma}(\mathbf{p},\varepsilon + |e|V + i0^+) A_{R\sigma}(\mathbf{k},\varepsilon) \right\}. \tag{2.298}$$

Making variable change $\varepsilon - |e|V = \varepsilon' \to \varepsilon$ in the first term, and considering

$$Im G_{R\sigma}(\mathbf{k},\varepsilon - i0^+) = \frac{A_{R\sigma}(\mathbf{k},\varepsilon)}{2}, \tag{2.299}$$

we obtain

$$I_p = |e| \sum_{\mathbf{kp}\sigma} |T_{\mathbf{kp}}|^2 \int_{-\infty}^{+\infty} \frac{d\varepsilon}{2\pi} A_{R\sigma}(\mathbf{k},\varepsilon) A_{L\sigma}(\mathbf{p},\varepsilon + |e|V) \left[f(\varepsilon) - f(\varepsilon + |e|V)\right]. \tag{2.300}$$

Considering Eqs. (2.182), (2.116), (2.117), and (2.299) gives

$$A_{R\sigma}(\mathbf{k},\varepsilon) = 2\pi \left[u_{\mathbf{k}}^2 \delta(\varepsilon - E_{\mathbf{k}}) + v_{\mathbf{k}}^2 \delta(\varepsilon + E_{\mathbf{k}})\right]. \tag{2.301}$$

From the nearly same reason as Eq. (2.301), we have

$$A_{L\sigma}(\mathbf{p},\varepsilon + |e|V) = 2\pi\delta[\varepsilon + |e|V - \xi_{\mathbf{p}}]. \tag{2.302}$$

Substituting Eqs. (2.301) and (2.302) into Eq. (2.300) and performing the spin summation, we obtain

$$I_p = 4\pi|e| \sum_{\mathbf{kp}} |T_{\mathbf{kp}}|^2 \left\{ u_{\mathbf{k}}^2 \delta(|e|V + E_{\mathbf{k}} - \xi_{\mathbf{p}}) [f(E_{\mathbf{k}}) - f(\xi_{\mathbf{p}})] \right.$$

$$+ v_{\mathbf{k}}^2\delta(|e|V - E_{\mathbf{k}} - \xi_{\mathbf{p}})[f(-E_{\mathbf{k}}) - f(\xi_{\mathbf{p}})]\}. \tag{2.303}$$

Considering $f(-E) = 1 - f(E)$, making the summation of wave vector, performing the integration of energy, and taking $T = 0$ K, we have

$$I_p = 4\pi|e|N_R(E_F)N_L(E_F)|T_{con}|^2 \int_{-\infty}^{+\infty} d\xi_{\mathbf{p}} \int_{-\infty}^{+\infty} d\xi_{\mathbf{k}} \{v_{\mathbf{k}}^2[1 - f(\xi_{\mathbf{p}})]\delta(|e|V - E_{\mathbf{k}} - \xi_{\mathbf{p}})$$

$$- u_{\mathbf{k}}^2 f(\xi_{\mathbf{p}})\delta(|e|V + E_{\mathbf{k}} - \xi_{\mathbf{p}})\}. \tag{2.304}$$

The limits of integration have been extended to infinity, which has no effect on the result since the main contribution to integration comes from a narrow interval around $\xi_{\mathbf{p}} = 0$.

We have taken $|e|V > 0$ (See Fig. 2.7.). $f(\xi_{\mathbf{p}}) \neq 0$ at $T = 0$ needs $\xi_{bfp} = 0$. Then $\delta(|e|V + E_{\mathbf{k}} - \xi_{\mathbf{p}}) = 0$, i.e., the second term of Eq. (2.304) is equal to zero. Let us evaluate the first term. At $T = 0$ K, the factor $[1 - f(\xi_{\mathbf{p}})]$ limits $\xi_{\mathbf{p}} > 0$, and thus the factor $\delta(|e|V - E_{\mathbf{k}} - \xi_{\mathbf{p}})$ limits $|e|V > \Delta$, i.e., there is a step function $\theta(|e|V - \Delta)$. Substituting Eq. (2.117) into Eq. (2.304) gives

$$I_p = 4\pi|e|N_R(E_F)N_L(E_F)|T_{con}|^2\theta(|e|V - \Delta) \int_0^{\infty} d\xi_{\mathbf{k}}[1 - f(|e|V - E_{\mathbf{k}})]. \tag{2.305}$$

Using the definition of $\sigma_\circ$ in Eq. (2.262) and making variable change from $\xi_{\mathbf{k}}$ to $E_{\mathbf{k}}$, we have

$$I_p = \frac{\sigma_\circ}{|e|}\theta(|e|V - \Delta) \int_{\Delta}^{\infty} dE_{\mathbf{k}} \frac{d\xi_{\mathbf{k}}}{dE_{\mathbf{k}}}[1 - f(|e|V - E_{\mathbf{k}})]. \tag{2.306}$$

Considering that at $T = 0$ K the factor $[1 - f(|e|V - E_{\mathbf{k}})]$ limits $|e|V > E_{\mathbf{k}}$, and $E_{\mathbf{k}} = \sqrt{\xi_{\mathbf{k}}^2 + \Delta^2} = E$, we have

$$I_p = \frac{\sigma_\circ}{|e|}\theta(|e|V - \Delta) \int_{\Delta}^{|e|V} dE \frac{E}{\sqrt{E^2 - \Delta^2}}$$

$$= \frac{\sigma_\circ}{|e|}\theta(|e|V - \Delta)\sqrt{(|e|V)^2 - \Delta^2}. \tag{2.307}$$

From Eq. (2.307) we may obtain the characteristic of I_p vs. V, which is shown in Fig. 2.12.

From Eq. (2.307) we obtain conductance dI_p/dV as a function of voltage V.

$$\frac{dI_p}{dV} = |e|V\sigma_\circ \frac{\theta(|e|V - \Delta)}{\sqrt{(|e|V)^2 - \Delta^2}} \tag{2.308}$$

The curve of dI_p/dV vs. V is shown in Fig. 2.13. $(dI_p/dV)/\sigma_\circ$ is the BCS density of states of quasiparticle exitation.

$I_p > 0$ indicates that the current is from the left side of insulator to the right side of insulator in case of $|e|V > \Delta$ (See Fig. 2.7.). If the tunneling carriers are holes, then the physical picture of tunneling is shown in Fig. 2.14.

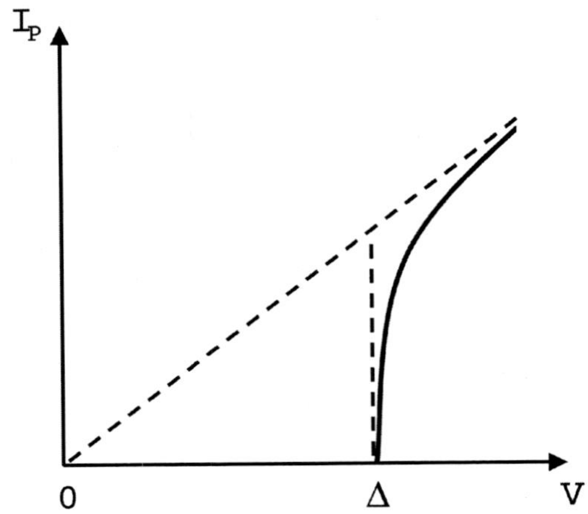

Figure 2.12. The characteristic of N-I-S tunneling junction at T = 0 K.

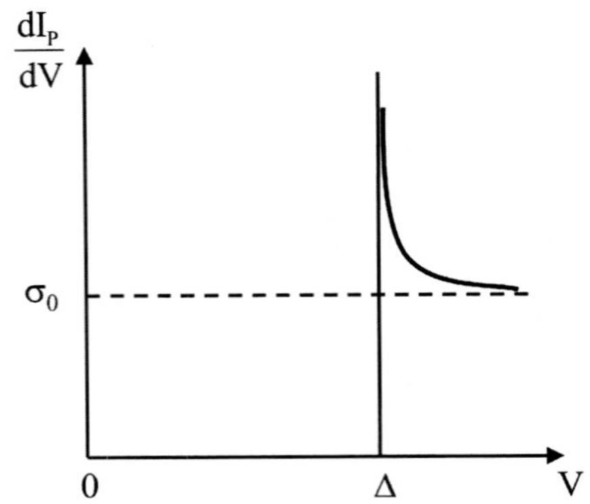

Figure 2.13. The curve of conductance vs. V of N-I-S tunneling junction at T = 0 K.

If $|e|V < -\Delta$, then $\delta(|e|V - E_{\mathbf{k}} - \xi_{\mathbf{p}}) = 0$ in the first term of Eq. (2.304), and the factor $\delta(|e|V + E_{\mathbf{k}} - \xi_{\mathbf{p}})$ in the second term limits $\xi_{\mathbf{p}} < 0$, and thus $f(\xi_{\mathbf{p}}) = 1$ at $T = 0$ K. In the case of $|e|V < -\Delta$, the second term is not equal to zero, and $I_p < 0$, which means that the current direction is from the right side to the left side. For $I_p < 0$, the physical picture of tunneling is shown in Fig. 2.15.

2.7.6. S-I-S Tunneling Junction

Here S and I represent superconductor in superconducting or pseudogap state and insulator, respectively. (Note that the S in Josephson tunneling and in Ref. [50] represents superconductor in superconducting state.) The single electron tunneling current formula Eq. (2.300) is still valid when the two sides of the insulator are superconductors in superconducting or pseudogap states. The following derivations are dependent on the existence of the gap or the pseudogap, independent of the long-range phase coherence of Cooper pairs, and are the

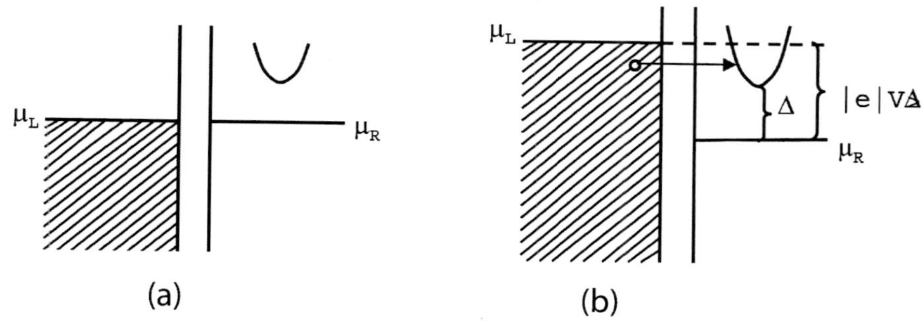

Figure 2.14. Hole tunneling between a normal metal at left side and a superconductor in superconducting or pseudogap state at the right side at T = 0 K. (a) No applied voltage. (b) For $|e|V > \Delta$, hole can tunnel from the normal metal and becomes a quasiparticle in the superconductor.

same as Ref. [50]. For S-I-S junction Eq. (2.302) becomes

$$A_{L\sigma}(\mathbf{p},\varepsilon+|e|V) = 2\pi\left[u_{\mathbf{p}}^2\delta(\varepsilon+|e|V-E_{\mathbf{p}})+v_{\mathbf{p}}^2\delta(\varepsilon+|e|V+E_{\mathbf{p}})\right]. \tag{2.309}$$

Substituting the spectral functions of superconducting or pseudogap state in Eqs. (2.309) and (2.301) into Eq. (2.300), we obtain the single electron tunneling current.

$$I_p = 4\pi|e|\sum_{\mathbf{kp}}|T_{\mathbf{kp}}|^2\left\{u_{\mathbf{k}}^2u_{\mathbf{p}}^2\delta(|e|V+E_{\mathbf{k}}-E_{\mathbf{p}})[f(E_{\mathbf{k}})-f(E_{\mathbf{p}})]\right.$$

$$+u_{\mathbf{k}}^2v_{\mathbf{p}}^2\delta(|e|V+E_{\mathbf{k}}+E_{\mathbf{p}})[f(E_{\mathbf{k}})-f(-E_{\mathbf{p}})]+v_{\mathbf{k}}^2u_{\mathbf{p}}^2\delta(|e|V-E_{\mathbf{k}}-E_{\mathbf{p}})[f(-E_{\mathbf{k}})-f(E_{\mathbf{p}})]$$

$$\left.+v_{\mathbf{k}}^2v_{\mathbf{p}}^2\delta(|e|V-E_{\mathbf{k}}+E_{\mathbf{p}})[f(-E_{\mathbf{k}})-f(-E_{\mathbf{p}})]\right\}. \tag{2.310}$$

The identity $f(-E) = 1-f(E)$ permits these to be grouped as follows.

$$I_p = 4\pi|e|\sum_{\mathbf{kp}}|T_{\mathbf{kp}}|^2\left\{[1-f(E_{\mathbf{p}})-f(E_{\mathbf{k}})]\right.$$

$$\left[v_{\mathbf{k}}^2u_{\mathbf{p}}^2\delta(|e|V-E_{\mathbf{p}}-E_{\mathbf{k}})-u_{\mathbf{k}}^2v_{\mathbf{p}}^2\delta(|e|V+E_{\mathbf{p}}+E_{\mathbf{k}})\right]$$

$$\left.+\left[f(E_{\mathbf{k}})-f(E_{\mathbf{p}})\right]\left[u_{\mathbf{k}}^2u_{\mathbf{p}}^2\delta(|e|V-E_{\mathbf{p}}+E_{\mathbf{k}})-v_{\mathbf{k}}^2v_{\mathbf{p}}^2\delta(|e|V+E_{\mathbf{p}}-E_{\mathbf{k}})\right]\right\} \tag{2.311}$$

Eq. (2.311) is the most general result, which is valid at finite temperatures. The variable change to $d\xi_{\mathbf{k}}d\xi_{\mathbf{p}}$ is done next. These integrals over the coherence factors u^2 and v^2 will eliminate their linear term in ξ. Then the variables can be changed to $E = E_{\mathbf{k}} = \sqrt{\xi_{\mathbf{k}}^2+\Delta_R^2}$ and $E' = E_{\mathbf{p}} = \sqrt{\xi_{\mathbf{p}}^2+\Delta_L^2}$, which brings us to the expression

$$I_p = 4\pi|e|N_L(E_F)N_R(E_F)|T_{con}|^2\int_{\Delta_R}^{\infty}dE\frac{d\xi_{\mathbf{k}}}{dE_{\mathbf{k}}}\int_{\Delta_L}^{\infty}dE'\frac{d\xi_{\mathbf{p}}}{dE_{\mathbf{p}}}$$

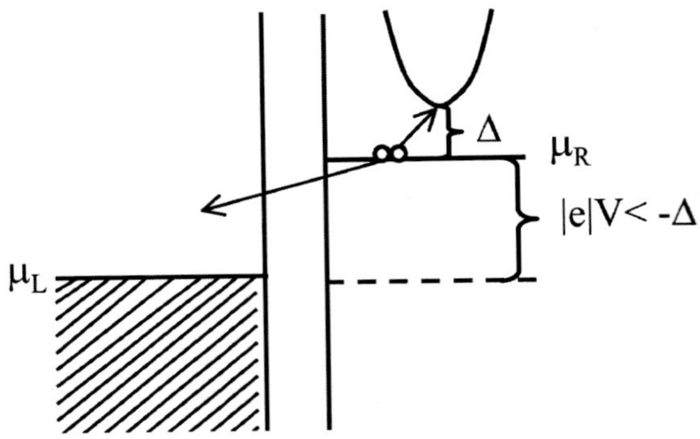

Figure 2.15. Hole tunneling between a normal metal at the left side and a superconductor in superconducting or pseudogap state at the right side at T = 0 K. For $|e|V < -\Delta$, a Cooper hole pair is broken in the superconductor, and one hole tunnels, while the other becomes a quasiparticle in the superconductor.

$$\{[1-f(E)-f(E')][\delta(|e|V-E-E')-\delta(|e|V+E+E')] + [f(E)-f(E')][\delta(|e|V+E-E')-\delta(|e|V+E'-E)]\}. \tag{2.312}$$

In the next, we discuss the simplest case, zero temperature and identical superconductors $\Delta_R = \Delta_L = \Delta$. At zero temperature, we can set to zero all the thermal factors $f(E)$ and $f(E')$.

$$I_p = \frac{\sigma_\circ}{|e|}\int_\Delta^\infty dE \frac{d\xi_{\mathbf{k}}}{dE_{\mathbf{k}}}\int_\Delta^\infty dE' \frac{d\xi_{\mathbf{p}}}{dE_{\mathbf{p}}}[\delta(|e|V-E-E')-\delta(|e|V+E+E')]. \tag{2.313}$$

Because we have taken $|e|V > 0$, the second term is zero. Thus only the first term needs to be done, where $E' = |e|V - E$ and $|e|V > 2\Delta$. Note that in this case the factor $[1-f(E)-f(E')]$ in Eq. (2.312) limits the integral over E to $|e|V-\Delta$ instead of infinity.

$$I_p = \frac{\sigma_\circ}{|e|}\theta(|e|V-2\Delta)\int_\Delta^{|e|V-\Delta} dE \frac{E(|e|V-E)}{\sqrt{E^2-\Delta^2}\sqrt{[(|e|V-E)^2-\Delta^2]}} \tag{2.314}$$

This integral is done in the following way: change variable of integration to x, where $2E = |e|V + x(|e|V-2\Delta)$. After some algebra, the integral is [50]

$$I_p = \frac{\sigma_\circ}{|e|}\frac{\theta(|e|V-2\Delta)}{|e|V+2\Delta}\int_{-1}^{+1} dx \frac{(|e|V)^2 - x^2(|e|V-2\Delta)^2}{\sqrt{(1-x^2)(1-\alpha^2x^2)}}, \tag{2.315}$$

$$\alpha = \frac{|e|V-2\Delta}{|e|V+2\Delta}. \tag{2.316}$$

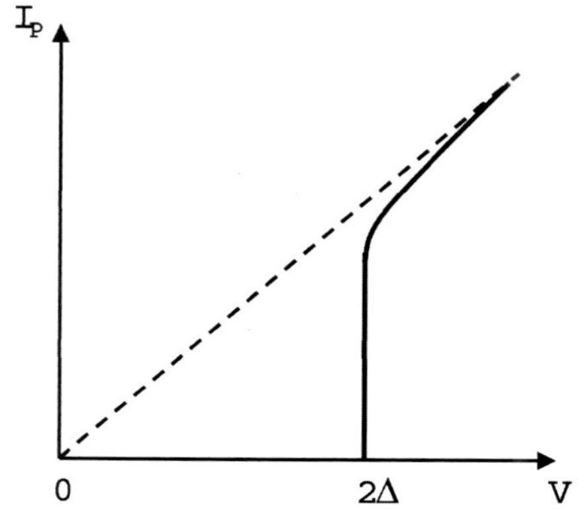

Figure 2.16. The characteristic of S-I-S tunneling junction at T = 0 K.

The dx integral is in the form of complete elliptic integrals [50].

$$I_p = \frac{\sigma_\circ}{|e|}\theta(|e|V-2\Delta)\left\{\frac{(|e|V)^2}{|e|V+2\Delta}K(\alpha)-(|e|V+2\Delta)[K(\alpha)-E(\alpha)]\right\}. \tag{2.317}$$

The tunneling current is zero for voltage less than twice the gap Δ. It is interesting to examine the threshold phenomenon, for voltage slightly larger than 2Δ. We have that $K(0) = E(0) = \pi/2$, so at $|e|V = 2\Delta^+$ the tunneling current is [50]

$$I_p = \frac{\sigma_\circ}{|e|}\frac{\Delta\pi}{2} = \sigma_\circ V\frac{\pi}{4}. \tag{2.318}$$

It has a finite value, so there is a discontinuous jump in the tunneling current at $|e|V = 2\Delta$. The tunneling current above threshold is $\pi/4$ of the value of a normal tunneling current at the same voltage. The characteristic of I_p vs. V at 0 K is shown in Fig. 2.16.

The I_p in Eq. (2.318) is positive, which means that the tunneling current is from the left to the right for $|e|V > 2\Delta$. For $I_p > 0$, the physical picture of tunneling is shown in Fig. 2.17.

2.8. On London Equations

The phenomenological London equations describing superconductivity are as follows (See section 1.2.1.).

$$a\mathbf{E} = \frac{\partial}{\partial t}\mathbf{J}_s, \tag{2.319}$$

$$\frac{a}{c}\mathbf{B} = -(\text{curl})\mathbf{J}_s, \tag{2.320}$$

$$a = \frac{n_s e^{*2}}{m^{**}}, \tag{2.321}$$

where $\mathbf{J}_s$ is supercurrent density of Cooper pairs with long-range phase coherence, m^{**} is the mass of a Cooper pair, e^* is the charge of a Cooper pair, and n_s is number density

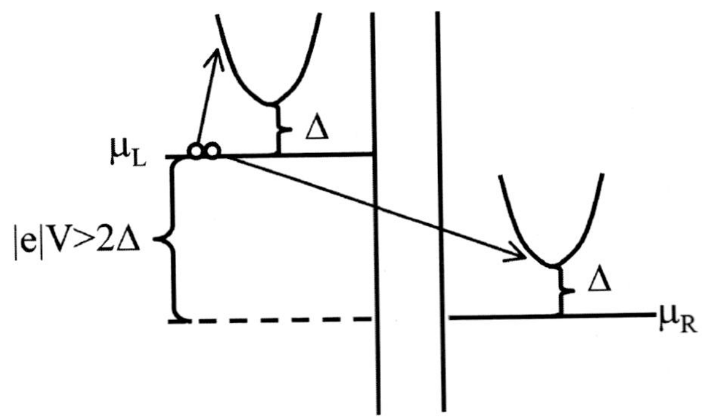

Figure 2.17. Hole tunneling between a superconductor and a superconductor. For $|e|V > 2\Delta$, a Cooper hole pair breaks on the left side, one hole tunnels, and both holes become excited quasiparticles in the two sides. Note that the superconductors can be in superconducting or pseudogap state.

of supercurrent Cooper pairs. Eqs. (2.319) and Eq. (2.320) describe zero-resistance and Meissner effect, respectively.

The London equations can be derived by BCS microscopic pairing theory as follows. The Cooper pairs with or without the long-range phase coherence are in bound states consisting of two quasiparticles. For small disturbance the Cooper pairs are free. The canonical momentum $\mathbf{q}$ of mass-center of every Cooper pair with or without long-range phase coherence should be same. All Cooper pairs with or without long-range phase coherence condense into the level of chemical potential μ [35]. Let us at first discuss superconducting state. For the static superconducting state the phases of wavefunctions of all Cooper pairs are same, which is represented by ϕ. The wavefunction of Cooper pairs with long-range phase coherence in supercurrent-carrying state is thus

$$\psi_{oN}(\mathbf{r}_1,\cdots,\mathbf{r}_N) = \hat{A}\,\phi(\mathbf{r}_1-\mathbf{r}_2)\exp(i\phi)\exp(i\mathbf{q}\cdot\mathbf{R}_1 - 2\mu t/\hbar)\uparrow_1\downarrow_2\cdots$$
$$\phi(\mathbf{r}_{N-1}-\mathbf{r}_N)\exp(i\phi)\exp(i\mathbf{q}\cdot\mathbf{R}_{N/2} - 2\mu t/\hbar)\uparrow_{N-1}\downarrow_N, \tag{2.322}$$

where $\hat{A}$ is antisymmetric operator, $\mathbf{r}_i$ is the position vector of the i-th electron, $\mathbf{R}_i = (\mathbf{r}_{2i-1} + \mathbf{r}_{2i})/2$ is the position vector of mass-center of the i-th Cooper pair, and the arrows denote electron spins. In an external applied field [58]

$$\mathbf{q} = m^{**}\mathbf{V}_s + \frac{e^*}{c}\mathbf{A}, \tag{2.323}$$

where $\mathbf{A}$ is the vector potential, and $\mathbf{V}_s$ is the velocity of a Cooper pair with long-range phase coherence. Considering supercurrent density

$$\mathbf{J}_s = n_s e^* \mathbf{V}_s, \tag{2.324}$$

we have

$$\mathbf{J}_s = \frac{n_s e^*}{m^{**}}\mathbf{q} - \frac{n_s e^{*2}}{cm^{**}}\mathbf{A}. \tag{2.325}$$

Eq. (2.325) comes from BCS microscopic pairing theory Eq. (2.322). Taking the time derivative of both sides yields Eq. (2.319) and taking the curl leads to Eq. (2.320). Therefore, Eq. (2.325) can be called general London equation.

The phases of wavefunctions of Cooper pairs without long-range phase coherence in the static pseudogap state are arbitrary. So, the wavefunction of Cooper pairs without long-range phase coherence in current-carrying state is

$$\psi_{oN}(\mathbf{r}_1,\cdots,\mathbf{r}_N) = \hat{A}\,\phi(\mathbf{r}_1-\mathbf{r}_2)\exp(i\phi_{1,arbitrary})\exp(i\mathbf{q}\cdot\mathbf{R}_1-2\mu t/\hbar)\uparrow_1\downarrow_2\cdots$$
$$\phi(\mathbf{r}_{N-1}-\mathbf{r}_N)\exp(i\phi_{N/2,arbitrary})\exp(i\mathbf{q}\cdot\mathbf{R}_{N/2}-2\mu t/\hbar)\uparrow_{N-1}\downarrow_N. \tag{2.326}$$

In an external applied field [58]

$$\mathbf{q} = m^{**}\mathbf{V}_n + \frac{e^*}{c}\mathbf{A}, \tag{2.327}$$

where $\mathbf{V}_n$ is the velocity of Cooper pair without long-range phase coherence. Considering

$$\mathbf{J}_n = ne^*\mathbf{V}_n, \tag{2.328}$$

where n is the number density of Cooper pairs without long-range phase coherence, we have

$$\mathbf{J}_n = \frac{ne^*}{m^{**}}\mathbf{q} - \frac{ne^{*2}}{cm^{**}}\mathbf{A}. \tag{2.329}$$

Eq. (2.329) shows that Cooper pairs without long-range phase coherence obey the general London equation, which is called microscopic superconductivity. However, Eq. (2.329) does not mean zero-resistance and Meissner effect. The reason is as follows. The wavefunction of any Cooper pair without long-range phase coherence has an arbitrary phase factor such as $\exp(i\phi_{1,arbitrary})$. The average value of wavefunctions of many Cooper pairs without long-range phase coherence at any location is zero. Therefore, exactly speaking, that London equations describe zero-resistance and Meissner effect needs an additional condition, i.e., the current-carrying Cooper pairs must be in long-range phase coherence.

2.9. On Shapiro Steps

The Josephson tunneling current given by Eq. (2.280) comes from Eq. (2.279) under condition $\mu_R - \mu_L = -|e|V = 0$. So, Eq. (2.280) can not be used when $\mu_R \neq \mu_L$. Eq. (2.322) tells us that the phase of wavefunction of a Cooper pair is a function of chemical potential. Josephson demonstrated that in cases of $\mu_R \neq \mu_L$, and μ_R and μ_L time-dependent, Josephson tunneling current is [59]

$$I_{pp} = iK(\Delta_L\Delta_R^* - \Delta_R\Delta_L^*). \tag{2.330}$$

Note that gap Δ is proportional to the wavefunction of Cooper pair. Taking $\Delta_L = |\Delta_L|\exp\{i\phi_L[\mu_L(t)]\}$ and $\Delta_R = |\Delta_R|\exp\{i\phi_R[\mu_R(t)]\}$, we obtain from Eq. (2.330)

$$I_{pp} = I_c\sin\{\phi_R[\mu_R(t)] - \phi_L[\mu_L(t)]\}. \tag{2.331}$$

From Eq. (2.322) we have

$$\frac{\partial\phi_R[\mu_R(t)]}{\partial t} = -\frac{2[\mu_R(t)+\mu']}{\hbar}, \tag{2.332}$$

and

$$\frac{\partial\phi_L[\mu_L(t)]}{\partial t} = -\frac{2[\mu_L(t)+\mu']}{\hbar}, \tag{2.333}$$

where μ' denotes term whose magnitude depends on the degree of departure from equilibrium. The leading contributions to μ' had been discussed in some papers [60, 61]. Usually one takes $\mu' = 0$ [59].

When a Josephson junction is irradiated with radiation of angular frequency ω, the response of the supercurrent gives rise to constant voltage Shapiro steps in the dc I-V curve at voltage $V_n = n\hbar\omega/2|e|$ [62]. In the following we give a mathematical demonstration for the Shapiro steps. Suppose we have a Josephson junction with a dc voltage V in a microwave field $|e|v\cos(\omega t+\theta)$. From Eqs. (2.332) and (2.333) we have

$$\begin{aligned}\frac{\partial}{\partial t}\{\phi_R[\mu_R(t)]-\phi_L[\mu_L(t)]\} &= -\frac{2[\mu_R(t)-\mu_L(t)]}{\hbar}\\ &= \frac{2|e|[V+v\cos(\omega t+\theta)]}{\hbar}.\end{aligned} \tag{2.334}$$

Eq. (2.334) gives

$$\phi_R[\mu_R(t)]-\phi_L[\mu_L(t)] = \frac{2|e|V}{\hbar}t + \frac{2|e|v}{\hbar\omega}\sin(\omega t+\theta)+S_\circ, \tag{2.335}$$

where $S_\circ$ is a constant of integration for the phase difference.

Substituting Eq. (2.335) into Eq. (2.331), we obtain an expression of I_{pp}. The resulting expression can be analysed with the aid of the identity

$$e^{iY\sin x} = \sum_{n=-\infty}^{\infty} J_n(Y)e^{inx}. \tag{2.336}$$

Here $J_n(Y)$ is the Bessel function of order n. We find for the supercurrent across the Josephson junction

$$I_{pp} = I_c\sum_{-\infty}^{\infty}(-1)^n J_n\left(\frac{2|e|v}{\hbar\omega}\right)\sin\left[\left(\frac{2|e|V}{\hbar}-n\omega\right)t-n\theta+S_\circ\right]. \tag{2.337}$$

The most important feature of this expression is that there is a dc supercurrent when

$$2|e|V = n\hbar\omega. \tag{2.338}$$

Eq. (2.338) is the condition of tunneling with absorption of n photons simultaneously. The magnitude is

$$I_{pp,n} = (-1)^n I_c J_n\left(\frac{2|e|v}{\hbar\omega}\right)\sin(S_\circ-n\theta). \tag{2.339}$$

This is like the zero-voltage supercurrent, in that the phase difference $S_\circ$ can adjust to match the applied dc current over a certain range. We have a constant-voltage Shapiro supercurrent steps, of magnitude

$$I_{pp,n}^{step} = 2I_c J_n\left(\frac{2|e|v}{\hbar\omega}\right). \tag{2.340}$$

In addition, the zero-voltage supercurrent is depressed:

$$I_{pp,\circ}^{step} = 2I_c J_\circ\left(\frac{2|e|v}{\hbar\omega}\right). \tag{2.341}$$

We have discussed P-I-P junction and the pseudogap state in sections 1.1.6 and 2.7.4. Here P represents right and left electrode in the pseudogap state where the Cooper pairs have no long-range phase coherence. Eq. (2.326) gives the wavefunction of Cooper pairs without long-range phase coherence, which shows that the random phases of Cooper pairs are dependent on the chemical potential μ. Let $\phi_{arbitrary}(\mu)$ represent the phase of any Cooper pair without long-range phase coherence. Then, we have

$$\frac{\partial\phi_{R,arbitrary}[\mu_R(t)]}{\partial t} = -\frac{2[\mu_R(t)+\mu']}{\hbar}, \tag{2.342}$$

and

$$\frac{\partial\phi_{L,arbitrary}[\mu_L(t)]}{\partial t} = -\frac{2[\mu_L(t)+\mu']}{\hbar}. \tag{2.343}$$

From the derivations of Eq. (2.330) in Ref. [59] we know that the formula of tunneling current across the barrier layer is still correct for the Cooper pairs without long-range phase coherence. For the Cooper pairs without long-range phase coherence, we have

$$\Delta_L = |\Delta_L|\exp\{i\phi_{L,arbitrary}[\mu_L(t)]\} \tag{2.344}$$

and

$$\Delta_R = |\Delta_R|\exp\{i\phi_{R,arbitrary}[\mu_R(t)]\}. \tag{2.345}$$

Substituting Eqs. (2.344) and (2.345) into Eq. (2.330), we obtain the tunneling supercurrent across the barrier layer

$$I_{pp} = I_c\sin\{\phi_{R,arbitrary}[\mu_R(t)] - \phi_{L,arbitrary}[\mu_L(t)]\}. \tag{2.346}$$

If we have a P-I-P junction with a dc voltage V in a microwave field $|e|v\cos(\omega t+\theta)$, then

$$\frac{\partial}{\partial t}\{\phi_{R,arbitrary}[\mu_R(t)] - \phi_{L,arbitrary}[\mu_L(t)]\} = -\frac{2[\mu_R(t)-\mu_L(t)]}{\hbar} = \frac{2|e|[V+v\cos(\omega t+\theta)]}{\hbar}. \tag{2.347}$$

Eq. (2.347) gives

$$\phi_{R,arbitrary}[\mu_R(t)] - \phi_{L,arbitrary}[\mu_L(t)] = \frac{2|e|V}{\hbar}t + \frac{2|e|v}{\hbar\omega}\sin(\omega t+\theta) + S_{\circ,arbitrari}, \tag{2.348}$$

where $S_{o,arbitrary}$ is a constant of integration for the arbitrary phase difference. Substituting Eq. (2.348) into Eq. (2.346), we obtain

$$I_{pp} = I_c \sum_{-\infty}^{\infty} (-1)^n J_n \left(\frac{2|e|v}{\hbar\omega} \right) \sin \left[\left(\frac{2|e|V}{\hbar} - n\omega \right) t - n\theta + S_{o,arbitrary} \right]. \tag{2.349}$$

The most important feature of Eq. (2.349) is that there is a dc supercurrent when

$$2|e|V = n\hbar\omega. \tag{2.350}$$

Eq. (2.350) is the condition of tunneling with absorption of n photons simultaneously. The magnitude is

$$I_{pp,n} = (-1)^n I_c J_n \left(\frac{2|e|v}{\hbar\omega} \right) \sin(S_{o,arbitrary} - n\theta). \tag{2.351}$$

Because $S_{o,arbitrary}$ is random, the contribution of all Cooper pairs without long-range phase coherence to the total dc supercurrent across the barrier layer is zero.

2.10. Probability of Cooper Pair

2.10.1. Constant Weak Magnetic Field

This section extends the corresponding discussions for Cooper pairs with long-range phase coherence in superconducting state of conventional superconductor in Ref. [47] to Cooper pairs without long-range phase coherence in pseudogap state of unconventional superconductor. The aim of this section is to find the probability of appearance of a Cooper pair at any temperature. The weak magnetic field in this section means $H \ll H_c$, where H represents applied constant magnetic field, and H_c represents thermodynamic critical magnetic field. Suppose that a superconductor with a plane surface occupies the half-space $z < 0$. $\mathbf{H}(\mathbf{r}) = \nabla \times \mathbf{A}(\mathbf{r})$. If $\mathbf{H} = H\mathbf{x}$, where $\mathbf{x}$ is unit vector in direction $\mathbf{x}$, then we can take

$$\mathbf{A} = (0, -Hz, 0). \tag{2.352}$$

Maxwell equation is

$$\nabla \times \mathbf{H} = 4\pi\mathbf{J} + \frac{\partial \mathbf{D}}{\partial t}. \tag{2.353}$$

If $\mathbf{D}$ is not a function of time, then the second term in Eq. (2.353) is zero. From Eq. (2.353) we have

$$\nabla \times \nabla \times \mathbf{A} = 4\pi\mathbf{J}. \tag{2.354}$$

Considering

$$\nabla \times \nabla \times \mathbf{A} = \nabla\nabla \cdot \mathbf{A} - \nabla^2 \mathbf{A} \tag{2.355}$$

and Eq. (2.352), we obtain

$$\nabla^2 \mathbf{A} = -4\pi\mathbf{J}. \tag{2.356}$$

Because the field is weak, we can assume $\mathbf{J} \propto \mathbf{A}$. The connection between current density and field must generally be of non-local form, i.e.,

$$\mathbf{J}(x) = -\int Q(x-y)\mathbf{A}(y)d^4y, \tag{2.357}$$

where $x = (\mathbf{x}, t)$, $y = (\mathbf{y}, t)$, and $Q(x-y)$ is a kernel of an integration equation. Making Fourier transformation, Eq. (2.356) becomes

$$\mathbf{J}(k) = -Q(k)\mathbf{A}(k). \tag{2.358}$$

The basic aim of section 2.10 is to expound the physical meaning of $Q(k)$. Our explanation for $Q(k)$ is different from the previous (See, for example, Ref. [47].).

2.10.2. Kernel Q(x-y)

The current density operator in second quantization approach is $(\hbar = c = k_B = 1.)$

$$\tilde{\mathbf{J}}_{sq}(\mathbf{r},t) = \frac{ie}{2m^*}(\nabla_{\mathbf{r}'} - \nabla_{\mathbf{r}})\Big|_{\mathbf{r}'\to\mathbf{r}} \tilde{\psi}^+_\alpha(\mathbf{r}',t)\,\tilde{\psi}_\alpha(\mathbf{r},t), \tag{2.359}$$

where the double spin subscripts mean spin summation. In Schrödinger representation

$$\mathbf{J}_{sq}(\mathbf{r}) = \frac{ie}{2m^*}(\nabla_{\mathbf{r}'} - \nabla_{\mathbf{r}})\Big|_{\mathbf{r}'\to\mathbf{r}} \psi^+_\alpha(\mathbf{r}')\psi_\alpha(\mathbf{r}). \tag{2.360}$$

In an external field

$$\nabla\psi^+_\alpha(\mathbf{r}) \longrightarrow (\nabla + ie\mathbf{A})\psi^+_\alpha. \tag{2.361}$$

$$\nabla\psi_\alpha(\mathbf{r}) \longrightarrow (\nabla - ie\mathbf{A})\psi_\alpha. \tag{2.362}$$

If external field $\mathbf{H} \neq 0$, then

$$\mathbf{J}_{sq}(\mathbf{r}) = \frac{i|e|}{2m^*}(\nabla_{\mathbf{r}'} - \nabla_{\mathbf{r}})|_{\mathbf{r}'\to\mathbf{r}}\psi^+_\alpha(\mathbf{r}')\psi_\alpha(\mathbf{r}) - \frac{e^2\mathbf{A}(\mathbf{r})}{m^*}\psi^+_\alpha(\mathbf{r})\psi_\alpha(\mathbf{r}). \tag{2.363}$$

The statistical average value of $\mathbf{J}_{sq}$ is

$$\overline{\mathbf{J}(\mathbf{r})} = \langle \mathbf{J}_{sq}(\mathbf{r})\rangle$$

$$= 2\left\{\frac{ie}{2m^*}(\nabla_{\mathbf{r}'} - \nabla_{\mathbf{r}})G(x,x') - \frac{e^2\mathbf{A}(\mathbf{r})}{m^*}G(x,x')\right\}_{\mathbf{r}'\to\mathbf{r},\tau'\to\tau+0^+}, \tag{2.364}$$

where the coefficient "2" comes from spin summation, and $x = (\mathbf{r}, \tau)$.

G and F^+ are functions of $\tau - \tau'$. Their Fourier components G_{ω_n} and $F^+_{\omega_n}$ satisfy the following equations [See Eq. (2.188).].

$$\left\{i\omega_n + \frac{1}{2m^*}[\nabla_{\mathbf{r}} - ie\mathbf{A}(\mathbf{r})]^2 + \mu\right\} G_{\omega_n}(\mathbf{r},\mathbf{r}') + \Delta(\mathbf{r})F^+_{\omega_n}(\mathbf{r},\mathbf{r}') = \delta(\mathbf{r}-\mathbf{r}'), \tag{2.365}$$

$$\left\{-i\omega_n + \frac{1}{2m^*}[\nabla_{\mathbf{r}} + ie\mathbf{A}(\mathbf{r})]^2 + \mu\right\} F^+_{\omega_n}(\mathbf{r},\mathbf{r}') - \Delta^*(\mathbf{r})G_{\omega_n}(\mathbf{r},\mathbf{r}') = 0. \tag{2.366}$$

Note that Eqs. (2.365) and (2.366) can be used to Cooper pairs with or without long-range phase coherence.

Write the Green functions G and F^+ as follows.

$$G = G_\circ + G^{(1)}, \tag{2.367}$$

$$F = F_\circ + F^{(1)}, \tag{2.368}$$

$$F^+ = F_\circ^+ + F^{+(1)}, \tag{2.369}$$

$$\Delta = \Delta_\circ + \Delta^{(1)}, \tag{2.370}$$

$$\Delta^* = \Delta_\circ^* + \Delta^{(1)*}, \tag{2.371}$$

where subscript "o" denotes that the corresponding quantity is in zero field, and superscript "(1)" denotes the corresponding quantity is linearly dependent on the field. To obtain the equations satisfied by $G^{(1)}$ and $F^{+(1)}$, we make the following four considerations. First, substitute Eqs. (2.367) - (2.371) into Eqs. (2.365) and (2.366). Second, keep only the terms linear in field. Third, use the equations satisfied by $G_{\circ\omega_n}$ and $F_{\circ\omega_n}$. Fourth, use $[\nabla_\mathbf{r} - ie\mathbf{A}(\mathbf{r})]^2 = \nabla_\mathbf{r}\cdot\nabla_\mathbf{r} - ie\mathbf{A}(\mathbf{r})\cdot\nabla_\mathbf{r} - ie\nabla_\mathbf{r}\cdot\mathbf{A}(\mathbf{r}) - e^2[\mathbf{A}(\mathbf{r})]^2$. We obtain

$$\begin{aligned}
&\left(i\omega_n + \frac{\nabla^2}{2m^*} + \mu\right) G^{(1)}_{\omega_n}(\mathbf{r},\mathbf{r}') + \Delta_\circ F^{+(1)}_{\omega_n}(\mathbf{r},\mathbf{r}') \\
&= -\Delta^{(1)}(\mathbf{r}) F^+_{\circ\omega_n}(\mathbf{r}-\mathbf{r}') + \frac{ie}{2m^*}(\nabla\cdot\mathbf{A} + \mathbf{A}\cdot\nabla) G_{\circ\omega_n}(\mathbf{r}-\mathbf{r}') \\
&= L_{11}(\mathbf{r},\mathbf{r}')
\end{aligned} \tag{2.372}$$

and

$$\begin{aligned}
&\left(-i\omega_n + \frac{\nabla^2}{2m^*} + \mu\right) F^{+(1)}_{\omega_n}(\mathbf{r},\mathbf{r}') - \Delta_\circ G^{(1)}_{\omega_n}(\mathbf{r},\mathbf{r}') \\
&= \Delta^{(1)*}(\mathbf{r}) G_{\circ\omega_n}(\mathbf{r}-\mathbf{r}') - \frac{ie}{2m^*}(\nabla\cdot\mathbf{A} + \mathbf{A}\cdot\nabla) F^+_{\circ\omega_n}(\mathbf{r}-\mathbf{r}') \\
&= -L_{21}(\mathbf{r},\mathbf{r}').
\end{aligned} \tag{2.373}$$

Let us show that

$$\mathbf{J}_\circ(\mathbf{r}) = \frac{ie}{m^*} T \sum_n (\nabla_{\mathbf{r}'} - \nabla_\mathbf{r})\Big|_{\mathbf{r}'\to\mathbf{r}} G_{\circ\omega_n}(\mathbf{r}-\mathbf{r}') = 0 \tag{2.374}$$

From Eq. (2.363)

$$\begin{aligned}
\mathbf{J}_\circ(\mathbf{r}) &= \frac{ie}{m^*} (\nabla_{\mathbf{r}'} - \nabla_\mathbf{r})|_{\mathbf{r}'\to\mathbf{r}} G_\circ(\mathbf{r}\tau, \mathbf{r}'\tau + 0^+) \\
&= -\frac{ie}{m^*} (\nabla_{\mathbf{r}'} - \nabla_\mathbf{r})|_{\mathbf{r}'\to\mathbf{r}} \langle \psi^+_\uparrow(\mathbf{r}')\psi_\uparrow(\mathbf{r})\rangle \\
&= -\frac{ie}{m^*} (\nabla_{\mathbf{r}'} - \nabla_\mathbf{r})|_{\mathbf{r}'\to\mathbf{r}} \langle \frac{1}{\sqrt{N}} \sum_{\mathbf{p}'} e^{-i\mathbf{p}'\cdot\mathbf{r}'} c^+_{\mathbf{p}'\uparrow} \frac{1}{\sqrt{N}} \sum_{\mathbf{p}} e^{i\mathbf{p}\cdot\mathbf{r}} c_{\mathbf{p}\uparrow}\rangle
\end{aligned}$$

$$= -\frac{2e}{Nm^*}\sum_{\mathbf{p}} \mathbf{p}n(\mathbf{p}) = 0. \tag{2.375}$$

Using Eq. (2.374), the current density $\mathbf{J}(\mathbf{r})$ in linear approximation of field is

$$\mathbf{J}(\mathbf{r}) = \frac{ie}{m^*}T\sum_n (\nabla_{\mathbf{r}'} - \nabla_{\mathbf{r}})|_{\mathbf{r}'\to\mathbf{r}} G^{(1)}_{\omega_n}(\mathbf{r},\mathbf{r}') - \frac{e^2\mathbf{A}(\mathbf{r})}{m^*}N(\mathbf{r}), \tag{2.376}$$

where $N(\mathbf{r}) = N$ is the number density of electrons in zero field.

$$N(\mathbf{r}) = \langle \psi^+(\mathbf{r})\psi(\mathbf{r})\rangle = G_\circ(\mathbf{r}\tau,\mathbf{r}\tau')|_{\tau'\to\tau+0^+}. \tag{2.377}$$

According to the following analyses we can demonstrate $\Delta^{(1)*} = 0$. First, $\Delta^{(1)*}$ depends linearly on $\mathbf{A}$, and $\Delta^{(1)*}$ is a scalar. Therefore, $\Delta^{(1)*}$ should be proportional to $\mathbf{r}\cdot\mathbf{A}$ or $\nabla\cdot\mathbf{A}$. Second, when there is no external field, the system is of uniformity in space. The space-dependence of $\Delta^{(1)*}$ should come from $\mathbf{A(r)}$. So, $\Delta^{(1)*} \propto \nabla\cdot\mathbf{A}(\mathbf{r})$. Third, we take London gauge $\nabla\cdot\mathbf{A} = 0$.

Let us show that $G^{(1)}_{\omega_n}(\mathbf{r},\mathbf{r}')$ satisfies the following equation.

$$G^{(1)}_{\omega_n}(\mathbf{r},\mathbf{r}') = \frac{ie}{m^*}\int d\mathbf{l}\left\{G_{\circ\omega_n}(\mathbf{r}-\mathbf{l})\left[\mathbf{A}(\mathbf{l})\cdot\nabla_{\mathbf{l}}\right]G_{\circ\omega_n}(\mathbf{l}-\mathbf{r}')\right.$$
$$\left. + F_{\circ\omega_n}(\mathbf{l}-\mathbf{r})\left[\mathbf{A}(\mathbf{l})\cdot\nabla_{\mathbf{l}}\right]F^+_{\circ\omega_n}(\mathbf{r}'-\mathbf{l})\right\}. \tag{2.378}$$

From Eq. (2.188) we have

$$\begin{pmatrix} i\omega_n + \frac{\nabla^2_{\mathbf{r}}}{2m^*} + \mu & \Delta_\circ \\ \Delta^*_\circ & i\omega_n - \frac{\nabla^2_{\mathbf{r}}}{2m^*} - \mu \end{pmatrix}\begin{pmatrix} G_{\circ\omega_n}(\mathbf{r}-\mathbf{r}') & F_{\circ\omega_n}(\mathbf{r}-\mathbf{r}') \\ F^+_{\circ\omega_n}(\mathbf{r}-\mathbf{r}') & -G_{\circ-\omega_n}(\mathbf{r}'-\mathbf{r}) \end{pmatrix}$$
$$= \begin{pmatrix} 1 & 0 \\ 0 & 1 \end{pmatrix}\delta(\mathbf{r}-\mathbf{r}'). \tag{2.379}$$

From Eq. 2.379) we have

$$\begin{pmatrix} G_{\circ\omega_n}(\mathbf{r}-\mathbf{r}') & F_{\circ\omega_n}(\mathbf{r}-\mathbf{r}') \\ F^+_{\circ\omega_n}(\mathbf{r}-\mathbf{r}') & -G_{\circ-\omega_n}(\mathbf{r}'-\mathbf{r}) \end{pmatrix} = \begin{pmatrix} i\omega_n + \frac{\nabla^2_{\mathbf{r}}}{2m^*} + \mu & \Delta_\circ \\ \Delta^*_\circ & i\omega_n - \frac{\nabla^2_{\mathbf{r}}}{2m^*} - \mu \end{pmatrix}^{-1}$$
$$\begin{pmatrix} 1 & 0 \\ 0 & 1 \end{pmatrix}\delta(\mathbf{r}-\mathbf{r}'). \tag{2.380}$$

Referring to the proof of Eqs. (2.372) and (2.373), we obtain

$$\begin{pmatrix} i\omega_n + \frac{\nabla^2_{\mathbf{r}}}{2m^*} + \mu & \Delta_\circ \\ \Delta^*_\circ & i\omega_n - \frac{\nabla^2_{\mathbf{r}}}{2m^*} - \mu \end{pmatrix}\begin{pmatrix} G^{(1)}_{\omega_n}(\mathbf{r},\mathbf{r}') & F^{(1)}_{\omega_n}(\mathbf{r},\mathbf{r}') \\ F^{+(1)}_{\omega_n}(\mathbf{r},\mathbf{r}') & -G^{(1)}_{-\omega_n}(\mathbf{r}',\mathbf{r}) \end{pmatrix}$$
$$= \begin{pmatrix} L_{11}(\mathbf{r},\mathbf{r}') & L_{12}(\mathbf{r},\mathbf{r}') \\ L_{21}(\mathbf{r},\mathbf{r}') & L_{22}(\mathbf{r},\mathbf{r}') \end{pmatrix}. \tag{2.381}$$

From Eq. (2.381) we obtain

$$\begin{pmatrix} G^{(1)}_{\omega_n}(\mathbf{r},\mathbf{r}') & F^{(1)}_{\omega_n}(\mathbf{r},\mathbf{r}') \\ F^{+(1)}_{\omega_n}(\mathbf{r},\mathbf{r}') & -G^{(1)}_{-\omega_n}(\mathbf{r}',\mathbf{r}) \end{pmatrix} = \begin{pmatrix} i\omega_n + \frac{\nabla^2_{\mathbf{r}}}{2m^*} + \mu & \Delta_\circ \\ \Delta^*_\circ & i\omega_n - \frac{\nabla^2_{\mathbf{r}}}{2m^*} - \mu \end{pmatrix}^{-1}$$

$$\begin{pmatrix} L_{11}(\mathbf{r},\mathbf{r}') & L_{12}(\mathbf{r},\mathbf{r}') \\ L_{21}(\mathbf{r},\mathbf{r}') & L_{22}(\mathbf{r},\mathbf{r}') \end{pmatrix}.$$

$$= \int d\mathbf{r}" \begin{pmatrix} i\omega_n + \frac{\nabla^2_{\mathbf{r}"}}{2m^*} + \mu & \Delta_\circ \\ \Delta^*_\circ & i\omega_n - \frac{\nabla^2_{\mathbf{r}"}}{2m^*} - \mu \end{pmatrix}^{-1} \begin{pmatrix} 1 & 0 \\ 0 & 1 \end{pmatrix} \delta(\mathbf{r}-\mathbf{r}")$$

$$\begin{pmatrix} L_{11}(\mathbf{r}",\mathbf{r}') & L_{12}(\mathbf{r}",\mathbf{r}') \\ L_{21}(\mathbf{r}",\mathbf{r}') & L_{22}(\mathbf{r}",\mathbf{r}') \end{pmatrix}$$

$$= \int d\mathbf{r}" \begin{pmatrix} G_{\circ\omega_n}(\mathbf{r}-\mathbf{r}") & F_{\circ\omega_n}(\mathbf{r}-\mathbf{r}") \\ F^+_{\circ\omega_n}(\mathbf{r}-\mathbf{r}") & -G_{\circ-\omega_n}(\mathbf{r}"-\mathbf{r}) \end{pmatrix} \begin{pmatrix} L_{11}(\mathbf{r}",\mathbf{r}') & L_{12}(\mathbf{r}",\mathbf{r}') \\ L_{21}(\mathbf{r}",\mathbf{r}') & L_{22}(\mathbf{r}",\mathbf{r}') \end{pmatrix}. \tag{2.382}$$

In the last step of Eq. (2.382) we considered Eq. (2.380). From Eq. (2.382) we have

$$G^{(1)}_{\omega_n}(\mathbf{r},\mathbf{r}') = \int d\mathbf{r}" \left[G_{\circ\omega_n}(\mathbf{r}-\mathbf{r}")L_{11}(\mathbf{r}",\mathbf{r}') + F_{\circ\omega_n}(\mathbf{r}-\mathbf{r}")L_{21}(\mathbf{r}",\mathbf{r}')\right] \tag{2.383}$$

To obtain Eq. (2.378) from Eq. (2.383), we should consider the following six points. First, Substitute Eqs. (2.372) and (2.373) into Eq. (2.382). Second, consider $\Delta^{(1)} = 0$. Third, consider $\nabla \cdot \mathbf{A} = 0$. Fourth, set $\mathbf{r}" = \mathbf{l}$. Fifth, $(\nabla \cdot \mathbf{A} + \mathbf{A} \cdot \nabla)F^+_{\circ\omega_n} = \nabla \cdot (\mathbf{A}F^+_{\circ\omega_n}) + \mathbf{A} \cdot \nabla F^+_{\circ\omega_n} = 2\mathbf{A} \cdot \nabla F^+_{\circ\omega_n}$. Sixth, $F_{\circ\omega_n}(\mathbf{r}-\mathbf{l})L_{21}(\mathbf{l},\mathbf{r}') = [-F_{\circ\omega_n}(\mathbf{l}-\mathbf{r})]\mathbf{A} \cdot \nabla_{\mathbf{l}}[-F^+_{\circ\omega_n}(\mathbf{r}'-\mathbf{l})]ie/m^* = F_{\circ\omega_n}(\mathbf{l}-\mathbf{r})[\mathbf{A} \cdot \nabla_{\mathbf{l}}]F^+_{\circ\omega_n}(\mathbf{r}'-\mathbf{l})ie/m^*$. Considering the above six points, we obtain Eq. (2.378) from Eq. (2.383).

Substituting Eq. (2.378) into Eq. (2.376), we obtain $\mathbf{J}(\mathbf{r})$ expressed by $G_{\circ\omega_n}$, $F_{\circ\omega_n}$, $F^+_{\circ\omega_n}$, and $\mathbf{A}$.

$$\mathbf{J}(\mathbf{r}) = \frac{e^2}{m^{*2}} T \sum_n (\nabla_{\mathbf{r}} - \nabla_{\mathbf{r}'}) \int d\mathbf{l} \{G_{\circ\omega_n}(\mathbf{r}-\mathbf{l})[\mathbf{A}(\mathbf{l}) \cdot \nabla_{\mathbf{l}}] G_{\circ\omega_n}(\mathbf{l}-\mathbf{r}')$$

$$+ F_{\circ\omega_n}(\mathbf{l}-\mathbf{r})[\mathbf{A}(\mathbf{l}) \cdot \nabla_{\mathbf{l}}] F^+_{\circ\omega_n}(\mathbf{r}'-\mathbf{l})\}_{\mathbf{r}'\to\mathbf{r}} - \frac{Ne^2}{m^*}\mathbf{A}(\mathbf{r}). \tag{2.384}$$

Let us look for the Fourier component of $\mathbf{J}(\mathbf{r})$. For example, let us make the Fourier transformation for the first term of Eq. (2.384). The first term can be written as

$$\mathbf{J}_1(\mathbf{r}) = \frac{e^2}{m^{*2}} T \sum_n (\nabla_{\mathbf{r}} - \nabla_{\mathbf{r}'})|_{\mathbf{r}'\to\mathbf{r}} \int d\mathbf{l} \int \frac{d\mathbf{p}}{(2\pi)^3} e^{i(\mathbf{r}-\mathbf{l})\cdot\mathbf{p}} G_{\circ\omega_n}(\mathbf{p})$$

$$\int \frac{d\mathbf{k}}{(2\pi)^3} e^{i\mathbf{l}\cdot\mathbf{k}} \mathbf{A}(\mathbf{k}) \cdot \nabla_{\mathbf{l}} \int \frac{d\mathbf{p}_1}{(2\pi)^3} e^{i(\mathbf{l}-\mathbf{r}')\cdot\mathbf{p}_1} G_{\circ\omega_n}(\mathbf{p}_1). \tag{2.385}$$

Finishing the operation $\nabla_{\mathbf{l}}$, the integration over $\mathbf{l}$, the operations $\nabla_{\mathbf{r}}$ and $\nabla_{\mathbf{r}'}$, and setting $\mathbf{r}' = \mathbf{r}$, we obtain

$$\mathbf{J}_1(\mathbf{r}) = \frac{e^2}{m^{*2}} T \sum_n \int \frac{d\mathbf{p}}{(2\pi)^3} \int \frac{d\mathbf{k}}{(2\pi)^3} \int \frac{d\mathbf{p}_1}{(2\pi)^3} (i\mathbf{p} + i\mathbf{p}_1)(2\pi)^3 \delta(-\mathbf{p} + \mathbf{k} + \mathbf{p}_1)$$

$$G_{o\omega_n}(\mathbf{p})\mathbf{A}(\mathbf{k})i\mathbf{p}_1 G_{o\omega_n}(\mathbf{p}_1)e^{i\mathbf{k}\cdot\mathbf{r}}$$

$$= -\frac{e^2}{m^{*2}}T\sum_n\int\frac{d\mathbf{p}}{(2\pi)^3}\int\frac{d\mathbf{k}}{(2\pi)^3}(\mathbf{p}+\mathbf{p}-\mathbf{k})G_{o\omega_n}(\mathbf{p})\mathbf{A}(\mathbf{k})\cdot(\mathbf{p}-\mathbf{k})G_{o\omega_n}(\mathbf{p}-\mathbf{k})e^{i\mathbf{k}\cdot\mathbf{r}} \quad (2.386)$$

Setting **p**-**k**/2=**p**$'$, and considering **A(k)**·**k**=0 [$\nabla\cdot\mathbf{A}(\mathbf{r})=0$], Eq. (2.386) becomes

$$\mathbf{J}_1(\mathbf{r}) = \frac{-2e^2T}{m^{*2}}\sum_n\int\frac{d\mathbf{p}'}{(2\pi)^3}\int\frac{d\mathbf{k}}{(2\pi)^3}\mathbf{p}'G_{o\omega}\left(\mathbf{p}'+\frac{\mathbf{k}}{2}\right)\mathbf{A}(\mathbf{k})\cdot\mathbf{p}'G_{o\omega_n}\left(\mathbf{p}'-\frac{\mathbf{k}}{2}\right)e^{i\mathbf{k}\cdot\mathbf{r}}. \quad (2.387)$$

Note

$$\mathbf{J}(\mathbf{r}) = \int\frac{d\mathbf{k}}{(2\pi)^3}e^{i\mathbf{k}\cdot\mathbf{r}}\mathbf{J}(\mathbf{k}). \quad (2.388)$$

So,

$$\mathbf{J}(\mathbf{k}) = -\frac{2e^2}{m^{*2}}T\sum_n\int\frac{d\mathbf{p}}{(2\pi)^3}\mathbf{p}[\mathbf{p}\cdot\mathbf{A}(\mathbf{k})]\left[G_{o\omega_n}(\mathbf{p}_+)G_{o\omega_n}(\mathbf{p}_-)+F_{o\omega_n}(\mathbf{p}_+)F^+_{o\omega_n}(\mathbf{p}_-)\right]$$

$$-\frac{e^2}{m^*}\mathbf{A}(\mathbf{k})N. \quad (2.389)$$

$$\mathbf{p}_\pm = \mathbf{p}\pm\frac{\mathbf{k}}{2}. \quad (2.390)$$

Because there is a factor **p[p·A(k)]** in Eq. (2.389), we have **J(k)**$\|$**A(k)**.

$$\mathbf{J}(\mathbf{k}) = -\frac{Ne^2}{m^*}\overline{Q}(\mathbf{k})\mathbf{A}(\mathbf{k}). \quad (2.391)$$

Choose the vector **k** as the poler axis of variable **p**. From $\mathbf{k}\cdot\mathbf{A}(\mathbf{k})=0$, we have $\mathbf{k}\perp\mathbf{A}(\mathbf{k})$. Suppose $\mathbf{A}(\mathbf{k})\|\mathbf{p}_x$. Then $\mathbf{p}[\mathbf{p}\cdot\mathbf{A}(\mathbf{k})]=p_x^2\mathbf{A}(\mathbf{k})$.

$$\overline{Q}(\mathbf{k}) = 1+\frac{2T}{m^*N}\sum_n\int\frac{d\mathbf{p}}{(2\pi)^3}p_x^2\left[G_{o\omega_n}(\mathbf{p}_+)G_{o\omega_n}(\mathbf{p}_-)+F_{o\omega_n}(\mathbf{p}_+)F^+_{o\omega_n}(\mathbf{p}_-)\right]. \quad (2.392)$$

Considering $d\xi = d(p^2/2m*-\mu) = pdp/m^*$, $d\mathbf{p} = p^2\sin\theta d\theta d\phi dp$, $p_x^2 = p^2\sin^2\theta\cos^2\phi$, and Eqs. (2.182) and (2.183), we obtain from Eq. (2.392)

$$\overline{Q}(\mathbf{k}) = 1+\frac{2T}{m^*N}\sum_n\frac{1}{(2\pi)^3}\int p^2\sin\theta d\theta d\phi\frac{m^*}{p}d\xi p^2\sin^2\theta\cos^2\phi$$

$$\frac{(i\omega_n+\xi_+)(i\omega_n+\xi_-)+|\Delta|^2}{(\omega_n^2+\xi_+^2+|\Delta|^2)(\omega_n^2+\xi_-^2+|\Delta|^2)}, \quad (2.393)$$

where Δ represents the Δ_o in Eq. (2.370). To continue derivation from Eq. (2.393), we make the following analyses. First, $p^3\approx p_F^3$, because there is ξ^4 in the denominator, and thus the integration near Fermi surface has important contribution. Second,

$$\int_0^\infty dp \longrightarrow \int_{-\mu}^\infty d\xi \approx \int_{-\infty}^{+\infty} d\xi. \quad (2.394)$$

Third, $p_F^3/(3\pi^2) = N =$ electron number per unit volume. Fourth,

$$\int_0^{2\pi} \cos^2\phi d\phi = \pi. \tag{2.395}$$

$$\overline{Q}(\mathbf{k}) = 1 + \frac{3T}{4}\sum_n \int_0^{\pi} \sin^3\theta d\theta \int_{-\infty}^{+\infty} d\xi \frac{(i\omega_n + \xi_+)(i\omega_n + \xi_-) + |\Delta|^2}{(\omega_n^2 + \xi_+^2 + |\Delta|^2)(\omega_n^2 + \xi_-^2 + |\Delta|^2)}. \tag{2.396}$$

Considering $|\mathbf{k}| \ll p_F$ and $p \approx p_F$, we have

$$\xi_\pm = \frac{\left(\mathbf{p} \pm \frac{\mathbf{k}}{2}\right)^2 - p_F^2}{2m^*} \approx \xi \pm \frac{\mathbf{v}_F \cdot \mathbf{k}}{2}, \tag{2.397}$$

where $\mathbf{v}_F = \mathbf{p}_F/m^*$.

Is the order of integration and summation in Eq. (2.396) arbitrary? To answer this question, let us look at the corresponding expression of normal metal ($\Delta = 0$).

$$E_R = \frac{3T}{4}\sum_n \int_0^{\pi} \sin^3\theta d\theta \int_{-\infty}^{+\infty} d\xi \frac{1}{(i\omega_n - \xi_-)(i\omega_n - \xi_+)}. \tag{2.398}$$

At first, let us try to perform the integration with respect to ξ. Because the singular points are in the same side of real axis, we have

$$\int_{-\infty}^{+\infty} d\xi \frac{1}{(i\omega_n - \xi_+)(i\omega_n - \xi_-)} = 0. \tag{2.399}$$

If we do the summation over ω_n at first, then after summation we obtain

$$E_R = \frac{3}{8}\int_0^{\pi} \sin^3\theta d\theta \lim_{a\to\infty} \int_{-a}^{+a} d\xi \frac{1}{\mathbf{v}_F \cdot \mathbf{k}} \left[\tanh\frac{\xi_-}{2T} - \tanh\frac{\xi_+}{2T}\right]. \tag{2.400}$$

Noting $\int \tanh x dx = \ln(\mathrm{ch} x)$ and $\xi_\pm = \xi \pm \mathbf{v}_F \cdot \mathbf{k}/2$, we obtain

$$E_R = -1. \tag{2.401}$$

The cause of the non-unique-value of E_R is as follows. The ξ integration in Eq. (2.400) has large value only near the Fermi surface, i.e., $\xi \approx 0$. On the contrary, the ξ integration in Eq. (2.398) has large value even for large value of ξ. We should do the ω_n summation at first because we have used the small ξ approximation in Eq. (2.397). The ω_n summation in Eq. (2.396) is very difficult. To avoid this difficulty, we use Eq. (2.401) to substitute for the "1" in Eq. (2.396). Thus we obtain

$$\overline{Q}(\mathbf{k}) = \frac{3T}{4}\sum_n \int_0^{\pi} \sin^3\theta d\theta \int_{-\infty}^{+\infty} d\xi \left[\frac{(i\omega_n + \xi_+)(i\omega_n + \xi_-) + |\Delta|^2}{(\omega_n^2 + \xi_+^2 + |\Delta|^2)(\omega_n^2 + \xi_-^2 + |\Delta|^2)}\right.$$

$$-\left.\frac{(i\omega_n+\xi_+)(i\omega_n+\xi_-)}{(\omega_n^2+\xi_+^2)(\omega_n^2+\xi_-^2)}\right]. \tag{2.402}$$

In Eq. (2.402) the integrated function is significant only near the Fermi surface. So, we may perform the ξ integration at first. Reducing the two fractions to a common denominator, finishing the integration over ξ, and using

$$\int_0^{\pi} d\theta \sin^3\theta = \int_{-1}^{+1}(1-\mu^2)d\mu, \tag{2.403}$$

we obtain

$$\overline{Q}(\mathbf{k}) = \frac{3T}{4}\sum_n \int_{-1}^{+1} d\mu(1-\mu^2)\pi\frac{1}{\sqrt{\omega_n^2+|\Delta|^2}}\frac{|\Delta|^2}{\omega_n^2+|\Delta|^2+v_F^2k^2\mu^2/4}. \tag{2.404}$$

2.10.3. Physical Meaning of $\overline{Q}$

We calculate $\overline{Q}(\mathbf{k})$ in Eq. (2.404) under two limiting cases: $v_Fk \ll T^*$ and $v_Fk \gg T^*$. The physical meaning of $v_Fk \ll T^*$ is as follows. We know from Eq. (2.21) that the coherence length, i.e., the size of a Cooper pair, $\xi_\circ \approx v_F/|\Delta(0)|$. Here, $\Delta(0)$ is the gap at $T = 0$ K. $|\Delta(0)| > T^*$. So, $\xi_\circ < v_F/T^*$. Considering $v_Fk \ll T^*$, we have $\xi_\circ \ll 1/k \approx \lambda \approx \lambda_L$. Here, λ is the wave length of current wave, and λ_L is the London penetration depth. The $\xi_\circ \ll \lambda_L$, i.e., coherence length is much less than the London penetration depth, means that the field variation in the range of $\xi_\circ$ is not large. So, the relation between current of Cooper pairs and field is local. $\xi_\circ \ll \lambda_L$ may be called London case. In London case

$$\overline{Q}(\mathbf{k}) \cong \overline{Q} = \frac{3\pi T|\Delta|^2}{4}\sum_n \int_{-1}^{+1} d\mu\frac{1-\mu^2}{(\omega_n^2+|\Delta|^2)^{3/2}} = \pi T|\Delta|^2\sum_n\frac{1}{(\omega_n^2+|\Delta|^2)^{3/2}}. \tag{2.405}$$

From Eq. (2.405) we know that $\overline{Q}(\mathbf{k})$ does not depend on $\mathbf{k}$. Substituting Eq. (2.405) into Eq. (2.391), we obtain

$$\mathbf{J}(\mathbf{r}) = -\frac{Ne^2}{m^*}\overline{Q}\mathbf{A}(\mathbf{r}), \tag{2.406}$$

where N is the number density of electrons. Introducing $e^* = 2e$ and $m^{**} = 2m^*$, Eq. (2.406) becomes

$$\mathbf{J}(\mathbf{r}) = -\frac{|e|^{*2}N/2}{m^{**}}\overline{Q}\mathbf{A}(\mathbf{r}). \tag{2.407}$$

In appropriate gauge Eq. (2.329) becomes ($c = 1$)

$$\mathbf{J}(\mathbf{r}) = -\frac{e^{*2}n}{m^{**}}\mathbf{A}(\mathbf{r}), \tag{2.408}$$

where n is the number density of Cooper pairs without long-range phase coherence. Making comparison between Eq. (2.407) and (2.408), we have

$$\overline{Q} = \frac{n}{N/2}, \tag{2.409}$$

i.e., $\overline{Q}$ represents the probability of formation of one Cooper pair without long-range phase coherence at $T_c < T < T^*$ for unconventional superconductor. In unconventional superconductor at $T < T_c$ there are Cooper pairs with and without long-range phase coherence. In this case $\overline{Q}$ represents the probability of formation of one Cooper pair with or without long-range phase coherence because Eq. (2.407) is correct for Cooper pairs with and without long-range phase coherence. Therefore, for unconventional superconductor at $T < T_c$

$$\overline{Q} = \frac{n+n_s}{N/2}, \tag{2.410}$$

where n_s represents the number density of Cooper pairs with long-range phase coherence.

At $T = 0$ K

$$\overline{Q} = \frac{n+n_s}{N/2} = \pi|\Delta|^2 \int_{-\infty}^{+\infty} \frac{d\omega}{2\pi} \frac{1}{(\omega^2+|\Delta|^2)^{3/2}} = 1, \tag{2.411}$$

i.e., all electrons become Cooper pairs with and without long-range phase coherence.

At $T < T^*$ $(T \approx T^*)$, $|\Delta| \ll T$.

$$\overline{Q} = \frac{n}{N/2} \cong 2\pi T|\Delta|^2 \sum_{n\geq 0} \frac{1}{[(2n+1)\pi T]^3}$$

$$= \frac{2|\Delta|^2}{\pi^2 T^2} \frac{2^3-1}{2^3} \zeta(3). \tag{2.412}$$

Substituting Eq. (2.146) into Eq. (2.412), we have

$$\overline{Q} = 2\left(1 - \frac{T}{T^*}\right). \tag{2.413}$$

For conventional superconductor we have at $T = 0$ K

$$\overline{Q} = \frac{n_s}{N/2} = 1. \tag{2.414}$$

Considering that for the conventional superconductor T_c is very nearly equal to T^*,we have at $T < T_c$ $(T \approx T_c)$ $(|\Delta| \ll T)$

$$\overline{Q} = \frac{n_s}{N/2} = 2\left(1 - \frac{T}{T_c}\right). \tag{2.415}$$

Let us discuss the case $v_F k \gg T^*$ in Eq. (2.404). In this case $\xi_o \gg \lambda_L$, which means that the field variation in the range of ξ_o is large. So, the relation between current of Cooper pairs and the field is non-local. The $v_F k \gg T^*$ may be called Pippard case. In Eq. (2.404) the integrated expression has poles at $v_F k\mu/2 = \pm i\sqrt{\omega_n^2+|\Delta|^2}$. Because $\sqrt{\omega_n^2+|\Delta|^2}$ is always larger than T^*, at poles $|\mu| < T^*/v_F k \ll 1$. So, the μ^2 in numerator can be omitted. Eq. (2.404) becomes

$$\overline{Q}(\mathbf{k}) = \frac{3\pi T}{4} \sum_n \int_{-1}^{+1} d\mu \frac{1}{\sqrt{\omega_n^2+|\Delta|^2}} \frac{|\Delta|^2}{\omega_n^2+|\Delta|^2+v_F^2 k^2\mu^2/4}$$

$$
\begin{aligned}
&= \frac{3\pi^2 T}{2v_F k}\sum_n \frac{|\Delta|^2}{\omega_n^2+|\Delta|^2} \\
&= \frac{3\pi^2}{4v_F k}|\Delta|\tanh\frac{|\Delta|}{2T}.
\end{aligned}
\tag{2.416}
$$

From Eq. (2.416) we know that $\overline{Q}(\mathbf{k})$ depends on $|\mathbf{k}|$. Therefore, the $Q(x-y)$ in Eq. (2.357) is space-dependent, and the relation between current of Cooper pairs and field is non-local.

In chapters 1 and 2 we introduce the general phenomenological and microscopic theories for conventional and unconventional superconductors. For conventional superconductors there are many excellent books to interpret special phenomenological and microscopic theories [47], [63]– [72]. We suggest readers consulting them.

Chapter 3

Two Local Spin-Mediated Interaction

3.1. Introduction

The field of high-T_c superconductivity will approach its twenty-first birthday. The first publication in this area of research appeared in September, 1986 [5]. At the beginning of this discovery Anderson clearly pointed out that the strong correlation between electrons causes the high T_c [73]. The effective attractive interaction between two carriers mediated by two local spins, i.e., two local spin-mediated interaction (TLSMI), proposed in this book and Refs. [74, 75], is a kind of interactions caused by the strong correlation. The physical picture of TLSMI is as follows. There are both Kondo and Heisenberg Hamiltonian at the same time in TLSMI system. TLSMI is constituted by the following three interactions, i.e., the interaction between $\mathbf{k}\uparrow$ and the first local spin, the interaction between the first and the second local spins in antiferromagnetic or ferromagnetic orders, and the interaction between $-\mathbf{k}\downarrow$ and the second local spin. The essence of TLSMI is a multiple exchange interaction. Taking TLSMI as the Cooper pairing potential, using BCS gap equation, and considering the long-range phase coherence condition of the wavefunctions of Cooper pairs, this chapter points out the six generic features of the superconductivity and the pseudogap in the high-T_c cuprates, chapter 6 points out the six generic features in many other systems, and chapter 4 explains a series of observed phenomena, and emphasizes that the verified symmetry of the gap in the high-T_c cuprates is just our composite d-wave in this chapter rather than the recent popular simple d-wave in many different theories in Refs. [76] – [80].

Before this book there were already some popular theories on the high-T_c and the pseudogap. The differences between our theory and the other main theories are as follows. For TLSMI the spin and the charge do not separate, so our mechanism of the superconductivity and the pseudogap in the high-T_c cuprates is different from the recent popular spin-charge separation theory in Refs. [76, 77]. The single local spin flip in antiferromagnetic and ferromagnetic regions possessing long- or short-range orders causes large energy increment, so the single spin flip in these regions is not stable. In our derivation of TLSMI we think that the single local spin flip in these regions is prohibited when a single local spin scatters the carrier. The crucial conceptual difference between our mechanism and the magnon-mediated mechanism [78, 79], therefore, is that the spin fluctuation is not required in our mechanism. There has been considerable attention focused on Zhang's SO(5) theory [80], which attempts to unify the d-wave superconductivity and antiferromagnetism. In this theory, the antiferromagnetic Neel vector and the complex superconducting order parameter form the components of a five-dimensional "superspin" vector. The SO(5) rotational symmetry of the superspin is explicitly broken by anisotropy which favors antiferromagnetism for undoped or lightly doped high-T_c cuprates and superconductivity for more heavily doped high-T_c cuprates. The transition from antiferromagnetism to d-wave superconductivity is analogous to spin-flop transition of a uniaxial Heisenberg antiferromagnet in a parallel magnetic field. We think that the SO(5) theory is still a phenomenological theory in essence. However, the superconductivity caused by TLSMI has to coexist with, for example, antiferromagnetism, and can be both d- and p-wave. The review papers on the high-T_c superconductivity are Refs. [81, 82].

This chapter is organized as follows. In section 3.2 we derive the expressions of TLSMI for 2-, 3-, and 1-D (dimension) system with 1/2 and high spin, discuss the solution of BCS gap equation, and point out the six generic features of the solutions. In section 3.3 we discuss the granular model of the high-T_c cuprates, give the condition of long-range phase coherence of the wavefunctions of Cooper pairs, and give T_c equation. Section 3.3 demonstrates that the small size of Cooper pairs decides the high-T_c cuprates different from conventional superconductors and belonging to the unconventional superconductors. Here, T_c is the temperature of the long-range phase coherence of Cooper pairs. i.e., the superconducting transition temperature.

3.2. Expression of TLSMI and Gap Equation

3.2.1. 2-D and Spin 1/2

The TLSMI in this case has the simplest form. The formulas in this section can be directly applied to the high-T_c cuprates. Considering the CuO_2 plane of the high-T_c cuprates, we take the system as a 2-D lattice. The CuO_2 plane is a square lattice, where the Cu^{++} ions are on the four corners of a square, and the O^{--} ions are in the middle of two nearest-neighbor Cu^{++} ions. The schematic plot of the CuO_2 plane is shown in Fig. 3.1. The Ref. [83] quoted seventeen papers for the d-p_σ model of the CuO_2 plane. The d-p_σ model is constituted by Cu:$d_{x^2-y^2}$ and O:p_σ orbits in the CuO_2 plane. The strong correlated Hamiltonian is [83]

$$H = t_\circ \sum_{i,\alpha,s} (d_{is}^{+} p_{\alpha s} + h.c) + U_d \sum_i n_{di\uparrow} n_{di\downarrow} + U_p \sum_\alpha n_{p\alpha\uparrow} n_{p\alpha\downarrow}$$
$$+ V_\circ \sum_{i,\alpha} n_{di} n_{p\alpha} + \varepsilon_d \sum_i n_{di} + \varepsilon_p \sum_\alpha n_{p\alpha}, \tag{3.1}$$

where d_{is}^{+} and $p_{\alpha s}^{+}$ create holes on the Cu:$d_{x^2-y^2}$ and O:p_σ orbits at sites i and α with spin s, respectively; n_{di} ($n_{p\alpha}$) is the number operator of d (p) holes, $n_{di} \equiv n_{di\uparrow} + n_{di\downarrow}$, $n_{p\alpha} \equiv n_{p\alpha\uparrow} + n_{p\alpha\downarrow}$; $-t_\circ$ is the hopping integral for the holes between adjacent Cu:$d_{x^2-y^2}$ and O:p_σ orbits, and U_p, U_d, and $V_\circ$ are intra- and interatomic Coulomb repulsion on O:p_σ orbits, Cu:$d_{x^2-y^2}$ orbits, and between both orbits, respectively. Hereafter, the site index with an alphabetic letter stands for Cu site and that with a Greek letter stands for the O site in the

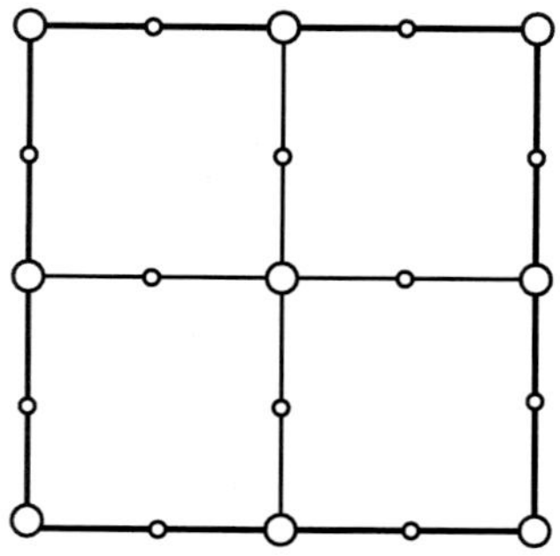

Figure 3.1. The schematic plot of the CuO_2 plane of the high-T_c cuprates. The large circle indicates Cu^{++} ion. The small circle indicates O^{--} ion.

CuO_2 plane. The phases of the orbits are chosen so that the hopping integral, $-t_\circ$, is a negative constant. It should be noted that the sign of hopping integral for holes is opposite to that for electrons. The $t_\circ$ is given as a constant about 1 eV [84]– [87]. We use x to stand for the hole number in one unit cell of Cu lattice in the CuO_2 plane. It is generally assumed that at half-filling, $x = 0$, there exists one hole per Cu site, i.e., Cu^{++}; and for $x > 0$ extra holes go into O:p_σ orbits and Cu^{++} is stable under doping. Hence, each Cu site has a spin whose magnitude is equal to 1/2. The conditions that Cu^{++} is stable under doping are [83]

$$\varepsilon_p - \varepsilon_d - V_\circ > 0, \tag{3.2}$$

$$U_d - \varepsilon_p + \varepsilon_d - 2V_\circ > 0, \tag{3.3}$$

$$U_d - \varepsilon_p + \varepsilon_d - V_\circ - U_p > 0. \tag{3.4}$$

We will see in chapter 5 that these conditions are satisfied in case of the high-T_c cuprates. The mobile carrier in the CuO_2 plane is $O_{p\sigma}$ hole. By treating the first term in Eq. (3.1) as a perturbation, many authors derived an effective Hamiltonian consisted of free $O_{p\sigma}$ hole Hamiltonian, Kondo Hamiltonian, and Heisenberg Hamiltonian [83]. This book adopts the form in Ref. [83], which is

$$H = -\sum_{i\alpha\beta s} T_{\alpha\beta} p^+_{\alpha s} p_{\beta_s} + J_K \sum_{i\alpha\beta ss'} \widehat{\mathbf{S}}_i \cdot \vec{\sigma}_{ss'} p^+_{\alpha s} p_{\beta s'} + J \sum_{ij(i<j)} \widehat{\mathbf{S}}_i \cdot \widehat{\mathbf{S}}_j, \tag{3.5}$$

where the summation over α and β stands for the oxygen sites around i-th Cu^{++} site; $p_{\alpha s}$ annihilates $O_{p\sigma}$ hole with spin s at site α; $\widehat{\mathbf{S}}_i$ is the local spin operator of Cu^{++} at site i; $\vec{\sigma}$ is Pauli matrix vector; and i and j are the nearest neighbors. Expand $p_{\alpha s}$ in $\mathbf{k}$ space. Here $\mathbf{k}$ is the wave vector in Brillouin zone of the oxygen lattice in the CuO_2 plane. The second term in Eq. (3.5) is Kondo Hamiltonian, H_K, which implies that the $O_{p\sigma}$ holes with $\mathbf{k}\uparrow$ and $-\mathbf{k}\downarrow$ can have interactions with the local spins of Cu^{++} at sites i and j, respectively. The third term in Eq. (3.5), on the other hand, is Heisenberg interaction, H_H, between the two nearest-neighbor local spins at sites i and j. We call formally the effective interaction between the

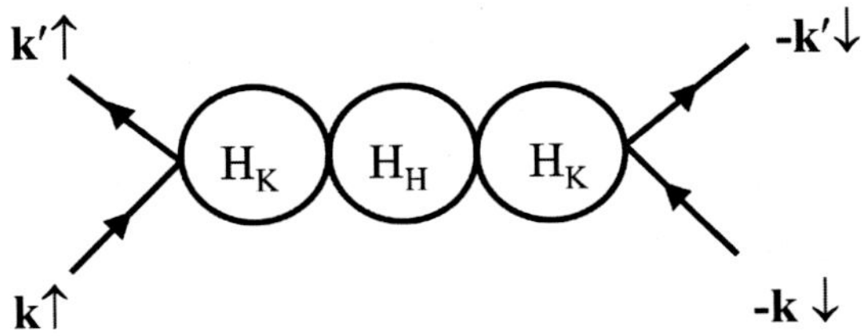

Figure 3.2. The visual picture of TLSMI between **k** ↑ and -**k** ↓.

two $O_{p\sigma}$ holes with **k** ↑ and $-$**k** ↓, mediated by $H_K - H_H - H_K$, TLSMI (two local spin-mediated interaction). The TLSMI between holes with **k** ↑ and $-$**k** ↓ can be shown visually by Fig. 3.2. To write down the expression of TLSMI we extend the Abrikosov pseudo-fermion method for the local spin operator of magnetic impurity to the Hamiltonian in Eq. (3.5). In view of the importance and universality of our extended Abrikosov pseudo-fermion method in general magnetic problems, this method is given in chapter 7. We can conveniently write down the expressions for any process describing by the Hamiltonian in Eq. (3.5) using our extended Abrikosov pseudo-fermion method. We transform $p_{\alpha s}$ into **k** space, and use the pseudo-fermion representation, Eq. (7.1), for the local spin operator, $\widehat{\mathbf{S}}_i$.

The space vectors appearing in this transformation are shown in Fig. 3.3. The expansion of $p_{\beta s}$ in **k** space is

$$p_{\beta s} = \frac{1}{\sqrt{2N_{Cu}}} \sum_{\mathbf{k}} p_{\mathbf{k}s} e^{i\mathbf{k}\cdot(\mathbf{R}_i + \mathbf{a}_\beta)}, \tag{3.6}$$

where N_{Cu} is the number of unit cell of Cu lattice in the CuO_2 plane, $2N_{Cu} = N_O$, N_O is the number of unit cell of O lattice in the CuO_2 plane, the definition of $\mathbf{a}_\beta$ is given by Fig. 3.3, and $\mathbf{R}_i$ is the position vector of Cu^{++} ion. Substituting Eq. (3.6) into the first term of Eq. (3.5) yields

$$-\sum_{i\alpha\beta s} T_{\alpha\beta} p^+_{\alpha s} p_{\beta s} = -\sum_{i\alpha\beta s} T_{\alpha\beta} \frac{1}{\sqrt{2N_{Cu}}} \sum_{\mathbf{k}} e^{-i\mathbf{k}\cdot(\mathbf{R}_i + \mathbf{a}_\alpha)} p^+_{\mathbf{k}s}$$
$$\frac{1}{\sqrt{2N_{Cu}}} \sum_{\mathbf{k}'} e^{i\mathbf{k}'\cdot(\mathbf{R}_i + \mathbf{a}_\beta)} p_{\mathbf{k}'s}. \tag{3.7}$$

Performing the summation over i yields

$$-\sum_{i\alpha\beta s} T_{\alpha\beta} p^+_{\alpha s} p_{\beta s} = -\frac{1}{2} \sum_{\mathbf{k}s} \sum_{\alpha\beta} T_{\alpha\beta} e^{-i\mathbf{k}\cdot(\mathbf{a}_\alpha - \mathbf{a}_\beta)} p^+_{\mathbf{k}s} p_{\mathbf{k}s}. \tag{3.8}$$

Performing the summation over α and β, we obtain 16 terms. Considering $T_{11} = T_{22} = T_{33} = T_{44}$, $T_{12} = T_{23} = T_{34} = T_{41} = T_{21} = T_{32} = T_{43} = T_{14}$, and $T_{42} = T_{24} = T_{13} = T_{31}$, we obtain

$$-\sum_{i\alpha\beta s} T_{\alpha\beta} p^+_{\alpha s} p_{\beta s} = -2T_{11} \sum_{\mathbf{k}s} p^+_{\mathbf{k}s} p_{\mathbf{k}s} - 2T_{12} \sum_{\mathbf{k}s} p^+_{\mathbf{k}s} p_{\mathbf{k}s} [\cos(k_x a) + \cos(k_y a)]$$

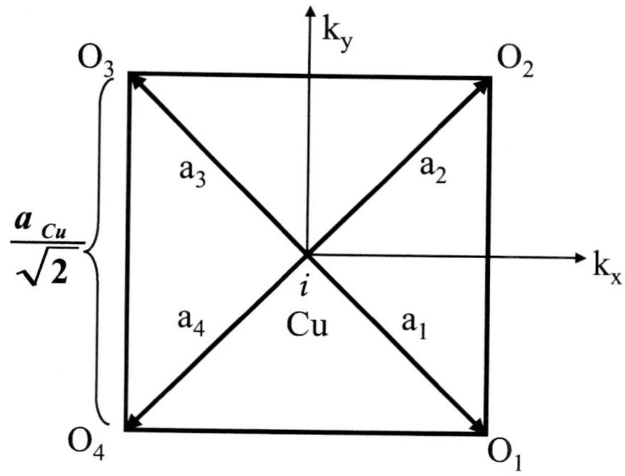

Figure 3.3. k_x and k_y are along the O-O bond direction. a_{Cu} is the nearest-neighbor distance between two Cu^{++} ions. O_1, O_2, O_3, and O_4 are the four O^{--} ions around the i-th Cu^{++} ion.

$$-2T_{13}\sum_{\mathbf{k}s} p^+_{\mathbf{k}s} p_{\mathbf{k}s} \cos(k_x a)\cos(k_y a), \tag{3.9}$$

where a $(=a_{Cu}/\sqrt{2})$ is the nearest-neighbor distance between two O^{--} ions in the CuO_2 plane. Substituting Eq. (3.6) and the pseudo-fermion representation for the local spin operator $\widehat{\mathbf{S}}_i$, Eq. (7.1), into the second term of Eq. (3.5) yields

$$H_K = J_K \sum_{i\delta\gamma ss'\alpha\beta} \mathbf{S}_{i\delta\gamma} \cdot \vec{\sigma}_{ss'} d^+_{i\delta} d_{i\gamma}$$

$$\frac{1}{\sqrt{2N_{Cu}}}\sum_{\mathbf{k}} p^+_{\mathbf{k}s} e^{-i\mathbf{k}\cdot(\mathbf{R_i}+\mathbf{a}_\alpha)} \frac{1}{\sqrt{2N_{Cu}}}\sum_{\mathbf{k}'} p_{\mathbf{k}'s'} e^{i\mathbf{k}'\cdot(\mathbf{R_i}+\mathbf{a}_\beta)}. \tag{3.10}$$

Let us calculate the summation over α and β in Eq. (3.10).

$$\sum_{\alpha\beta} e^{i(\mathbf{k}'\cdot\mathbf{a}_\beta-\mathbf{k}\cdot\mathbf{a}_\alpha)} = e^{i(\mathbf{k}'\cdot\mathbf{a}_1-\mathbf{k}\cdot\mathbf{a}_1)} + e^{i(\mathbf{k}'\cdot\mathbf{a}_1-\mathbf{k}\cdot\mathbf{a}_2)} + e^{i(\mathbf{k}'\cdot\mathbf{a}_1-\mathbf{k}\cdot\mathbf{a}_3)}$$

$$+e^{i(\mathbf{k}'\cdot\mathbf{a}_1-\mathbf{k}\cdot\mathbf{a}_4)} + e^{i(\mathbf{k}'\cdot\mathbf{a}_2-\mathbf{k}\cdot\mathbf{a}_1)} + e^{i(\mathbf{k}'\cdot\mathbf{a}_2-\mathbf{k}\cdot\mathbf{a}_2)}$$

$$+e^{i(\mathbf{k}'\cdot\mathbf{a}_2-\mathbf{k}\cdot\mathbf{a}_3)} + e^{i(\mathbf{k}'\cdot\mathbf{a}_2-\mathbf{k}\cdot\mathbf{a}_4)} + e^{i(\mathbf{k}'\cdot\mathbf{a}_3-\mathbf{k}\cdot\mathbf{a}_1)}$$

$$+e^{i(\mathbf{k}'\cdot\mathbf{a}_3-\mathbf{k}\cdot\mathbf{a}_2)} + e^{i(\mathbf{k}'\cdot\mathbf{a}_3-\mathbf{k}\cdot\mathbf{a}_3)} + e^{i(\mathbf{k}'\cdot\mathbf{a}_3-\mathbf{k}\cdot\mathbf{a}_4)}$$

$$+e^{i(\mathbf{k}'\cdot\mathbf{a}_4-\mathbf{k}\cdot\mathbf{a}_1)} + e^{i(\mathbf{k}'\cdot\mathbf{a}_4-\mathbf{k}\cdot\mathbf{a}_2)} + e^{i(\mathbf{k}'\cdot\mathbf{a}_4-\mathbf{k}\cdot\mathbf{a}_3)}$$

$$+e^{i(\mathbf{k}'\cdot\mathbf{a}_4-\mathbf{k}\cdot\mathbf{a}_4)}. \tag{3.11}$$

The vectors $\mathbf{a}_1$, $\mathbf{a}_2$, $\mathbf{a}_3$, and $\mathbf{a}_4$ can be expressed as

$$\mathbf{a}_1 = \frac{a}{2}\mathbf{e}_x + \frac{-a}{2}\mathbf{e}_y, \tag{3.12}$$

$$\mathbf{a}_2 = \frac{a}{2}\mathbf{e}_x + \frac{a}{2}\mathbf{e}_y, \tag{3.13}$$

$$\mathbf{a}_3 = \frac{-a}{2}\mathbf{e}_x + \frac{a}{2}\mathbf{e}_y, \tag{3.14}$$

$$\mathbf{a}_4 = \frac{-a}{2}\mathbf{e}_x + \frac{-a}{2}\mathbf{e}_y, \tag{3.15}$$

where $\mathbf{e}_x$ and $\mathbf{e}_y$ are unit vectors without dimension in the k_x and k_y directions, respectively. Substituting Eqs. (3.12) - (3.15) into Eq. (3.11) yields

$$\sum_{\alpha\beta} e^{i(\mathbf{k}'\cdot\mathbf{a}_\beta - \mathbf{k}\cdot\mathbf{a}_\alpha)} = 16\cos(k_x a/2)\cos(k_y a/2)\cos(k_x' a/2)\cos(k_y' a/2). \tag{3.16}$$

Substituting Eqs. (7.1), (3.9), (3.10), and (3.16) into Eq. (3.5) yields

$$H = \sum_{\mathbf{k}s} \varepsilon(\mathbf{k}) p^+_{\mathbf{k}s} p_{\mathbf{k}s}$$

$$+ \frac{J_K}{2N_{Cu}} \sum_{i\delta\gamma\mathbf{k}\mathbf{k}'ss'} \mathbf{S}_{i\delta\gamma} \cdot \vec{\sigma}_{ss'} d^+_{i\delta} d_{i\gamma} p^+_{\mathbf{k}s} p_{\mathbf{k}'s'} \exp[i(\mathbf{k}' - \mathbf{k}) \cdot \mathbf{R}_i]$$

$$16\cos(k_x a/2)\cos(k_y a/2)\cos(k_x' a/2)\cos(k_y' a/2)$$

$$+J \sum_{i<j\delta\delta'\gamma\gamma'} \mathbf{S}_{i\delta\delta'} \cdot \mathbf{S}_{j\gamma\gamma'} d^+_{i\delta} d_{i\delta'} d^+_{j\gamma} d_{j\gamma'}$$

$$\equiv H_\circ + H_K + H_H, \tag{3.17}$$

$$\varepsilon(\mathbf{k}) = -2T_{11} - 2T_{12}[\cos(k_x a) + \cos(k_y a)] - 2T_{13}\cos(k_x a)\cos(k_y a), \tag{3.18}$$

where a is the lattice constant of the oxygen lattice in the CuO_2 plane, $\mathbf{R}_i$ the position vector of Cu^{++}, d_{is} the pseudo-fermion annihilation operator of spin s at site i, N_{Cu} the number of Cu^{++}, and $\mathbf{S} = \vec{\sigma}/2$. The simplest Feynman diagram of TLSMI is shown in Fig. 3.4. The renormalized Feynman diagrams of TLSMI are shown in Figs. 3.5 and 3.6. In magnetic order region a single local spin flip is prohibited due to the energy minimum principle. Therefore for our aim we should only consider the z-component of local spin operator. To obtain the expression of TLSMI, we use the Matsubara temperature Green function [88]. For the Hamiltonian in Eq. (3.17), the rules for constructing Feynman diagram are as follows. (a) Each internal $O_{p\sigma}$ hole line (solid line) associates with the free $O_{p\sigma}$ Green function, $1/\{i\omega_n - [\varepsilon(\mathbf{k}) - E_F]\}$, where $\omega_n = (2n+1)\pi T$ and called Matsubara frequency. ω_n has energy dimension with taking $k_B = 1$ and $\hbar = 1$. (b) Each internal pseudo-fermion line (dashed line) associates with the free pseudo-fermion Green function, $1/(i\omega_n)$ [89]. (c) Each circle associates with a vertex of H_K, $[J_K/(2N_{Cu})]16\cos(k_x a/2)\cos(k_y a/2)\cos(k_x' a/2)\cos(k_y' a/2))\sigma^z_{ss'}S^z_{i\delta\delta}\exp[i(\mathbf{k}' - \mathbf{k}) \cdot \mathbf{R}_i]$. (d) Each wave line associates with a vertex of H_H, $JS^z_{i\delta\delta}S^z_{j\gamma\gamma}$. (e) Make summations over all internal degrees of freedoms: momentum, Matsubara frequency, spin, and position of Cu^{++} in the CuO_2 plane ($i < j$). (f) Multiply the expression by $(-T)^m(-1)^F$, where F is the total number of closed fermion and pseudo-fermion loops, and m the total number of J and J_K. m represents the m order interaction. The simplest effective interaction between two $O_{p\sigma}$ holes with $\mathbf{k}\uparrow$ and $-\mathbf{k}\downarrow$ is given by Fig. 3.4. Assume that after the scattering between $\mathbf{k}\uparrow$ and one local spin and between $-\mathbf{k}\downarrow$ and another nearest-neighbor local spin, the $\mathbf{k}\uparrow$

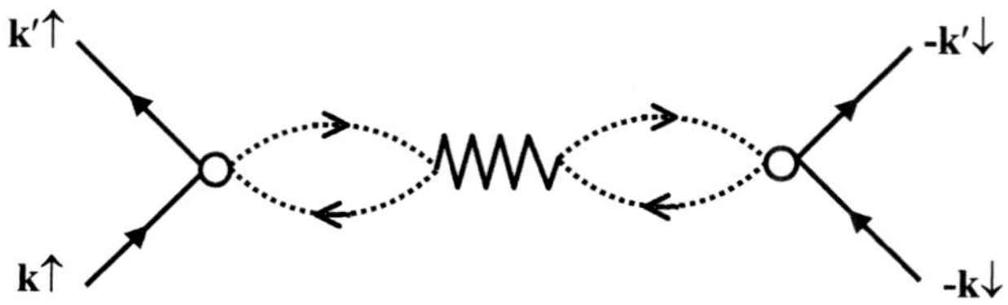

Figure 3.4. The simplest Feynman diagram of TLSMI between two $O_{p\sigma}$ holes with $\mathbf{k}\uparrow$ and -$\mathbf{k}\downarrow$. The dashed line represents the free pseudo-fermion Green function, the circle represents the vertex of H_K, and the wave line represents the vertex of H_H. The four external lines are not contained in the Feynman diagram.

and $-\mathbf{k}\downarrow$ become $\mathbf{k}'\uparrow$ and $-\mathbf{k}'\downarrow$, respectively. The expression of the simplest effective interaction in Fig. 3.4, I_3, is

$$I_3 = w(J/T)(-T)^3(-1)^2\frac{1}{-T}$$

$$\frac{J_K}{2N_{Cu}}\sum_{i\delta} S^z_{i\delta\delta}\sigma^z_{\uparrow\uparrow}e^{i(\mathbf{k}-\mathbf{k}')\cdot\mathbf{R}_i}16\cos(k_xa/2)\cos(k_ya/2)\cos(k'_xa/2)\cos(k'_ya/2)$$

$$J\sum_{j(j>i)\delta'} S^z_{i\delta\delta}S^z_{j\delta'\delta'}$$

$$\frac{J_K}{2N_{Cu}}S^z_{j\delta'\delta'}\sigma^z_{\downarrow\downarrow}e^{i(\mathbf{k}'-\mathbf{k})\cdot\mathbf{R}_j}16\cos(k_xa/2)\cos(k_ya/2)\cos(k'_xa/2)\cos(k'_ya/2)$$

$$\left(\sum_{nn'=-\infty}^{\infty}\frac{1}{i\omega_n}\frac{1}{i\omega_{n'}}\right)^2. \tag{3.19}$$

If the coupling between $\widehat{S}^z_i$ and $\widehat{S}^z_j$ is antiferromagnetic, i.e., $J>0$, then $\delta'=-\delta$. In Eq. (3.19) $w(J/T)$ is the transformation factor of pseudo-fermion method, which is given by Eqs. (7.41) and (7.66) of chapter 7, $(-T)^3$ arises from two J_K and one J, $(-1)^2$ from two pseudo-fermion closed loops, and $1/(-T)$ is a factor transforming the expression of a Feynman diagram into effective interaction. $\sigma^z_{\uparrow\uparrow}=1$. $\sigma^z_{\downarrow\downarrow}=-1$.

$$I_1 = \sum_{nn'=-\infty}^{\infty}\frac{1}{i\omega_n}\frac{1}{i\omega_{n'}} = \sum_{n=-\infty}^{\infty}\left[\frac{1}{i(2n+1)\pi T}\right]^2. \tag{3.20}$$

From Eq. (3.20) we see that the ω_n and $\omega_{n'}$ possessed by the two pseudo-fermion lines of a pseudo-fermion bubble should be equal to each other. Eq. (3.20) can be expressed by a contour integral.

$$I_1 = \frac{1}{2\pi i}\oint dz\frac{1}{z^2}\frac{1}{e^{z/T}+1}\frac{-1}{T}. \tag{3.21}$$

The integral contour is shown in Fig. 3.7. This integral contour can be transformed into the contour shown in Fig. 3.8, and I_1 becomes

$$I_1 = \frac{1}{2\pi i}\frac{1}{T}\oint dz\frac{1}{z^2}\frac{1}{e^{z/T}+1}. \tag{3.22}$$

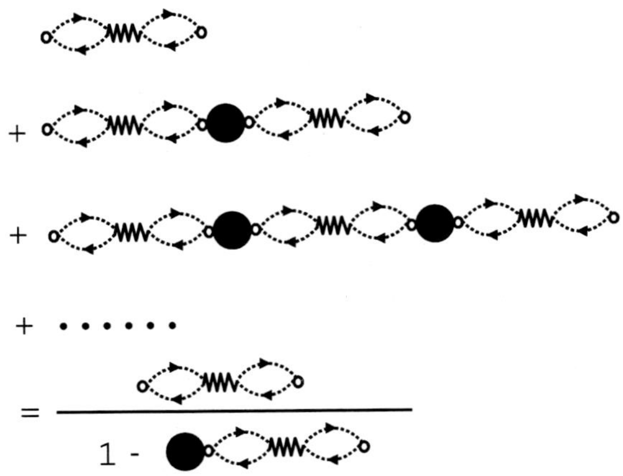

Figure 3.5. The renormalized Feynman diagram of TLSMI. The large solid circle is shown in detail in Fig. 3.6.

The $z = 0$ is a second order pole point of the integrand in Eq. (3.22). Applying the residue theorem yields

$$I_1 = -\frac{1}{4T^2}. \tag{3.23}$$

From Eqs. (3.20) and (3.23) we realize that every bubble of the pseudo-fermion contributes a factor $-1/(4T^2)$ besides -1.

The spin summation in Eq. (3.19) is

$$\sum_{\delta(\delta'=-\delta)} S^z_{i\delta\delta} S^z_{i\delta\delta} S^z_{j\delta'\delta'} S^z_{j\delta'\delta'} = 2\left(\frac{1}{2}\right)^4. \tag{3.24}$$

Next, we treat the summations over i and j in Eq. (3.19).

$$I_2 = \sum_{ij(i<j)} e^{i(\mathbf{k}-\mathbf{k}')\cdot(\mathbf{R}_i-\mathbf{R}_j)}. \tag{3.25}$$

Note that here i and j represent the nearest neighbors of Cu^{++} ions in the CuO_2 plane. The summation $\sum_{ij(i<j)}$ represents the total number of the nearest-neighbor pairs of Cu^{++} ions in all regions with antiferromagnetic short-range order of the CuO_2 plane. We only consider the z-component of spin operator $\widehat{S}$, which corresponds to that i and j are in the same cluster possessing antiferromagnetic short-range order. Assume that the CuO_2 plane consists of these magnetic short-rang order clusters. The sizes of clusters are, for simplicity, equal to each other. Let N' represent the number of Cu^{++} ions in one cluster. Let N'' represent the number of the nearest-neighbor pairs of Cu^{++} ions in the same cluster. The number of the clusters in the CuO_2 plane is N_{Cu}/N'. Roughly,

$$I_2 \approx \frac{N_{Cu}}{N'} N'' e^{i(\mathbf{k}-\mathbf{k}')\cdot(\mathbf{R}_i-\mathbf{R}_j)}, \tag{3.26}$$

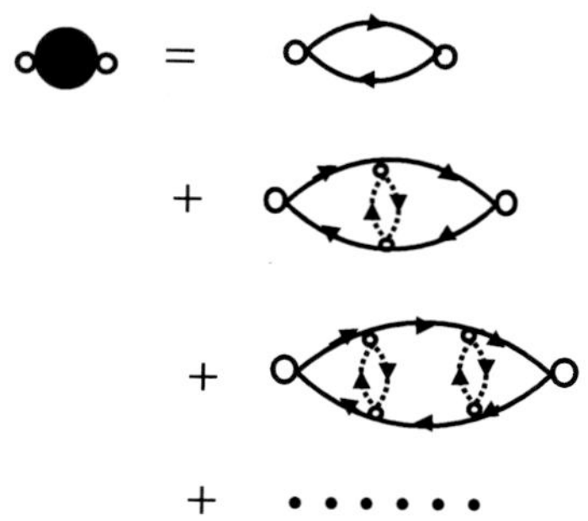

Figure 3.6. The large solid circle in Fig. 3.5. The solid line represents the Green function of mobile $O_{p\sigma}$ hole, the dashed line represents the Green function of free pseudo-fermion, and the small circle the vertex of H_K.

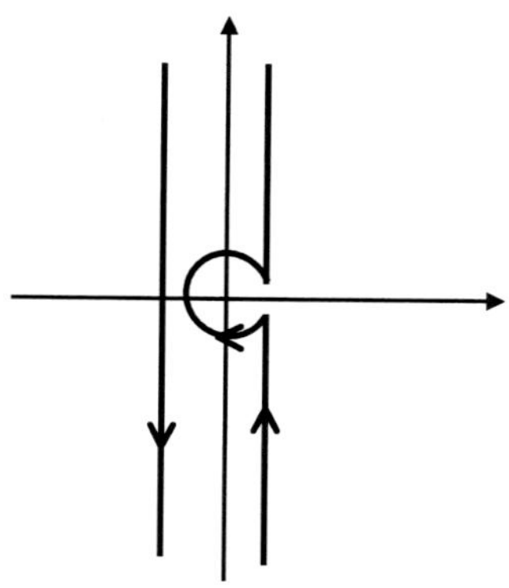

Figure 3.7. The integral contour in complex z-plane for Eq. (3.21).

where $\mathbf{R}_i$ and $\mathbf{R}_j$ are position vectors of some nearest-neighbor pair of Cu^{++} ions. Actually, for a specific i there are four nearest neighbors in the CuO_2 plane. Different nearest neighbor has different contribution to I_2 in Eq. (3.26). Therefore, I_2 is exactly

$$I_2 = \frac{N_{Cu}}{N'} N'' \frac{1}{4} \sum_j e^{i(\mathbf{k}-\mathbf{k}')\cdot(\mathbf{R}_i-\mathbf{R}_j)}, \tag{3.27}$$

where the summation over j is for the four nearest neighbors of i. Let us refer to Fig. 3.9 to calculate I_2. In Fig. 3.9 $\mathbf{k} = k_x\mathbf{e}_x + k_y\mathbf{e}_y$ and $\mathbf{A}_1$, for example, is $\mathbf{A}_1 = a\mathbf{e}_x - a\mathbf{e}_y$.

$$I_2 = N_{Cu}\frac{N''}{N'}\frac{1}{4}\left[e^{i(\mathbf{k}'-\mathbf{k})\cdot\mathbf{A}_1} + e^{i(\mathbf{k}'-\mathbf{k})\cdot\mathbf{A}_2} + e^{i(\mathbf{k}'-\mathbf{k})\cdot\mathbf{A}_3} + e^{i(\mathbf{k}'-\mathbf{k})\cdot\mathbf{A}_4}\right]$$

$$= N_{Cu}\frac{N''}{N'}\cos[(k'_x - k_x)a]\cos[(k'_y - k_y)a]$$

$$= N_{Cu}\frac{N''}{N'}[\cos(k'_x a)\cos(k_x a) + \sin(k'_x a)\sin(k_x a)]$$

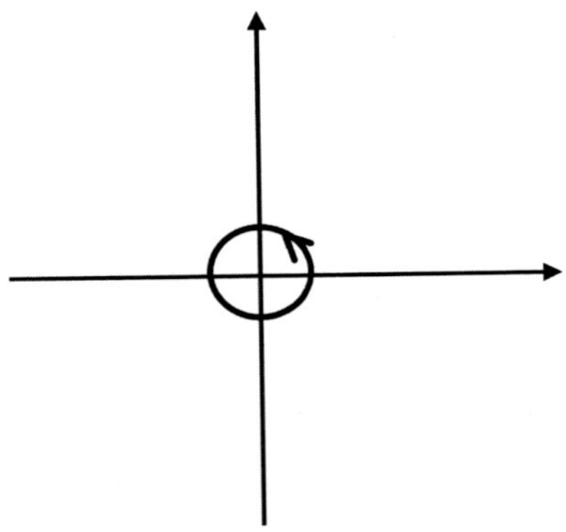

Figure 3.8. The integral contour in complex z-plane for Eq. (3.22).

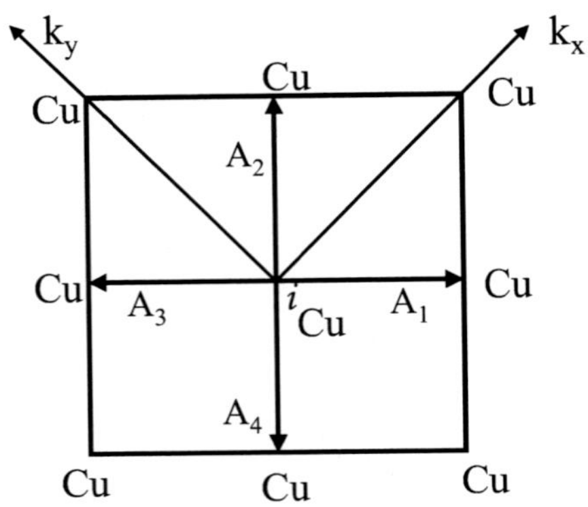

Figure 3.9. The schematic plot of vectors $\mathbf{A}_1$, $\mathbf{A}_2$, $\mathbf{A}_3$, and $\mathbf{A}_4$.

$$[\cos(k'_y a)\cos(k_y a) + \sin(k'_y a)\sin(k_y a)]. \tag{3.28}$$

Substituting Eqs. (3.28), (3.25), (3.24), and (3.20) into Eq. (3.19) yields

$$I_3 = -w(J/T)\frac{1}{2N_{Cu}}\frac{1}{T^2}JJ_K^2\frac{N''}{N'}F_{\mathbf{kk}'}. \tag{3.29}$$

$$F_{\mathbf{kk}'} = \sum_{i=1}^{4} F_i(\mathbf{k})F_i(\mathbf{k}')F_5(\mathbf{k})F_5(\mathbf{k}'), \tag{3.30}$$

$$F_1(\mathbf{k}) = \cos(k_x a)\cos(k_y a), \tag{3.31}$$

$$F_2(\mathbf{k}) = \sin(k_x a)\sin(k_y a), \tag{3.32}$$

$$F_3(\mathbf{k}) = \sin(k_x a)\cos(k_y a), \tag{3.33}$$

$$F_4(\mathbf{k}) = \cos(k_x a)\sin(k_y a), \tag{3.34}$$

$$F_5(\mathbf{k}) = \cos^2(k_x a/2)\cos^2(k_y a/2). \tag{3.35}$$

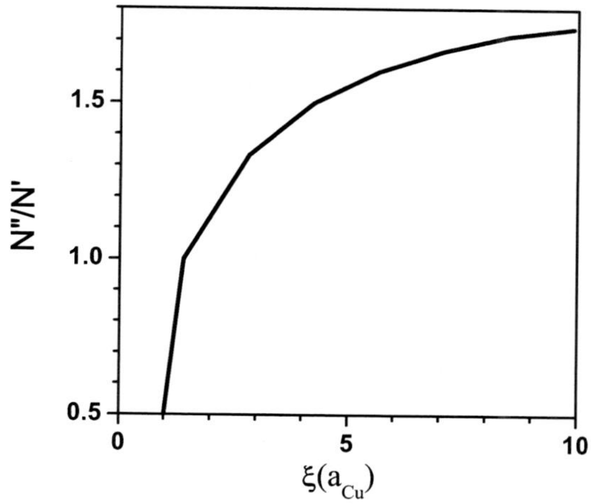

Figure 3.10. The curve of N''/N' vs. ξ. The ξ is the coherence length of antiferromagnetic short-range order, and a_{Cu} the nearest-neighbor distance between two Cu^{++} ions in the CuO_2 plane.

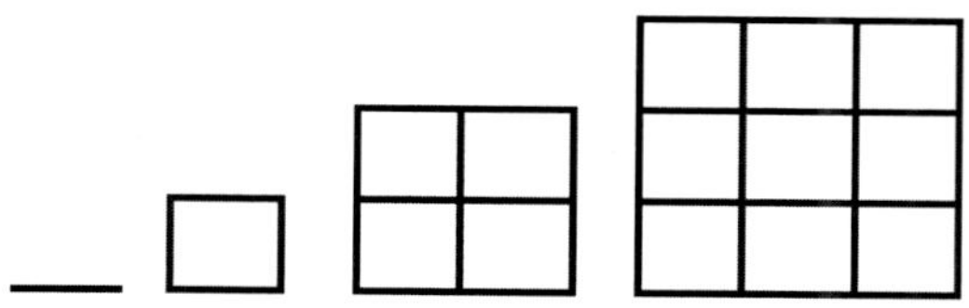

Figure 3.11. The schematic plots to obtain the Fig. 3.10. The cross point and the end point are the sites of Cu^{++} ions. The length of the line segment is a_{Cu}. Every plot represents a cluster possessing antiferromagnetic short-range order.

From the definitions of N'' and N' the curve of N''/N' vs. the coherence length of antiferromagnetic or ferromagnetic short-range order, ξ, is easily obtained and shown in Fig. 3.10. Let us explain how to obtain the Fig. 3.10 from Fig. 3.11. In Fig. 3.11, plot 1 gives $N''/N' = 1/2 = 0.5$ and $\xi = a_{Cu}$; plot 2 gives $N''/N' = 4/4 = 1$ and $\xi = \sqrt{2}a_{Cu}$; plot 3 gives $N''/N' = 12/9 = 1.333$ and $\xi = 2\sqrt{2}a_{Cu}$; and plot 4 gives $N''/N' = 24/16 = 1.5$ and $\xi = 3\sqrt{2}a_{Cu}$. Here a_{Cu} is the nearest-neighbor distance between two Cu^{++} ions. In terms of the simple forms of the clusters possessing different coherence lengths of antiferromagnetic short-range order we can obtain the curve in Fig. 3.10.

The TLSMI given by Eq. (3.29) is the simple third order interaction in H_K and H_H. To get a renormalized TLSMI we should consider sixth order interaction, I_6', in H_K and H_H, and much higher order interactions. The sixth order interaction, I_6', in H_K and H_H is shown in Fig. 3.12.

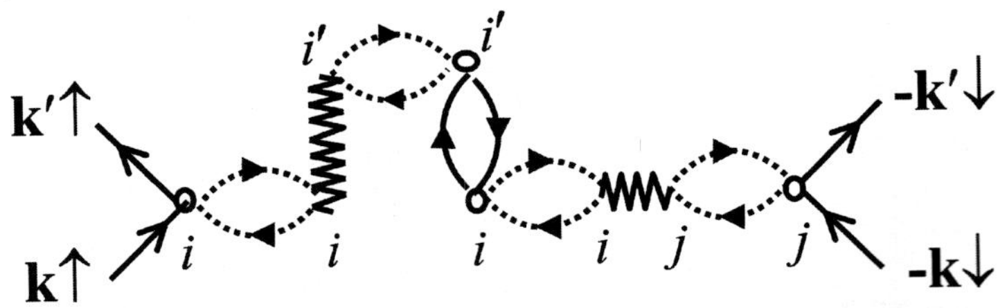

Figure 3.12. The sixth order interaction in H_K and H_H. The i' and j are the nearest neighbors of i. The four external lines do not belong to the sixth order interaction.

$$I_6' = w(J/T)(-T)^6(-1)^5\left(\frac{-1}{4T^2}\right)^4\frac{1}{-T}$$

$$\frac{J_K}{2N_{Cu}}\sum_{i\delta}S^z_{i\delta\delta}\sigma^z_{\uparrow\uparrow}e^{i(\mathbf{k}-\mathbf{k}')\cdot\mathbf{R}_i}16\cos\frac{k_xa}{2}\cos\frac{k_ya}{2}\cos\frac{k'_xa}{2}\cos\frac{k'_ya}{2}$$

$$J\sum_{i'(i'>i)}S^z_{i\delta\delta}S^z_{i'-\delta-\delta}$$

$$\frac{J_K}{2N_{Cu}}\sum_{s_1\mathbf{k}_3\mathbf{k}'_3\omega_m}S^z_{i'-\delta-\delta}\sigma^z_{s_1s_1}e^{i(\mathbf{k}_3-\mathbf{k}'_3)\cdot\mathbf{R}_{i'}}16\cos\frac{k_{3x}a}{2}\cos\frac{k_{3y}a}{2}\cos\frac{k'_{3x}a}{2}\cos\frac{k'_{3y}a}{2}$$

$$\frac{1}{i\omega_m-\xi(\mathbf{k}_3)}\frac{1}{i\omega_m-\xi(\mathbf{k}'_3)}$$

$$\frac{J_K}{2N_{Cu}}\sum_{\delta'}S^z_{i\delta'\delta'}\sigma^z_{s_1s_1}e^{i(\mathbf{k}'_3-\mathbf{k}_3)\cdot\mathbf{R}_i}16\cos\frac{k_{3x}a}{2}\cos\frac{k_{3y}a}{2}\cos\frac{k'_{3x}a}{2}\cos\frac{k'_{3y}a}{2}$$

$$J\sum_{j(j>i)}S^z_{i\delta'\delta'}S^z_{j-\delta'-\delta'}$$

$$\frac{J_K}{2N_{Cu}}S^z_{j-\delta'-\delta'}\sigma^z_{\downarrow\downarrow}e^{i(\mathbf{k}'-\mathbf{k})\cdot\mathbf{R}_j}16\cos\frac{k_xa}{2}\cos\frac{k_ya}{2}\cos\frac{k'_xa}{2}\cos\frac{k'_ya}{2}, \tag{3.36}$$

where $(-T)^6$ comes from sixth order interaction in H_K and H_H, $(-1)^5$ from four pseudo-fermion closed loops and one fermion closed loop, $[-1/(4T^2)]^4$ from four pseudo-fermion bubbles, 1/(-T) from the transformation of Feynman diagram into effective interaction, and $\xi(\mathbf{k}_3)=\varepsilon(\mathbf{k}_3)-E_F$. I_6' can be rewritten in a simple form.

$$I_6' = w(J/T)(-T)^6(-1)^5\left(\frac{-1}{4T^2}\right)^4\frac{1}{-T}J^2\left(\frac{J_K}{2N_{Cu}}\right)^4\sum_{ij(i<j)}e^{i(\mathbf{k}-\mathbf{k}')\cdot(\mathbf{R}_i-\mathbf{R}_j)}$$

$$\left(16\cos\frac{k_xa}{2}\cos\frac{k_ya}{2}\cos\frac{k'_xa}{2}\cos\frac{k'_ya}{2}\right)^2$$

$$\sum_{\delta i'(i'>i)}S^z_{i\delta\delta}S^z_{i\delta\delta}S^z_{i'-\delta-\delta}S^z_{i'-\delta-\delta}\sum_{\delta'}S^z_{i\delta'\delta'}S^z_{i\delta'\delta'}S^z_{j-\delta'-\delta'}S^z_{j-\delta'-\delta'}\sigma^z_{\uparrow\uparrow}\sigma^z_{\downarrow\downarrow}\sum_{s_1}\sigma^z_{s_1s_1}\sigma^z_{s_1s_1}$$

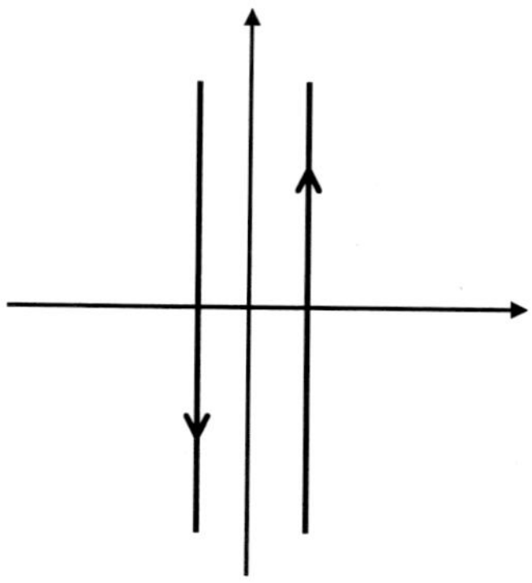

Figure 3.13. The first integral contour on the complex z-plane for Eq. (3.39).

$$\sum_{\mathbf{k}_3\mathbf{k}_3'\omega_m} e^{i(\mathbf{k}_3-\mathbf{k}_3')\cdot(\mathbf{R}_{i'}-\mathbf{R}_i)} \frac{1}{i\omega_m-\xi(\mathbf{k}_3)} \frac{1}{i\omega_m-\xi(\mathbf{k}_3')}$$

$$\left(16\cos\frac{k_{3x}a}{2}\cos\frac{k_{3y}a}{2}\cos\frac{k_{3x}'a}{2}\cos\frac{k_{3y}'a}{2}\right)^2. \tag{3.37}$$

Considering that there are also four nearest neighbors of j, the exact sixth order interaction I_6 is

$$I_6 = 2I_6'. \tag{3.38}$$

Let us calculate some factors in Eq. (3.37).

$$\sum_{\omega_m} \frac{1}{i\omega_m-\xi(\mathbf{k}_3)} \frac{1}{i\omega_m-\xi(\mathbf{k}_3')} = \frac{1}{2\pi i}\oint dz \frac{1}{e^{z/T}+1} \frac{1}{z-\xi(\mathbf{k}_3)} \frac{1}{z-\xi(\mathbf{k}_3')} \frac{-1}{T}. \tag{3.39}$$

The integral contour of Eq. (3.39) is shown in Fig. 3.13. Changing the integral contour from Fig. 3.13 to Fig. 3.14, the integral is easily calculated by residue theorem. The result is

$$\sum_{\omega_m} \frac{1}{i\omega_m-\xi(\mathbf{k}_3)} \frac{1}{i\omega_m-\xi(\mathbf{k}_3')} = \frac{1}{T}\frac{f(\mathbf{k}_3)-f(\mathbf{k}_3')}{\xi(\mathbf{k}_3)-\xi(\mathbf{k}_3')}, \tag{3.40}$$

where $f(\mathbf{k}_3)$ is Fermi distribution function. The spin summation in Eq. (3.37) is

$$\left(\sum_{\delta} S^z_{i\delta\delta} S^z_{i\delta\delta} S^z_{i'-\delta-\delta} S^z_{i'-\delta-\delta}\right)^2 \sum_{s_1} \sigma^z_{s_1s_1}\sigma^z_{s_1s_1} = \left(\frac{1}{2}\right)^5. \tag{3.41}$$

Substituting Eqs. (3.41), (3.40), and (3.38) into Eq. (3.37) yields

$$I_6 = I_3 \frac{32}{T^2}\frac{JJ_K^2}{(2N_{Cu})^2} \sum_{\mathbf{k}_3\mathbf{k}_3'} \frac{f(\mathbf{k}_3)-f(\mathbf{k}_3')}{\xi(\mathbf{k}_3)-\xi(\mathbf{k}_3')} e^{i(\mathbf{k}_3-\mathbf{k}_3')\cdot(\mathbf{R}_{i'}-\mathbf{R}_i)}$$

$$\left(\cos\frac{k_{3x}a}{2}\cos\frac{k_{3y}a}{2}\cos\frac{k_{3x}'a}{2}\cos\frac{k_{3y}'a}{2}\right)^2, \tag{3.42}$$

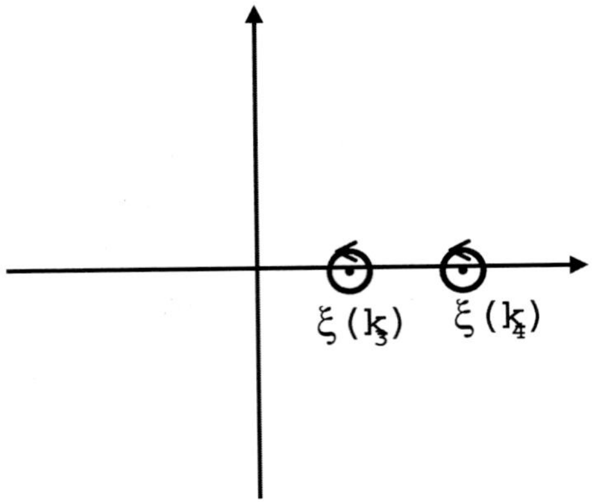

Figure 3.14. The second integral contour on the complex z-plane for Eq. (3.39).

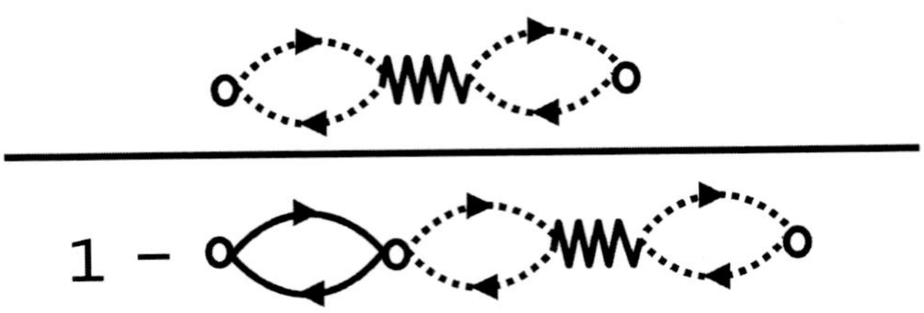

Figure 3.15. The simplest renormalized TLSMI.

where i' is any nearest neighbor of i. In chapter 5 we will see that the I_6 does not depend on i'. Imitating the method of obtaining I_6, we can obtain the expression for ninth order interaction in H_K and H_H. Actually, we do not need to calculate the ninth order interaction. The simplest renormalized TLSMI, I, is shown in Fig. 3.15. According to this diagram the expression of the denominator of Fig. 3.15 can be obtained from the sixth order interaction. The expression I is

$$I = \frac{I_3}{1 - \frac{32JJ_K^2}{T^2} g}, \tag{3.43}$$

$$g = \left(\frac{1}{2N_{Cu}}\right)^2 \sum_{\mathbf{k}_3 \mathbf{k}_3'} \frac{f(\mathbf{k}_3) - f(\mathbf{k}_3')}{\xi(\mathbf{k}_3) - \xi(\mathbf{k}_3')} e^{i(\mathbf{k}_3 - \mathbf{k}_3')\cdot(\mathbf{R}_{i'} - \mathbf{R}_i)}$$

$$\left(\cos\frac{k_{3x}a}{2}\cos\frac{k_{3y}a}{2}\cos\frac{k_{3x}'a}{2}\cos\frac{k_{3y}'a}{2}\right)^2.$$

To obtain a better expression for TLSMI, we make the renormalization for the fermion bubble in Fig. 3.15. The fermion bubble is shown in Fig. 3.16. The diagram of the next order is shown in Fig. 3.17.

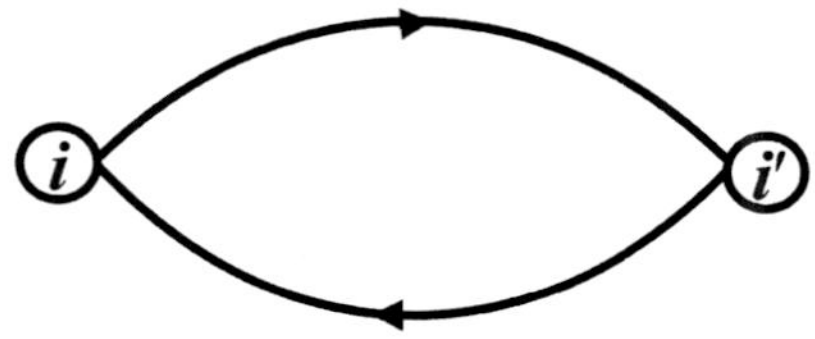

Figure 3.16. The fermion bubble in Fig.3.15.

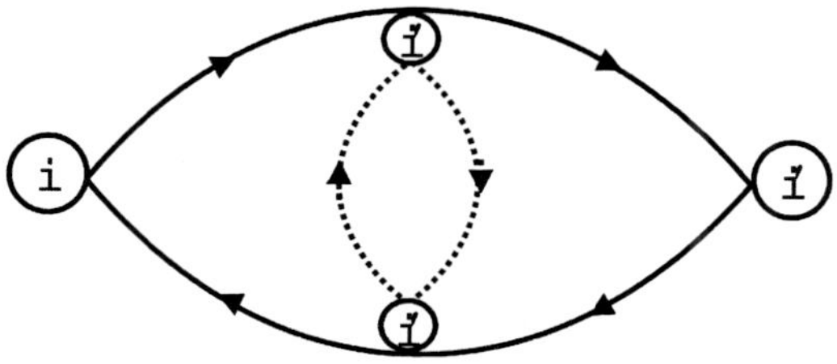

Figure 3.17. The Feynman diagram of the next order for the fermion bubble in Fig. 3.16.

The expression I_4 of the diagram in Fig. 3.17 is

$$I_4 = (-T)^2(-1)\left(\frac{J_K}{2N_{Cu}}\right)^2 \sum_{ii'\omega_m \mathbf{k}_4\mathbf{k}'_4 s_2} e^{i(\mathbf{k}_4-\mathbf{k}'_4)\cdot(\mathbf{R}_i-\mathbf{R}_{i'})} \frac{1}{i\omega_m - \xi(\mathbf{k}_4)} \frac{1}{i\omega_m - \xi(\mathbf{k}'_4)}$$

$$\sigma^z_{s_2s_2}\sigma^z_{s_2s_2}\sigma^z_{s_2s_2}\sigma^z_{s_2s_2}\left(16\cos\frac{k_{4x}a}{2}\cos\frac{k_{4y}a}{2}\cos\frac{k'_{4x}a}{2}\cos\frac{k'_{4y}a}{2}\right)^2$$

$$(-T)^2(-1)\left(\frac{-1}{4T^2}\right)\left(\frac{J_K}{2N_{Cu}}\right)^2 \sum_{\delta_8\mathbf{k}_3\mathbf{k}'_3} S^z_{i'\delta_8\delta_8}S^z_{i'\delta_8\delta_8}$$

$$\left(16\cos\frac{k_{3x}a}{2}\cos\frac{k_{3y}a}{2}\cos\frac{k'_{3x}a}{2}\cos\frac{k'_{3y}a}{2}\right)^2 \frac{1}{i\omega_m - \xi(\mathbf{k}_3)} \frac{1}{i\omega_m - \xi(\mathbf{k}'_3)}. \tag{3.44}$$

It is easy to show that

$$\sum_{s_2}\sigma^z_{s_2s_2}\sigma^z_{s_2s_2}\sigma^z_{s_2s_2}\sigma^z_{s_2s_2} = \sum_{s_2}\sigma^z_{s_2s_2}\sigma^z_{s_2s_2}. \tag{3.45}$$

The summation over ω_m in Eq. (3.44) can be calculated approximately.

$$\sum_{\omega_m}\frac{1}{i\omega_m - \xi(\mathbf{k}_3)}\frac{1}{i\omega_m - \xi(\mathbf{k}'_3)}\frac{1}{i\omega_m - \xi(\mathbf{k}_4)}\frac{1}{i\omega_m - \xi(\mathbf{k}'_4)}$$

$$= \sum_{\omega_m}\frac{1}{i\omega_m - \xi(\mathbf{k}_4)}\frac{1}{i\omega_m - \xi(\mathbf{k}'_4)}\frac{1}{i\omega' - \xi(\mathbf{k}_3)}\frac{1}{i\omega'' - \xi(\mathbf{k}'_3)}, \tag{3.46}$$

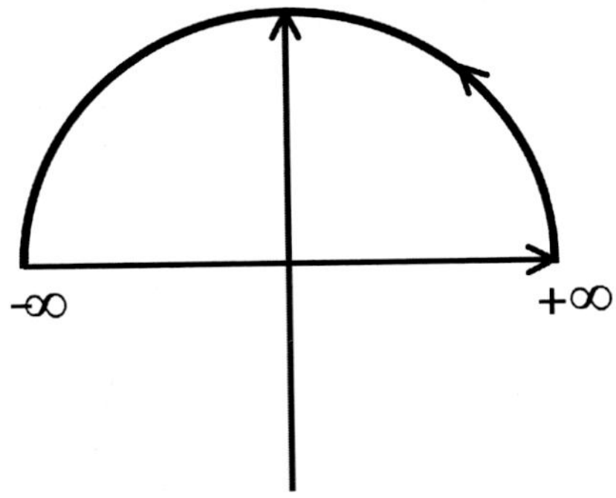

Figure 3.18. The integral contour on the complex ξ_3-plane for Eq. (3.50).

where ω' and ω'' are fixed unknown real constants. Substituting Eqs. (3.46) and (3.45) into Eq. (3.44) yields

$$I_4 = I_2(-T)^2(-1)\left(\frac{-1}{4T^2}\right)\left(\frac{J_K}{2N_{Cu}}\right)^2 \sum_{\delta_8 \mathbf{k}_3 \mathbf{k}'_3} S^z_{i'\delta_8\delta_8} S^z_{i'\delta_8\delta_8}$$

$$\left(16\cos\frac{k_{3x}a}{2}\frac{k_{3y}a}{2}\frac{k'_{3x}a}{2}\frac{k'_{3y}a}{2}\right)^2 \frac{1}{i\omega' - \xi(\mathbf{k}_3)}\frac{1}{i\omega'' - \xi(\mathbf{k}'_3)}, \tag{3.47}$$

where I_2 is the expression of the fermion bubble in Fig. 3.16. Eq. (3.47) can be rewritten as

$$I_4 = I_2\frac{8J_K^2}{N_{Cu}^2}\sum_{\mathbf{k}_3}(-1)\frac{\xi(\mathbf{k}_3)+i\omega'}{\xi^2(\mathbf{k}_3)+\omega'^2}\left(\cos\frac{k_{3x}a}{2}\cos\frac{k_{3y}a}{2}\right)^2$$

$$\sum_{\mathbf{k}'_3}(-1)\frac{\xi(\mathbf{k}'_3)+i\omega''}{\xi^2(\mathbf{k}'_3)+\omega''^2}\left(\cos\frac{k'_{3x}a}{2}\cos\frac{k'_{3y}a}{2}\right)^2. \tag{3.48}$$

For the triangular functions in Eq. (3.48) we take their averages on Fermi surface approximately. Then we transform the summation over $\mathbf{k}$ into integrations over ξ_3 and ξ'_3. Eq. (3.48) becomes approximately

$$I_4 = I_2\frac{8J_K^2}{N_{Cu}^2}\left[N'(E_F)\int_{-\infty}^{+\infty} d\xi_3\frac{-i\omega'}{\xi_3^2+\omega'^2}\right]\left[N'(E_F)\int_{-\infty}^{+\infty} d\xi'_3\frac{-i\omega''}{\xi_3'^2+\omega''^2}\right]$$

$$\overline{\left(\cos\frac{k_{3x}a}{2}\cos\frac{k_{3y}a}{2}\right)^4}, \tag{3.49}$$

where $N'(E_F)$ is the density of states of oxygen lattice per single spin. $N(E_F) = N'(E_F)/(2N_{Cu})$. The integral in Eq. (3.49) can be transformed into a contour integral shown in Fig. 3.18.

$$\int_{-\infty}^{+\infty} d\xi_3\frac{-i\omega'}{\xi_3+\omega'^2} = \oint d\xi_3\frac{-i\omega'}{(\xi_3+i\omega')(\xi_3-i\omega')} = -i\pi. \tag{3.50}$$

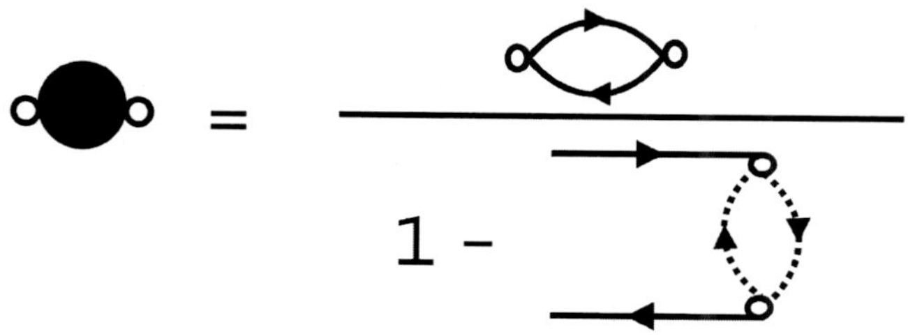

Figure 3.19. The equivalent diagram of Fig. 3.6.

Substituting Eq. (3.50) into Eq. (3.49) yields

$$I_4 = I_2(-1)32\pi^2 J_K^2 [N(E_F)]^2 \overline{\left(\cos\frac{k_{3x}a}{2}\cos\frac{k_{3y}a}{2}\right)^4}. \tag{3.51}$$

The bar represents the average on Fermi surface. Making the approximation in Eq. (3.46), the summation in Fig. 3.6 can be transformed into Fig. 3.19. The expression I_r of the renormalized fermion bubble is

$$I_r = \frac{I_2}{1+32\pi^2 J_K^2 [N(E_F)]^2 \overline{\left(\cos\frac{k_{3x}a}{2}\cos\frac{k_{3y}a}{2}\right)^4}}. \tag{3.52}$$

Considering the renormalization of the fermion bubble in Eq. (3.43), we obtain the expression $U_{\mathbf{kk}'}$ of the renormalized TLSMI shown in Fig. 3.5.

$$U_{\mathbf{kk}'} = -A(T)F_{\mathbf{kk}'}, \tag{3.53}$$

$$A(T) = \frac{1}{2N_{Cu}} \frac{JJ_K^2(1+C)N''/N'w(J/T)}{T^2 + 64JJ_K^2 \sum_{\mathbf{kp}} g(\mathbf{k},\mathbf{p})/\{1+32\pi^2 J_K^2 [N(E_F)]^2 \overline{h(\mathbf{q})}\}}, \tag{3.54}$$

$$g(\mathbf{k},\mathbf{p}) = \left(\frac{1}{2N_{Cu}}\right)^2 \frac{f[\varepsilon(\mathbf{k})-E_F]F_5(\mathbf{k})F_5(\mathbf{p})}{\varepsilon(\mathbf{p})-\varepsilon(\mathbf{k})} e^{i(\mathbf{k}-\mathbf{p})\cdot(\mathbf{R}_{i'}-\mathbf{R}_i)}, \tag{3.55}$$

$$h(\mathbf{q}) = [\cos(q_x a/2)\cos(q_y a/2)]^4, \tag{3.56}$$

where $f(x)$ is Fermi distribution, $\varepsilon(\mathbf{k})$ energy of $O_{p\sigma}$ holes, E_F Fermi energy, $N(E_F)$ density of states per single spin and per O^{--} ion, C weak tunneling coupling between two CuO_2 planes if the cuprates have two CuO_2 planes in a unit cell. In our numerical calculations in chapter 5, $C = 0.05$. Actually, C can be neglected. $w(J/T) = 1 + f_1/f_2$ is the transformation factor in Eqs. (7.41) and (7.66) of the chapter 7. The expression of f_1 and f_2 are given by Eqs. (7.44) and (7.45).

If $J > 0$, then TLSMI between two $O_{p\sigma}$ holes with $\mathbf{k}\uparrow$ and $-\mathbf{k}\downarrow$, $U_{\mathbf{kk}'}$, is negative, which means that TLSMI can become the Cooper pairing potential of the two $O_{p\sigma}$ holes in the regions with, at least, antiferromagnetic short-range order. Note that if $J < 0$, which means ferromagnetic coupling, then the two spins of the Cooper pair have to be parallel, otherwise TLSMI is a repulsion.

3.2.2. Gap Equation

In this section we will give the general solution of the BCS gap equation for the CuO_2 plane of the high-T_c cuprates. There are also the phonon-mediated interaction, V_{ep}, and the effective Coulomb interaction, V_c^*, between the two $O_{p\sigma}$ holes with $\mathbf{k}\uparrow$ and $-\mathbf{k}\downarrow$ besides TLSMI. Considering all the interactions, the BCS gap equation is

$$\Delta'(T,\mathbf{k}) = -\sum_{\mathbf{k}'} \frac{V_{\mathbf{kk}'}\Delta'(T,\mathbf{k}')}{2m(T,\mathbf{k}')} \tanh\frac{m(T,\mathbf{k}')}{2T}, \tag{3.57}$$

$$m(T,\mathbf{k}) = \sqrt{[\varepsilon(\mathbf{k}) - E_F]^2 + \Delta'(T,\mathbf{k})^2}. \tag{3.58}$$

$$V_{\mathbf{kk}'} = \begin{cases} V_{ep} + V_c^*, \\ 0 \le |\varepsilon(\mathbf{k}) - E_F|, |\varepsilon(\mathbf{k}') - E_F| \le E_D; \\ U_{\mathbf{kk}'}, \\ 0 \le E_F - \varepsilon(\mathbf{k}), E_F - \varepsilon(\mathbf{k}') \le E_F; \\ 0, \\ \text{otherwise}, \end{cases} \tag{3.59}$$

where $\Delta'(T,\mathbf{k})$ is a gap function in the CuO_2 plane, and E_D the Debye energy. Note that E_F can be less than E_D in some cases, and we take that the zero point of $\varepsilon(\mathbf{k})$ is at the top of conduction band ($E_F > 0$ for hole.). Considering that $\varepsilon(\mathbf{k})$ generally is an even function of k_x and k_y, V_{ep} and V_c^* are constants, $F_1(\mathbf{k})$ and $F_5(\mathbf{k})$ are even function of k_x and k_y, $F_2(\mathbf{k})$ is an odd function of k_x and k_y, $F_3(\mathbf{k})$ is an odd function of k_x, and $F_4(\mathbf{k})$ is an odd function of k_y, the BCS gap equation can be separated into following three decoupled equations and one decoupled equation set.

$$\Delta_1 = -\sum_{\mathbf{k}'} \frac{(V_{ep} + V_c^*)[\Delta_1 + \Delta_2 F_1(\mathbf{k}')F_5(\mathbf{k}')]}{2\sqrt{[\varepsilon(\mathbf{k}') - E_F]^2 + [\Delta_1 + \Delta_2 F_1(\mathbf{k}')F_5(\mathbf{k}')]^2}}$$

$$\tanh\frac{\sqrt{[\varepsilon(\mathbf{k}') - E_F]^2 + [\Delta_1 + \Delta_2 F_1(\mathbf{k}')F_5(\mathbf{k}'))]^2}}{2T}, \tag{3.60}$$

$$\Delta_2 F_1(\mathbf{k})F_5(\mathbf{k}) = -\sum_{\mathbf{k}'} \frac{-A(T)F_1(\mathbf{k})F_5(\mathbf{k})F_1(\mathbf{k}')F_5(\mathbf{k}')[\Delta_1 + \Delta_2 F_1(\mathbf{k}')F_5(\mathbf{k}')]}{2\sqrt{[\varepsilon(\mathbf{k}') - E_F]^2 + [\Delta_1 + \Delta_2 F_1(\mathbf{k}')F_5(\mathbf{k}')]^2}}$$

$$\tanh\frac{\sqrt{[\varepsilon(\mathbf{k}') - E_F]^2 + [\Delta_1 + \Delta_2 F_1(\mathbf{k}')F_5(\mathbf{k}'))]^2}}{2T}, \tag{3.61}$$

$$\Delta_3 F_2(\mathbf{k})F_5(\mathbf{k}) = -\sum_{\mathbf{k}'} \frac{-A(T)F_2(\mathbf{k})F_5(\mathbf{k})F_2(\mathbf{k}')F_5(\mathbf{k}')[\Delta_3 F_2(\mathbf{k}')F_5(\mathbf{k}')]}{2\sqrt{[\varepsilon(\mathbf{k}') - E_F]^2 + [\Delta_3 F_2(\mathbf{k}')F_5(\mathbf{k}')]^2}}$$

$$\tanh\frac{\sqrt{[\varepsilon(\mathbf{k}') - E_F]^2 + [\Delta_3 F_2(\mathbf{k}')F_5(\mathbf{k}'))]^2}}{2T}, \tag{3.62}$$

$$\Delta_4 F_3(\mathbf{k})F_5(\mathbf{k}) = -\sum_{\mathbf{k}'} \frac{-A(T)F_3(\mathbf{k})F_5(\mathbf{k})F_3(\mathbf{k}')F_5(\mathbf{k}')[\Delta_4 F_3(\mathbf{k}')F_5(\mathbf{k}')]}{2\sqrt{[\varepsilon(\mathbf{k}') - E_F]^2 + [\Delta_4 F_3(\mathbf{k}')F_5(\mathbf{k}')]^2}}$$

$$\tanh\frac{\sqrt{[\varepsilon(\mathbf{k}') - E_F]^2 + [\Delta_4 F_3(\mathbf{k}')F_5(\mathbf{k}'))]^2}}{2T}, \tag{3.63}$$

$$\Delta_5F_4(\mathbf{k})F_5(\mathbf{k}) = -\sum_{\mathbf{k}'}\frac{-A(T)F_4(\mathbf{k})F_5(\mathbf{k})F_4(\mathbf{k}')F_5(\mathbf{k}')[\Delta_5F_4(\mathbf{k}')F_5(\mathbf{k}')]}{2\sqrt{[\varepsilon(\mathbf{k}')-E_F]^2+[\Delta_5F_4(\mathbf{k}')F_5(\mathbf{k}')]^2}}$$

$$\tanh\frac{\sqrt{[\varepsilon(\mathbf{k}')-E_F]^2+[\Delta_5F_4(\mathbf{k}')F_5(\mathbf{k}'))]^2}}{2T}. \quad (3.64)$$

Eqs. (3.60) and (3.61) are coupled equation set. Eqs. (3.60) – (3.64) have the following four solutions.

$$\Delta'_s(T,\mathbf{k}) = \Delta_1(T)+\Delta_2(T)F_1(\mathbf{k})F_5(\mathbf{k}), \quad (3.65)$$

$$\Delta'_d(T,\mathbf{k}) = \Delta_3(T)F_2(\mathbf{k})F_5(\mathbf{k}), \quad (3.66)$$

$$\Delta'_{p1}(T,\mathbf{k}) = \Delta_4(T)F_3(\mathbf{k})F_5(\mathbf{k}), \quad (3.67)$$

$$\Delta'_{p2}(T,\mathbf{k}) = \Delta_5(T)F_4(\mathbf{k})F_5(\mathbf{k}). \quad (3.68)$$

Now we discuss the six generic features of the solutions. The first feature is about the form of $\Delta'_d(T,\mathbf{k})$ in Eq. (3.66). In a coordinate system $\mathbf{k}'$ which is 45°-rotating from the system $\mathbf{k}$, $F_2(\mathbf{k}') = [\cos(k'_x a_{Cu}) - \cos(k'_y a_{Cu})]/2$, i.e., $F_2(\mathbf{k}')$ is the recent popular simple d-wave. The reason to call the $F_2(\mathbf{k}')$ the d-wave is as follows. $F_2(\mathbf{k}')$ has the same symmetry property as $k_x^2 - k_y^2$ in D_{4h}. For this reason, $F_2(\mathbf{k}')$ is referred to the d-wave, i.e., the relative angular momentum of Cooper pair is $L = 2$, although the classification with respect to angular momentum has no meaning under the discrete crystal field symmetry. Because $\Delta'_d(T,\mathbf{k})$ contains $F_5(\mathbf{k})$ besides $F_2(\mathbf{k})$, we call $\Delta'_d(T,\mathbf{k})$ the composite d-wave. For our composite d-wave the maximum value of the gap is not always in the Cu-O bond direction of the CuO_2 plane, which will be analyzed in section 4.1.

The second feature is about the form of the spin part of the wavefunction of Cooper pair. Δ'_s and Δ'_d have even parity, and the same symmetries as the space wavefunctions of Cooper pairs [90]. The spin part of the corresponding wavefunction of Cooper pairs should be in spin-singlet state, i.e., $[\uparrow\downarrow - \downarrow\uparrow]/\sqrt{2}$. For spin-singlet state, TLSMI is negative only if $J > 0$. The gap function can also be constituted by the linear combination of Δ'_s and Δ'_d. $\Delta'_s + i\Delta'_d$ is more stable than $\Delta'_s + \Delta'_d$ [91, 92]. Note that the former corresponds to time-reversal symmetry breaking. The p-waves, Δ'_{p1} and Δ'_{p2}, have odd parity. In case of $J > 0$, i.e., antiferromagnetic coupling, we can also have p-wave superconductivity or pseudogap if the spin part of the Cooper pair is $[\uparrow\downarrow + \downarrow\uparrow]/\sqrt{2}$. The another two spin functions in the spin-triplet states, $\uparrow\uparrow$ and $\downarrow\downarrow$, do not appear because the TLSMI given by Eqs. (3.29) and (3.53) is positive. For the ferromagnetic long- or short- range order $J < 0$, the TLSMI in Eqs. (3.29) and (3.53) is negative for the two holes with spin-triplet states. In this case the space part of the wavefunction of Cooper pair should have odd parity, which means that the gap caused by TLSMI in ferromagnetic long- or short-range order regions should be, for example, p-wave.

The third feature is the zero influence of the Coulomb interaction on the d- or p-wave superconductivity and pseudogap. This feature is easily seen from BCS gap equation. For example, Δ'_d is determined by the equation

$$\Delta_3(T)F_2(\mathbf{k})F_5(\mathbf{k}) = -\sum_{\mathbf{k}'}\frac{U_{\mathbf{k}\mathbf{k}'d}\Delta_3(T)F_2(\mathbf{k}')F_5(\mathbf{k}')}{2m(T,\mathbf{k}')}\tanh\frac{m(T,\mathbf{k}')}{2T}, \quad (3.69)$$

$$m(T,\mathbf{k}) = \sqrt{[\varepsilon(\mathbf{k})-E_F]^2+[\Delta_3(T)F_2(\mathbf{k})F_5(\mathbf{k})]^2}, \quad (3.70)$$

$$U_{\mathbf{kk}'d} = -A(T)F_2(\mathbf{k})F_2(\mathbf{k}')F_5(\mathbf{k})F_5(\mathbf{k}'). \tag{3.71}$$

If we take $(U_{\mathbf{kk}'d} + V_c^*)$ as the Cooper pairing potential instead of $U_{\mathbf{kk}'d}$ alone, then the solution remains unchanged. This feature can explain why TLSMI can cause high-T_c superconductivity and high temperature pseudogap under the condition of existence of very large Coulomb interaction. The single local spin-mediated interaction consisting of second order of H_K ia also a constant independent of **k**. Therefore, its influence on the d- and p-wave superconductivity and pseudogap is zero.

The fourth feature of the solution is its complex dependence on temperature and concentration of carriers, coming from the complexity of TLSMI in Eq. (3.54). The TLSMI in Eq. (3.54) depends on the temperature through T^2, $w(J/T)$, and the Fermi distribution, and on the concentration of carriers through E_F, N''/N', and J [93]– [95]. These dependences of TLSMI are the basic reasons to have the complex temperature-dependences of gap and pseudogap, and the complex doping-dependences of gap, pseudogap, T^*, and T_c, which will be seen clearly from the numerical calculations in chapters 4 and 5. On the contrary, the phonon-mediated interaction and the Coulomb interaction depend on the concentration of carriers very weakly, and are nearly independent of the temperature.

The fifth feature is about the BCS equation. BCS gap equation comes from the mean field approximation. Therefore, it can not determine the phase of the wavefunction of Cooper pair. If we use zero-gap condition to determine a temperature, then this temperature is T^* rather than T_c. T^* is the temperature, at which the gap, determined by the BCS gap equation, begins to appear. T^* is called pairing temperature. T_c is the temperature, at which the superconductivity begins to appear. T_c is called superconducting transition temperature or temperature of long-range phase coherence. For the conventional superconductor, the BCS coherence length ξ_o, i.e., the average size of a Cooper pair, is between 10^3 and 10^4 Å. Therefore, for conventional superconductor the size of a Cooper pair is large compared to the average distance between two conduction electrons, which is about a few Å. Within the space of a single pair there exist about 10^1 electrons, each being correlated in Cooper pairs, when the temperature goes down to a temperature just a little less than T_c. The Cooper pairs overlap strongly at $T < T_c$, which leads to strong coupling of the Cooper pairs, and further leads to the unified phase of all Cooper pairs, i.e., superconductivity. For conventional superconductor $T_c = T^*$ approximately (See sections 1.1.1 and 3.3 for the exact discussion on the relation between T_c and T^* in conventional superconductor.). Note that the long-range phase coherence contains two forms: unified phase of all Cooper pairs and phase difference locking between any two nearest-neighbor superconducting units or for the high-T_c cuprates between any two nearest-neighbor units (See section 1.1.1.). There is only one single Cooper pair in the unit of the high- T_c cuprates. The size of a Cooper pair in the high-T_c cuprates is only 20 - 30 Å. Thus the Cooper pairs in the high-T_c cuprates are separated into many single Cooper pairs without overlap. The long-range phase coherence of the Cooper pairs, i.e., the long-range phase difference locking between any two nearest-neighbor Cooper pairs in the high-T_c cuprates, needs another condition besides the non-zero gap condition. Therefore, T^* and T_c are not approximately equal to each other for the high-T_c cuprates, which is very different from conventional superconductors. The gap at $T_c < T < T^*$ is called pseudogap. For the high-T_c cuprates we use the Doniach-Inui granular model to determine T_c in section 3.3.

The sixth feature is about the inherent inhomogeneities of the gap, the pseudogap, T^*, and T_c in the CuO_2 plane. The reason is as follows. The gap, the pseudogap, T^*, and T_c are related with the TLSMI in Eqs. (3.53) and (3.54). The TLSMI is proportional to N''/N'. From Fig. 3.10 we know that N''/N' is determined by the coherence length of antiferromagnetic short-range order, ξ. The statistical inhomogeneous distribution of ξ in the CuO_2 plane gives rise to the inherent inhomogeneities of the gap, the pseudogap, T^*, and T_c in the CuO_2 plane, which will be shown numerically in chapters 4 and 5.

3.2.3. 2-D and High Local Spin

The Hamiltonian of the system is still given by Eq. (3.5) formally for the high local spin. The chapter 7 shows that to write down the expression of a Feynman diagram the high local spin operator $\widehat{S}_i^z$ should be decomposed into series of low local spin operators with spin 1/2, $\widehat{S}_{i\gamma}^z$. If S = n/2, then

$$\widehat{S}_i^z = \widehat{S}_{i1}^z + \widehat{S}_{i2}^z + \cdots + \widehat{S}_{in}^z. \tag{3.72}$$

Every closed pseudo-fermion loop in Fig. 3.4 is formed by two local spin operators at one site. According to Eq. (3.72) there are n! possibilities to form the closed pseudo-fermion loop with spin 1/2 at every local spin site i. Therefore the expression of Fig. 3.4 for high local spin has an extra factor $(n!)^2$ in comparison with the expression in Eq. (3.29). The renormalized TLSMI in case of high local spin is thus

$$U'_{\mathbf{kk}'} = -A'(T)F_{\mathbf{kk}'}, \tag{3.73}$$

$$A'(T) = \frac{(n!)^2 JJ_K^2(1+C)N''/N'w'(J/T)}{T^2 + (n!)^2 64JJ_K^2 \sum_{\mathbf{kp}} g(\mathbf{k},\mathbf{p})/\{1+n!32\pi^2 J_K^2[N(E_F)]^2\overline{h(\mathbf{q})}\}}, \tag{3.74}$$

$$w'(J/T) = 1 + \frac{F_1}{F_2}, \tag{3.75}$$

where the F_1 and F_2 are given by Eqs. (7.95) and (7.96). TLSMI of high local spin operator is larger than TLSMI of low local spin operator with spin 1/2.

3.2.4. 3-D

At first we consider the local spin operator with spin 1/2. To make the derivation more concrete we take a 3-D cubic (bcc) lattice in which the local spin, such as Cu^{++}, is in the center of a unit cell, and the eight O^{--} ions are situated at the eight corners. The distances between any two corners are of three kinds. The distances between the nearest neighbors, the second, and the third neighbors are a, $\sqrt{2}a$, and $\sqrt{3}a$, respectively. a is the lattice constant. The Hamiltonian in Eq. (3.5) can still describe the system formally. Transforming into **k** space, we have

$$H_\circ = -\frac{1}{2}\sum_{\mathbf{k}s} p_{\mathbf{k}s}^+ p_{\mathbf{k}s}\{8T_{11} + 8T_n[\cos(k_x a) + \cos(k_y a) + \cos(k_z a)]$$
$$+8T_{nn}[\cos(k_x a)\cos(k_y a) + \cos(k_x a)\cos(k_z a) + \cos(k_y a)\cos(k_z a)]$$
$$+8T_{nnn}\cos(k_x a)\cos(k_y a)\cos(k_z a)\}, \tag{3.76}$$

where T_n, T_{nn}, and T_{nnn} mean that the α and β of $T_{\alpha\beta}$ are in the nearest neighbors, the second, and the third neighbors, respectively.

$$H_K = \frac{J_K}{N_O} \sum_{i\delta\gamma\mathbf{k}\mathbf{k}'ss'} \mathbf{S}_{\delta\gamma} \cdot \vec{\sigma}_{ss'} d^+_{i\delta} d_{i\gamma} p^+_{\mathbf{k}s} p_{\mathbf{k}'s'} \exp[i(\mathbf{k}' - \mathbf{k}) \cdot \mathbf{R}_i]$$

$$64\cos(k_x a/2)\cos(k_y a/2)\cos(k_z a/2)\cos(k'_x a/2)\cos(k'_y a/2)\cos(k'_z a), \tag{3.77}$$

where N_O is the unit cell number of the 3-D lattice. We write down the expression of TLSMI in a region with, at least, magnetic short-range order, and thus we can consider only the z-component of the local spin operator. The simplest Feynman diagram in 3-D case is the same as Fig. 3.4. Using the similar diagram rules as Eq. (3.29), the expression of the renormalized TLSMI corresponding to Fig. 3.5 is

$$U_{\mathbf{k}\mathbf{k}'} = -A(T) F_{\mathbf{k}\mathbf{k}'}, \tag{3.78}$$

$$A(T) = \frac{1}{N_O} \frac{16/3JJ_K^2 N''/N' w(J/T)}{T^2 + 1536JJ_K^2 \sum_{\mathbf{kp}} g(\mathbf{k},\mathbf{p})/\{1 + 32\pi^2 J_K^2 [N(E_F)]^2 \overline{h(\mathbf{q})}\}}, \tag{3.79}$$

$$g(\mathbf{k},\mathbf{p}) = \left(\frac{1}{N_O}\right)^2 \frac{f[\varepsilon(\mathbf{k}) - E_F] F_7(\mathbf{k}) F_7(\mathbf{p})}{\varepsilon(\mathbf{p}) - \varepsilon(\mathbf{k})} e^{i(\mathbf{k}-\mathbf{p})\cdot(\mathbf{R}_{i'} - \mathbf{R}_i)}, \tag{3.80}$$

$$h(\mathbf{q}) = [\cos(q_x a/2)\cos(q_y a/2)\cos(q_z a/2)]^4, \tag{3.81}$$

$$F_{\mathbf{k}\mathbf{k}'} = \sum_{i=1}^{6} F_i(\mathbf{k}) F_i(\mathbf{k}') F_7(\mathbf{k}) F_7(\mathbf{k}'), \tag{3.82}$$

$$F_1(\mathbf{k}) = \cos(k_x a), \tag{3.83}$$

$$F_2(\mathbf{k}) = \cos(k_y a), \tag{3.84}$$

$$F_3(\mathbf{k}) = \cos(k_z a) \tag{3.85}$$

$$F_4(\mathbf{k}) = \sin(k_x a), \tag{3.86}$$

$$F_5(\mathbf{k}) = \sin(k_y a), \tag{3.87}$$

$$F_6(\mathbf{k}) = \sin(k_z a), \tag{3.88}$$

$$F_7(\mathbf{k}) = \cos^2(k_x a/2)\cos^2(k_y a/2)\cos^2(k_z a/2). \tag{3.89}$$

The factor $\sum_{i=1}^{6} F_i(\mathbf{k}) F_i(\mathbf{k}')$ comes from the summation of $3\sum_{ij(i<j)} \exp[i(\mathbf{k} - \mathbf{k}') \cdot (\mathbf{R}_i - \mathbf{R}_j)]$. Note that the f_1 and f_2 in $w(J/T)$ should be given by Eqs. (7.50) and (7.51), respectively. As a Cooper pairing potential in BCS equation, the terms containing the sine functions in Eq. (3.78) are not coupled with the Coulomb repulsion, therefore TLSMI in 3-D case can also have possibility to cause high temperature superconductivity. If the local spin is high spin with S = n/2, then Eq. (3.79) has only a little correction such as from Eq. (3.54) to Eq. (3.74). The possible symmetries of the gap and the pseudogap for a cubic (bcc) lattice are s- and p-wave, which is different from the high-T_c cuprates. For a lattice other than cubic (bcc) the expression in Eq. (3.82) may change largely. The six generic features of the superconductivity and the pseudogap in 2-D case remain unchanged in principle in this 3-D case. Because the large factor "1536" of the denominator in Eq. (3.79), the TLSMI in 3-D system is generally smaller than the TLSMI in 2-D system for the same magnitude of local spin.

3.2.5. 1-D

We take that the local spin operator has spin 1/2, and the oxygen ion is in the middle between two nearest-neighbor local spins such as Cu^{++} ions. The lattice constant is a. The Hamiltonian of this system has the same form as Eq. (3.5). Transforming $p_{\alpha s}$ into $\mathbf{k}$ space, we have

$$H_\circ = -2T_{11}\sum_{\mathbf{k}s} p^+_{\mathbf{k}s}p_{\mathbf{k}s} - 2T_n\sum_{\mathbf{k}s} p^+_{\mathbf{k}s}p_{\mathbf{k}s}\cos(k_xa), \tag{3.90}$$

$$H_K = \frac{J_K}{N_O}\sum_{i\delta\gamma\mathbf{k}\mathbf{k}'ss'} \mathbf{S}_{\delta\gamma}\cdot\overrightarrow{\sigma}_{ss'}d^+_{i\delta}d_{i\gamma}p^+_{\mathbf{k}s}p_{\mathbf{k}'s'}\exp[i(\mathbf{k}'-\mathbf{k})\cdot\mathbf{R}_i]$$

$$4\cos(k_xa/2)\cos(k'_xa/2), \tag{3.91}$$

where N_O is the unit cell number, and T_n means that α and β of $T_{\alpha\beta}$ are the nearest neighbors. The expression of TLSMI in the region with, at least, magnetic short-range order is

$$U_{\mathbf{k}\mathbf{k}'} = -A(T)F_{\mathbf{k}\mathbf{k}'}, \tag{3.92}$$

$$A(T) = \frac{1}{N_O}\frac{JJ_K^2N''/N'w(J/T)/8}{T^2+2JJ_K^2\sum_{\mathbf{k}\mathbf{p}}g(\mathbf{k},\mathbf{p})/\{1+32\pi^2J_K^2[N(E_F)]^2\overline{h(\mathbf{q})}\}}, \tag{3.93}$$

$$g(\mathbf{k},\mathbf{p}) = \left(\frac{1}{N_O}\right)^2\frac{f[\varepsilon(\mathbf{k})-E_F]F_3(\mathbf{k})F_3(\mathbf{p})}{\varepsilon(\mathbf{p})-\varepsilon(\mathbf{k})}e^{i(\mathbf{k}-\mathbf{p})\cdot(\mathbf{R}_{i'}-\mathbf{R}_i)}, \tag{3.94}$$

$$h(\mathbf{q}) = [\cos(q_xa/2)]^4, \tag{3.95}$$

$$F_{\mathbf{k}\mathbf{k}'} = \sum_{i=1}^{2}F_i(\mathbf{k})F_i(\mathbf{k}')F_3(\mathbf{k})F_3(\mathbf{k}'), \tag{3.96}$$

$$F_1(\mathbf{k}) = \cos(k_xa), \tag{3.97}$$

$$F_2(\mathbf{k}) = \sin(k_xa), \tag{3.98}$$

$$F_3(\mathbf{k}) = \cos^2(k_xa/2), \tag{3.99}$$

where the term $\sum_{i=1}^2 F_i(\mathbf{k})F_i(\mathbf{k}')$ comes from the summation $\sum_{ij(i<j)}\exp[i(\mathbf{k}-\mathbf{k}')\cdot(\mathbf{R}_i-\mathbf{R}_j)]$. Note that the f_1 and f_2 in $w(J/T)$ should be given by Eqs. (7.52) and (7.53), respectively. As an interaction in BCS gap equation, the term containing the sine function in Eq. (3.92) does not couple with the Coulomb repulsion. Therefore TLSMI in 1-D case can also have possibility to cause high temperature superconductivity. Note that a real 1-D system is always quasi-1-D rather than exact 1-D. The phase transitions can occur in the quasi-1-D systems. If the local spin is the high spin with S = n/2, then Eq. (3.93) has only a little correction such as from Eq. (3.54) to Eq. (3.74). The possible symmetries of the wavefunction of Cooper pair for a 1-D lattice are s- and p-wave, which is different from the CuO_2 plane of the high-T_c cuprates. The six generic features of the superconductivity and the pseudogap in 2-D case remain unchanged in principle in this 1-D case. Because the factor "2" of the denominator in Eq. (3.93) is smaller than the corresponding factors of 2-D and 3-D systems, the TLSMI in 1-D system is generally larger than the TLSMI in 2-D and 3-D systems for the same magnitude of local spin.

3.3. Determination of T_c

We discuss the determination of T_c for the high-T_c cuprates. The appearance of the gap in a superconductor just means that there are Cooper pairs in the superconductor. Additional condition is needed by superconductivity, i.e., the long-range phase coherence of the wavefunctions of Cooper pairs.

The size of Cooper pair, $\xi_\circ$, is only 20 - 30 $\mathring{A}$ in the CuO_2 plane. The average distance between two centers of nearest-neighbor Cooper pairs at T_c is

$$a_\circ = \sqrt{\frac{a_{Cu}^2}{x/2}}, \tag{3.100}$$

where a_{Cu} is the nearest-neighbor distance between two Cu^{++} ions in the CuO_2 plane, and x is the average number of $O_{p\sigma}$ holes in a unit cell of Cu^{++} lattice in the CuO_2 plane. $a_{Cu} = 3.9\,\mathring{A}$. The $O_{p\sigma}$ holes are the conduction carriers in the CuO_2 plane. If $x = 0.15$, then $a_\circ = 14.24\,\mathring{A}$. For describing the Cooper pairs in the CuO_2 plane Doniach and Inui proposed a granular model [96]. Granular model of the high-T_c cuprates assumes that the Cooper pairs constitute a square lattice, every lattice point is the center of a Cooper pair, and the lattice constant is $a_\circ$. The granular model is shown in Fig. 3.20. The simplified granular model is shown in Fig. 3.21 in which every Cooper pair corresponds to a superconductor, and it is separated from the others by the barrier layers of the tunneling junctions. Considering that Cooper pair is a wave packet, the average distance between two nearest-neighbor Cooper pairs should be shorter than the distance between two nearest-neighbor wave lines in Fig. 3.20. In our numerical calculations of chapter 5 we take that the nearest-neighbor distance between two Cooper pairs, d, is equal to d'. The d' is the thickness of one layer of the CuO_2 plane. $d' = 2.6\,\mathring{A}$. The coupling between two Cooper pairs is the Josephson tunneling coupling which tends to keep the phase difference of the Cooper pairs on the two sides coupled. If the Josephson tunneling coupling between two Cooper pairs is larger than the zero-point oscillation of the phase difference and the thermal disturbance, then there is a phase difference locking between the two nearest-neighbor Cooper pairs. If there is phase difference locking between any two nearest-neighbor Cooper pairs, then there is the long-range phase difference locking, which belongs to the long-range phase coherence as well.

For the Josephson tunneling junction lattice the Hamiltonian is

$$H = \frac{2e^2}{C} \sum_{ij(i<j)} n_{i,j}^2 + \sum_{ij(i<j)} E_J[1 - \cos(\phi_i - \phi_j)], \tag{3.101}$$

where e is the charge of a free electron, C is the capacitance of the junction, $n_{i,j}$ is the number of Cooper pairs transferred across junction from a superconductor (unit) j to another superconductor (unit) i, ϕ_i and ϕ_j are the phases of two superconductors (units), i.e., two Cooper pairs i and j, respectively, and E_J is the Josephson coupling energy. The first term of Eq. (3.101) is the electrostatic energy associated with any difference in charge between the two sides of the junction. The second term is the energy associated with the phase difference of the two superconductors (units), i.e., the two Cooper pairs, i and j on the two sides of the barrier layer of the Josephson tunneling junction.

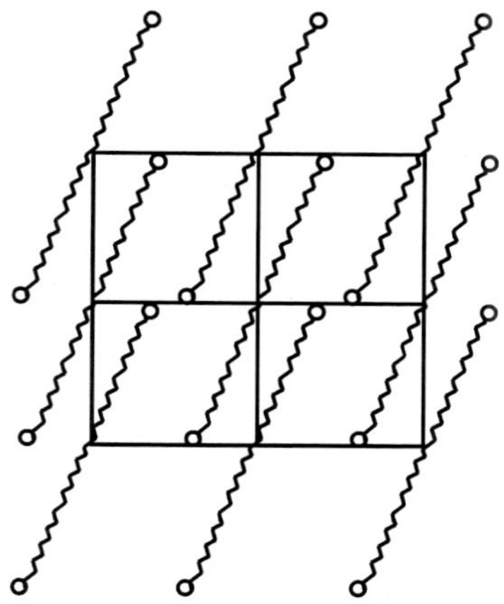

Figure 3.20. The schematic plot of the granular model for the CuO_2 plane of the high-T_c cuprates. The wave line represents a Cooper pair. The mass center of a Cooper pair is on a lattice point. The lattice constant is $a_\circ$.

Let us give a derivation for the second term in Eq. (3.101) [97]. The Josephson effect has the following two equations.

$$I_s = I_c \sin(\phi_i - \phi_j), \tag{3.102}$$

$$2|e|V = \hbar \frac{d}{dt}(\phi_i - \phi_j). \tag{3.103}$$

Consider a junction through which a constant current I_s is passing, the current having raised from zero to I_s over a time t. During the time of the current increasing there must be a variation rate of current dI_s/dt. Eq. (3.103) tells us that if the current through a junction is being changed, then a voltage will appear across it. During the time t that the current is being established there is a voltage V across the junction, and power I_sV is being delivered to it. So, work,

$$W = \int_0^t I_s V dt, \tag{3.104}$$

is performed in setting up the current and the consequent phase difference across the junction. W must be the increase in energy of the junction due to current passing through it. Eqs. (3.102) and (3.103) relate the voltage and current to the phase difference. Substituting Eqs. (3.102) and (3.103) into Eq. (3.104) gives the increase in junction energy as

$$\begin{aligned} W &= \frac{I_c\hbar}{2|e|}\int_0^t \sin(\phi_i - \phi_j)\frac{d}{dt}(\phi_i - \phi_j)dt \\ &= \frac{I_c\hbar}{2|e|}[1 - \cos(\phi_i - \phi_j)]. \end{aligned} \tag{3.105}$$

E_J is given by [98].

$$E_J = \frac{I_c\hbar}{2|e|}, \tag{3.106}$$

Figure 3.21. The schematic plot of the simplified granular model. The square constituted by four real lines represents a superconductor (unit), i.e., a Cooper pair. The distance between two nearest-neighbor squares is d. The nearest-neighbor distance between two centers of the squares is $a_\circ$.

where I_c is the maximum supercurrent of a Josephson tunneling junction. The I_c is [88,99]

$$I_c = \frac{\pi\overline{\Delta(T)}}{2|e|R_n}\tanh\left[\frac{\overline{\Delta(T)}}{2T}\right], \tag{3.107}$$

where R_n is the normal resistance at $T \approx T_c$, and $\overline{\Delta(T)}$ the average value of the gap at Fermi surface. Substituting Eq. (3.107) into Eq. (3.106) gives that E_J is a function of T.

$$E_J(T) = \frac{\pi\hbar\overline{\Delta(T)}}{4e^2R_n}\tanh\left[\frac{\overline{\Delta(T)}}{2T}\right]. \tag{3.108}$$

Considering that the Josephson tunneling junctions constitute a square lattice in the CuO_2 plane, the Hamiltonian in Eq. (3.101) can be rewritten as

$$H = \frac{2e^2}{C}2\sum_i n_{i,i+l}^2 + E_J 2\sum_i[1-\cos(\phi_i - \phi_{i+l})], \tag{3.109}$$

where $i+l$ indicates the nearest neighbor of superconductor (unit), i.e., Cooper pair, i. Because every junction in Fig. 3.21 is the same, Eq. (3.109) can be further simplified.

$$H = N\left\{\frac{4e^2}{C}n_{1,2}^2 + 2E_J[1-\cos(\phi_1-\phi_2)]\right\}, \tag{3.110}$$

where N is the number of Josephson tunneling junctions in the CuO_2 plane, and 1 and 2 are the nearest-neighbor superconductors (units), i.e., Cooper pairs. If we approximate the smaller part of the cosine curve by a parabola [100], then the H in Eq. (3.110) becomes

$$H = N\left[\frac{1}{2}\frac{8e^2}{C}n_{1,2}^2 + \frac{1}{2}8E_J\left(\frac{\phi_1-\phi_2}{2}\right)^2\right]. \tag{3.111}$$

Next, we show that $n_{1,2}$ is a variable operator conjugating to the phase difference $(\phi_1 - \phi_2)/2$ [101]. Assume that the number operators of Cooper pairs in superconductors (units) 1 and 2 are N_1 and N_2, respectively.

$$N_1 = \overline{N_\circ} + n_{1,2}, \tag{3.112}$$

$$N_2 = \overline{N_\circ} - n_{1,2}, \tag{3.113}$$

where $\overline{N_\circ}$ is the average number of Cooper pairs in every superconductor (unit) 1 or 2, and $n_{1,2}$ is the number operator of Cooper pairs transferred across junction from superconductor (unit) 2 to superconductor (unit) 1. We have

$$[\phi_1, N_1] = [\phi_1, n_{1,2}] = i, \tag{3.114}$$

$$[\phi_2, N_2] = [\phi_2, -n_{1,2}] = i. \tag{3.115}$$

Adding Eqs. (3.114) and (3.115), we obtain

$$\left[n_{1,2}, \frac{\phi_1 - \phi_2}{2}\right] = -i. \tag{3.116}$$

From Eq. (3.116) we know that

$$n_{1,2} = -i\frac{\partial}{\partial(\phi_1 - \phi_2)/2}. \tag{3.117}$$

The Hamiltonian of Eq. (3.111) is a Hamiltonian of N harmonic oscillators. The frequency of the oscillators, ω, is

$$\hbar\omega = \sqrt{\frac{8E_J}{C/(8e^2)}}. \tag{3.118}$$

The zero-point energy of the fluctuation of $(\phi_1 - \phi_2)$ destroys the locking of the phase difference $(\phi_1 - \phi_2)$ at

$$\frac{\hbar\omega}{2} = 8E_J. \tag{3.119}$$

The $8E_J$ in Eq. (3.119) comes from Eq. (3.111). Substituting Eq. (3.118) into Eq. (3.119) yields

$$E_J = \frac{e^2}{4C}, \tag{3.120}$$

or

$$8E_J = \frac{2e^2}{C}. \tag{3.121}$$

Eq. (3.121) tells us that $8E_J$ must be, at least, larger than $2e^2/C$, otherwise the zero-point fluctuation of $(\phi_1 - \phi_2)$ will destroy the phase difference locking between superconductor (unit) 1 and superconductor (unit) 2, i.e., between Cooper pair 1 and Cooper pair 2.

The thermal disturbance can also destroy the phase difference locking between ϕ_1 and ϕ_2. Doniach demonstrated that the condition, that the Josephson coupling energy is larger than the thermal disturbance, is [102]

$$8E_J(T) > 1.075T + 0.924\frac{T^2}{4E_J(T)}. \tag{3.122}$$

If the Josephson coupling energy can overcome both the thermal disturbance energy and zero-point fluctuation energy, then there will be the long-range phase difference locking in the high-T_c cuprates, and thus there will be a number density of supercurrent Cooper pairs, $n_s(T)$, in the system. The number density of supercurrent Cooper pairs is [102]

$$n_s(T) = B\left\{4E_J(T)\left[2 - 1.075\frac{T}{4E_J} - 0.924\left(\frac{T}{4E_J}\right)^2\right] - \frac{2e^2}{C}\right\}^{2/3}, \tag{3.123}$$

where B is a temperature-independent constant. The theoretical value of T_c, $T_{c,theo}$, is determined by

$$n_s(T_{c,theo}) = 0. \tag{3.124}$$

In actual numerical calculations we take

$$n_s(T) = B\left\{\alpha 4E_J(T)\left[2 - 1.075\frac{T}{4E_J} - 0.924\left(\frac{T}{4E_J}\right)^2\right] - \frac{2e^2}{C}\right\}^{2/3}, \tag{3.125}$$

where α is a number less than 1, which guarantees that the Josephson coupling energy is a little larger than the zero-point fluctuation energy [96].

Because T_c must satisfy Eq. (3.124), generally speaking, $T_c < T^*$ for the high-T_c cuprates. Even at $T < T_c$ ($T \approx T_c$) only partial Cooper pairs have the long-range phase coherence, and, at the same time, partial Cooper pairs still do not have the long-range phase coherence, because at $T = T_c$ the number density of supercurrent Cooper pairs is zero, and thus all Cooper pairs are non-supercurrent Cooper pairs. Therefore, we can say that at $T < T_c$ there are two sorts of Cooper pairs in the high-T_c cuprates.

In the GL theory of chapter 1, the amplitude of the order parameter ψ is

$$|\psi| = \sqrt{n_s(T)}. \tag{3.126}$$

From Eqs. (3.125) and (3.108) we know that $n_s(T)$ is a very complex function of the gap and thus the amplitude of the wavefunction of Cooper pair in the high-T_c cuprates.

We have to emphasize that there are two kinds of the long-range phase coherence. One is the unified phase for all Cooper pairs in the superconducting state. The other is the long-range phase difference locking between any two nearest-neighbor Cooper pairs in the superconducting state. The superconducting transition temperature T_c of the conventional superconductors means that the Cooper pairs possessing unified phase begin to appear. We explain that for the conventional superconductors at temperature interval $T_c < T < T_c + 10^{-3}$ K there should be another two characteristic temperatures T^* and $T_{c,1}$ (See section 1.1.1.). The estimation of 10^{-3} K comes from the specific heat jump experiment in Fig. 1.1 of chapter 1. When we use zero-gap condition to the BCS gap equation in case of conventional superconductors, we obtain pairing temperature T^* as well as in the case of unconventional superconductors because only the zero-gap condition can not guarantee long-range phase coherence. At $T \approx T^*$ only a few Cooper pairs begin to appear in the conventional superconductors. Although the size of Cooper pair in conventional superconductors is very large (10^4 $\AA$), there are only a few Cooper pairs, and thus the distance between two Cooper pairs is larger. As the temperature goes down from T^*, the Cooper pairs will appear more in the system, and thus the distance between Cooper pairs will become shorter. The shorter distance between Cooper pairs leads to the stronger Josephson tunneling coupling between two nearest-neighbor Cooper pairs because R_n in Eq. (3.108) becomes smaller, and leads to the weak zero-point fluctuation because the C in Eq. (3.123) becomes larger. At last, at $T = T_{c,1}$, where the Josephson tunneling coupling energy overcomes both the zero-point fluctuation energy and the thermal disturbance energy, the long-range phase difference locking begins to appear in the system, and thus leads to the finite number density of supercurrent Cooper pairs at $T < T_{c,1}$. The latter means superconductivity. We estimate in physics that at $T = T_{c,1}$ within the space of a Cooper pair there exist 2 - 3 other carriers, each being correlated in Cooper pairs. As the temperature continues to go down from $T_{c,1}$, the Cooper pairs will appear more and more in the system. At last at T_c the Cooper pairs strongly overlap, and thus all the Cooper pairs in the conventional superconductors have a unified phase. We estimate in physics that at $T = T_c$ within the space of a Cooper pair there exist at least 10 other carriers, each being correlated in Cooper pairs. The forms of the long-range phase coherence at T_c and $T_{c,1}$ are different. Both the two forms can cause superconductivity. It is very difficult to find experimentally all the three temperatures T^*, $T_{c,1}$, and T_c because all the three characteristic temperatures appear in a very small temperature interval about 10^{-3} K. The reason is as follows. When the temperature goes down from T^*, the gap, and thus the number of Cooper pairs, go from zero to finiteness very fast [See Eq. (2.146.)]. As we know, the size of the Cooper pair is very large in conventional superconductors, therefore, in a very small temperature interval from T^* the Cooper pairs in conventional superconductors can accomplish a transition from large distance between Cooper pairs to strong overlap of Cooper pairs. Because $T_c \approx T_{c,1} \approx T^*$ for conventional superconductors, we can say roughly that the second order phase transition and the zero-resistance occur at T_c.

In the specific heat experiment of conventional superconductors, exactly speaking, the first specific heat jump should occur at T^*, whose magnitude is given by Eq. (2.154), and the second specific heat jump should occur at $T_{c,1}$, where the number density of supercurrent Cooper pairs begins to appear due to the appearance of the long-range phase difference locking. We do not know that whether there is a second order phase transition at T_c, where the long-range phase difference locking between any two nearest-neighbor Cooper pairs becomes the same unified phase of all Cooper pairs. In the electric experiment of conventional superconductors, exactly speaking, the zero resistance occurs at $T = T_{c,1}$, where there is the long-range phase difference locking between any two nearest-neighbor Cooper pairs.

It is worth to mention Ogg's pair [103]– [106] here. Refs. [103]– [105] thought that the origin of superconductivity is due to the Bose-Einstein condensation of spatial local non-overlapping electron pairs (Ogg's pairs). Since the discovery of high temperature superconductivity in the high-T_c cuprates, in which the size of Cooper pair is only about 20 - 30 $\AA$, people begin to pay attention to the Ogg's pair [107, 108]. However, according to the theory in this section, the existence of condensed local carrier pairs is only one necessary condition to yield the superconductivity in the high- T_c cuprates. The other necessary condition is the long-range phase difference locking between any two nearest-neighbor local pairs, which is shown by Eq. (3.124).

The method to determine $T_{c,theo}$, given in this section, can also be used to determine the $T_{c,theo}$ of single Josephson tunneling junction and granular superconductor. The previous method to determine the $T_{c,theo}$ of granular superconductor is much different from the method in this section [101].

Chapter 4

Properties of High-T_c Cuprates

4.1. Features of Gap Anisotropy and Explanations

The energy gap anisotropy in the high-T_c cuprates, particularly in the a-b plane, can have significant implications about the quasiparticle pairing mechanisms [109, 110]. Because of the strong Coulomb interaction among conduction carriers and the related antiferromagnetic spin fluctuations, non-phonon-mediated pairing mechanisms with d-wave symmetry have been proposed to explain the superconductivity in the high-T_c cuprates [111]– [116] [74]. Most d-wave theories use a $d_{x^2-y^2}$, i.e., simple d-wave $\cos(k'_x a_{Cu})$ - $\cos(k'_y a_{Cu})$, to describe the gap [111]– [116]. The a_{Cu} is the lattice constant of the Cu lattice in the CuO_2 plane of the high-T_c cuprates. $\mathbf{k}'$ is a wave vector in the Brillouin zone corresponding to the Cu lattice in the CuO_2 plane. However, according to the d-wave theory in Ref. [74] and chapter 3 of this book, the $\mathbf{k}'$ dependence of the gap is $[\cos(k'_x a_{Cu}) - \cos(k'_y a_{Cu})][\cos(k'_x a_{Cu}/2) + \cos(k'_y a_{Cu}/2)]^2$ which is a composite d-wave. The $\mathbf{k}'$ dependence of the gap, which can be measured by angle-resolved photoemission spectroscopy (ARPES) experiment, offers an opportunity to test the d-wave theories. There are many ARPES experiments to test the $\mathbf{k}'$ dependence of the gap [117, 118]. The aim of this section is to summarize the experimental features of the gap anisotropy in the high-T_c cuprates, and to explain these experimental features in theory. This section and Ref. [119] show that the composite d-wave is better than the simple d-wave.

Shen *et al.* observed in the a-b plane of Bi2212 in ARPES that the gap anisotropy is at least an order of magnitude larger than that of the isotropic conventional superconductors [117]. The values of T_c for the samples # 1, # 2, and # 3, used in Ref. [117], are 78 K, 88 K, and 88 K, respectively. Shen *et al.* pointed out that the sample # 1 is overdoped. Dessau *et al.* pointed out that because of a less flat surface (as confirmed by a laser reflection), the sample # 2 was not as representative as the others [118]. According to the empirical T_c formula for Bi2212 and Y123, $T_c = 93[1 - 82.6(x - 0.16)^2]$ [120]. The x in this formula is the carrier ($O_{p\sigma}$ hole) number in one unit cell of the Cu lattice in the CuO_2 plane. The value of x, which is not given in the reference, can be obtained by the empirical T_c formula. Considering that the sample # 1 is overdoped, the values of x of the samples # 1 and # 3 are 0.204 and 0.134, respectively. The ARPES experiment in Ref. [117] concluded that the maximum gaps are along the Cu-O bond direction and the gaps vanish along line 45° away from the Cu-O bond direction for the overdoped and slightly underdoped samples. Ding *et al.* observed the gap anisotropy in the a-b plane of Bi2212 in ARPES [20]. There are three samples in Ref. [20]. The values of T_c for the samples # 1, # 2, and # 3 are 10 K, 83 K, and 87 K, respectively [20]. Using the empirical T_c formula, the values of x of the sample # 1, # 2, and # 3 are 0.056, 0.124. and 0.132, respectively. The gaps vanish along line 45° away from the Cu-O bond direction for all the three samples [20]. The gap reaches a maximum along the Cu-O bond direction for the slightly underdoped sample # 3. However, for more underdoped samples # 1 and # 2, the maximum gaps are along line about 15° away from the Cu-O bond direction. It is well known that the data from ARPES are quite scattered. We just choose the representative references. In many references the samples are not the much underdoped samples such as in Ref. [20]. The Ref. [20] is the only paper which gives systematic studies from slightly to much underdoped samples. The phenomenon that the direction of the maximum gap deviates from the Cu-O bond direction was also observed in Ref. [20, 118]. The directions of the maximum gaps for the samples # 6 and # 8 deviate from the Cu-O bond direction clearly [118, 121]. The directions of the maximum gap for the samples # 1, # 12, and # 13 deviate from the Cu-O bond direction also possibly because the values of the gaps at the Cu-O bond direction are not given in Refs. [118, 121]. For all the 14 samples in Refs. [118,121], the gaps vanish along line 45° away from the Cu-O bond direction. Thus, we summarize the three experimental features of the gap anisotropy in the high-T_c cuprates as follows. The gaps vanish along line 45° away from the Cu-O bond direction for any doping level of the high-T_c cuprates. The gaps reach maximum along the Cu-O bond direction for slightly underdoped and overdoped samples. The directions of the maximum gaps deviate from the Cu-O bond direction for much underdoped samples.

According to the simple d-wave, i.e., $\cos(k'_x a_{Cu})$ - $\cos(k'_y a_{Cu})$, the gap always reaches maximum along the Cu-O bond direction, and always vanishes along the line 45° away from the Cu-O bond direction. Therefore, the simple d-wave can not explain all the three experimental features of the gap anisotropy in the high-T_c cuprates.

We show in this paragraph that the composite d-wave can explain all the three experimental features of the gap anisotropy in the high-T_c cuprates. As we know, the gap opens at Fermi surface. The observed Fermi surface of Bi2212 sample with $T_c = 85$ K is $-1.456[\cos(k'_x a_{Cu}) - \cos(k'_y a_{Cu})] = -0.294$ eV [122]. From the empirical T_c formula in Ref. [120] the value of x is 0.192. The Fermi surface is shown in Fig. 4.1. When $x \neq 0.192$, we can use the rigid band approximation to obtain the corresponding Fermi surfaces. The concrete procedures are as follows. As we take the rigid band approximation, the form $-1.456[\cos(k'_x a_{Cu}) - \cos(k'_y a_{Cu})] = -E_F$ does not change for any values of x. The holes are in the range ABDC in Fig. 4.1. There are 1/4 holes in the range OAB for one unit cell of Cu lattice in the CuO_2 plane. The value of x is given by

$$x = \frac{\text{range(ABDC)}}{\text{range(OAB)}}. \tag{4.1}$$

From Eq. (4.1) and $-1.456[\cos(k'_x a_{Cu}) - \cos(k'_y a_{Cu})] = -E_F$, the obtained Fermi energies are $E_F = 0.057, 0.92, 0.13, 0.172, 0.214, 0.31, 0.361$, and 0.458 eV for $x = 0.05, 0.075$, 0.10, 0.125, 0.15, 0.20, 0.225, and 0.27, respectively. From the values of the expressions of the simple and the composite d-waves on Fermi surfaces, we can obtain relative values of the gaps of the simple and the composite d-waves on Fermi surfaces. Figs. 4.2 - 4.3 are the

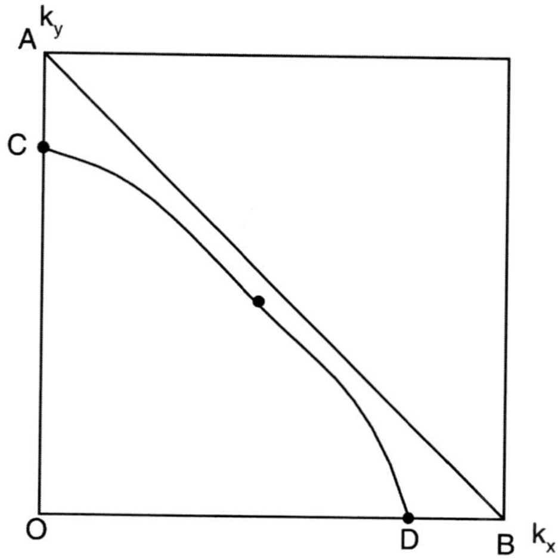

Figure 4.1. Fermi surface of Bi2212 in one quadrant. The line AB is the lower Hubbard band top. The line CD is the Fermi surface of Bi2212. The data are from Ref. [122].

curves of relative gap at 0 K vs. angle of the simple and the composite d-waves for $x = 0.05$, 0.075, 0.10, 0.125, 0.15, 0.20, 0.225, and 0.27, respectively. The units of the gaps in Figs. 4.2 - 4.4 are the values of the gaps at 0°, i.e., at the Cu-O bond direction. Using the empirical T_c formula, $T_c = 0$ for $x = 0.05$ and 0.27. From Figs. 4.2((a)(b)(c)) we can see clearly that the maximum gaps for $x = 0.05$, 0.075, and 0.10 in case of the composite d-wave are not at the Cu-O bond direction, and the maximum gaps in case of the simple d-wave are always at the Cu-O bond direction. Figs. 4.2 - 4.3 show that the composite d-wave can explain all the three experimental features of the gap anisotropy in the high-T_c cuprates. Fig. 4.4 is comparisons of the experimental data with the simple and the composite d-waves. Note that the values of x are not given in Ref. [20] directly. However, we can have the values of x from the observed values of T_c by the empirical T_c formula in Ref. [120]. From Fig. 4.4, we can see clearly that the composite d-wave fits the experimental data much better than the simple d-wave. As we know, the simple and the composite d-waves come from completely different quasiparticle pairing mechanism [111]– [116] [74]. Therefore, this section suggests that the quasiparticle pairing mechanism in Ref. [74] and chapter 3 of this book, which gives the composite d-wave, is correct probably, and that the quasiparticle pairing mechanisms in Refs. [111]– [116], which give the simple d-wave, is wrong probably. The content in this section can be found in Ref. [119] as well.

4.2. Essence of Pseudogap State

4.2.1. Introduction

One of the major challenges in condensed matter physics today is to understand the electronic phase diagram of the high-T_c cuprates. At the center of the debate is the nature of electronic states, which has attracted a great deal of attention because of the observation

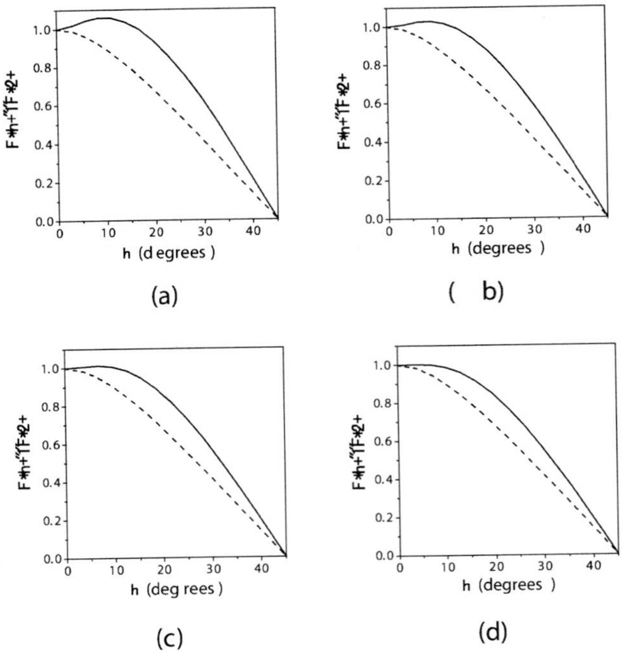

Figure 4.2. (a) Anisotropy of the gap at 0 K for $x = 0.05$. (b) Anisotropy of the gap at 0 K for $x = 0.075$. (c) Anisotropy of the gap at 0 K for $x = 0.10$. (d) Anisotropy of the gap at 0 K for $x = 0.125$. The value of x is the number of carriers in one unit cell of the Cu lattice in the CuO_2 plane. The solid line is the composite d-wave curve. The dashed line is the simple d-wave curve. 0° corresponds to the Cu-O bond direction.

of a pseudogap opened at Fermi surface [18, 19] [123] [124]– [129]. To understand the pseudogap state, more than ten mechanisms had been proposed and no conclusion has been reached [18, 19] [123] [124]– [129]. The reason for the situations is lacking direct calculation to show the essence of the pseudogap state and direct observation of Cooper pairs in the pseudogap state. In this section, we focus on solutions of the problems existing in understanding the pseudogap state (Also see Refs. [130, 131].). At first, we only use zero-gap condition for the gap equation of Cooper pairs to determine the pairing temperature T^*, at which the pseudogap is opened, and use the condition of zero-density of supercurrent Cooper pairs to determine the superconducting transition temperature T_c, at which the superconductivity appears, in the high-T_c cuprates. Thus, we demonstrate in mathematics that at $T_c < T < T^*$ the pseudogap is a gap of Cooper pairs without the long-range phase coherence and at $T < T_c$ there is the long-range phase coherence between some Cooper pairs. That there are Cooper pairs with and without the long-range phase coherence below and above T_c, respectively, might be observed by Refs. [132, 133]. Then, we show that there should be an oscillatory current, correlated with charge $2e$, induced by an applied constant bias V on the P-I-P junction and other weakly coupled systems. Here, P represents the high-T_c cuprate in the pseudogap state, where the Cooper pairs have no long-range phase coherence, and I represents a thin insulating barrier layer. The two electrodes of other weakly coupled systems consist of the high-T_c cuprate in the pseudogap state as well. At

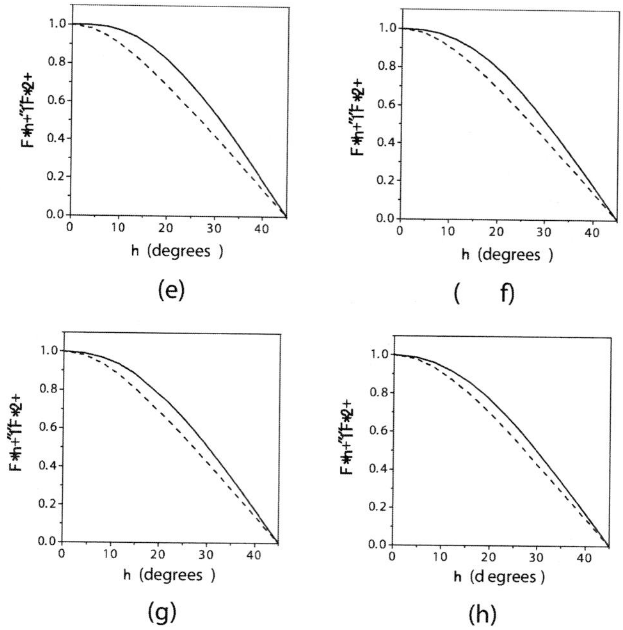

Figure 4.3. (e) Anisotropy of the gap at 0 K for $x = 0.15$. (f) Anisotropy of the gap at 0 K for $x = 0.20$. (g) Anisotropy of the gap at 0 K for $x = 0.225$. (h) Anisotropy of the gap at 0 K for $x = 0.27$.

last, we point out that there is a possibility to yield high temperature superconductivity in the heavily underdoped high-T_c cuprates.

4.2.2. Essence of Pseudogap State

The BCS gap equation for the CuO_2 plane in Bi2212 and Y123 is (See [2, 74], and chapter 3 of this book.)

$$\Delta'(T,\mathbf{k}) = -\sum_{\mathbf{k}'} \frac{U_{\mathbf{k}\mathbf{k}'}\Delta'(T,\mathbf{k}')}{2m(T,\mathbf{k}')} \tanh\frac{m(T,\mathbf{k}')}{2T}, \tag{4.2}$$

$$m(T,\mathbf{k}) = \sqrt{[\varepsilon(\mathbf{k}) - E_F]^2 + |\Delta'(T,\mathbf{k})|^2}, \tag{4.3}$$

where $\Delta'(T,\mathbf{k})$ is a gap function in the CuO_2 plane, $U_{\mathbf{k}\mathbf{k}'}$ a pairing potential, $\varepsilon(\mathbf{k})$ the dispersion relation of $O_{p\sigma}$ holes, and E_F the Fermi energy of $O_{p\sigma}$ holes. The form of $U_{\mathbf{k}\mathbf{k}'}$ of the high-T_c cuprates can be found in Eq. (3.53), which is called two local spin-mediated interaction (TLSMI). $U_{\mathbf{k}\mathbf{k}'}$ exists in antiferromagnetic or ferromagnetic long- or short-range coherence regions (See [74] and chapter 3 of this book.). When one only applies a zero-gap condition to Eq. (4.2), then the temperature obtained is T^* rather than T_c because the zero-gap condition can not give a guarantee that there is the long-range phase coherence between Cooper pairs. Because the gap is proportional to amplitude of the complex wave function of Cooper pair (See [134] and chapter 2 of this book.), we can say that T^* is only a temperature at which Cooper pairs without the long-range phase coherence begin to appear. Since 1957, when the famous BCS theory was published [2], people always think that the

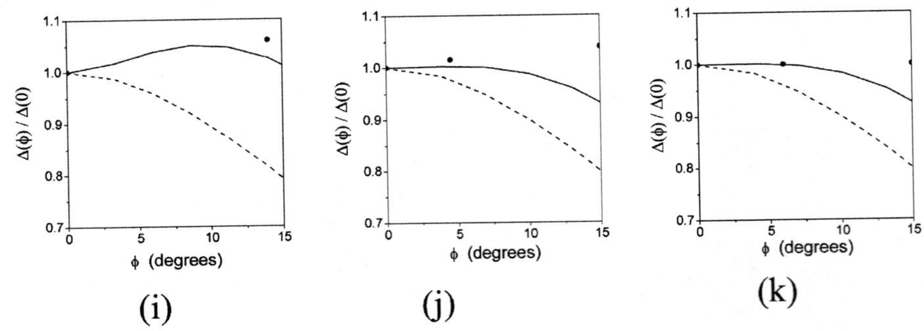

Figure 4.4. (i) Anisotropy of the gap at 0 K for $x = 0.056$. The experimental data are from Ref. [120]. (j) Anisotropy of the gap at 0 K for $x = 0.124$. The experimental data are from Ref. [120]. (k) Anisotropy of the gap at 0 K for $x = 0.132$. The experimental data are from Ref. [120].

solution of BCS gap equation under the zero-gap condition is T_c. Actually, we will see that the solution is T^* and $T_c \neq T^*$ for the high-T_c cuprates. Using TLSMI in Eq. (3.53) as the pairing potential $U_{\mathbf{kk}'}$ in Eq. (4.2), we know that the gap function is a composite d-wave in the CuO_2 plane of Bi2212 and Y123 (See [74] and chapter 3 of this book.). Using only zero-gap condition, the solution of Eq. (4.2) is T^* for a fixed x (See [74] and chapter 5 of this book.). The relation between T^* and x is shown in Fig. 4.5. Here, x represents the $O_{p\sigma}$ hole number in a unit cell of the Cu lattice in the CuO_2 plane. Fig. 4.5 shows clearly that $T_c < T^*$ for Bi2212 and Y123. The detailed numerical calculations to obtain the curve of T^* vs. x can be found in chapter 5.

If the Josephson coupling energy can overcome both the thermal disturbance energy and the zero-point fluctuation energy, then there will be a supercurrent density $n_s(T)$ in the system (See section 3.3.). The supercurrent density is given by Eq. (3.123)

$$n_s(T) = B\left\{4E_J(T)\left[2 - 1.075\frac{T}{4E_J} - 0.924\left(\frac{T}{4E_J}\right)^2\right] - \frac{2e^2}{C}\right\}^{2/3}, \tag{4.4}$$

where B is a temperature-independent constant. The theoretical value of T_c, $T_{c,theo}$, is determined by Eq. (3.124), i.e.,

$$n_s(T_{c,theo}) = 0. \tag{4.5}$$

Because $T_{c,theo}$ must satisfy Eq. (4.5), generally speaking, $T_c < T^*$.

The theoretical curve of $T_{c,theo}$ vs. x is shown in Fig. 4.6. The data in Fig. 4.6 are determined by the empirical formula of T_c in Refs. [120,135], $T_c = 93[1 - 82.6(x - 0.16)^2]$.

The gap at $T < T_c$ and the gap at $T_c < T < T^*$ are the same gap in essence for the high-T_c cuprates. They come from the same BCS gap equation in Eq. (4.2) and the same pairing potential in Eq. (3.53). The evolution from the pseudogap to the gap at T_c should be smooth

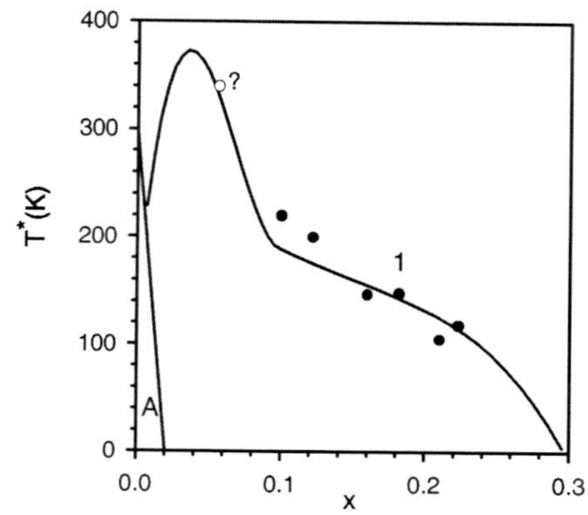

Figure 4.5. The relation of T^* vs. x. The x is the $O_{p\sigma}$ hole number of one unit cell of Cu^{++} lattice in the CuO_2 plane. The curve 1 is the theoretical curve. In the range A the samples of Bi2212 are in antiferromagnetic insulating state. The solid circles and the circles are taken from Refs. [94] and [20], respectively. There is a question mark for the circle because the actual datum in Ref. [20] is $T^* > 300$ K at $x = 0.056$.

as shown in Fig. 4.7. The global phase diagram of Bi2212 is shown in Fig. 4.8. From Fig. 4.8 we can see that the states in the part of $x < 0.05$ in region 2 are the pseudogap states, and these states do not have superconductivity at any temperature. Therefore, the materials in these states are called the pseudogap body.

4.2.3. Bias-Induced Oscillatory Current in P-I-P Junction and Other Weakly Coupled Systems

To sum up, the pseudogap state is a state of Cooper pairs without the long-range phase coherence. As far as we know, there is no direct experiment to show that the carriers with charge $2e$ exist in the pseudogap state. Sections 1.1.6 and 2.7.4 showed that the P-I-P junction and other weakly coupled systems can test it directly by the existence of bias-induced oscillatory current of frequency $\omega = 2|e|V/\hbar$ or the bias-induced external radiation of $\omega = 2|e|V/\hbar$ (Also see Refs. [130, 131].). Here, P represents the high-T_c cuprate in the pseudogap state, I the thin insulating barrier layer, and V the applied voltage.

The best materials to make the P in P-I-P junction and the electrodes in other weakly coupled systems are the underdoped high-T_c cuprates, such as Bi2212 and Y123. Fig. 4.9 plots the curve of $\Delta(T)$ vs. T of Bi2212 and Y123 with $x = 0.05$. Note that the $\Delta(T)$ represents the gap at the Cu-O bond direction in the CuO_2 plane if we do not give other definition. The numerical method to obtain the relation between $\Delta(T)$ and T is given in chapter 5. Fig. 4.9 shows that the pseudogap is equal to 26.6 meV even at 300 K. Note that $T_c = 0$ at $x = 0.05$ [120, 135]. Therefore, Bi2212 and Y123 with $x = 0.05$ are in the pseudogap state at $T \leq 300$ K. If this bias-induced oscillation current and external radiation can be verified by experiment, then many Josephson generators and mixers do not need to work at superconducting state. These devices can work in the pseudogap state at room temperature, and thus a superconducting electronics at room temperature can be found by using the high-T_c cuprates. This is only a possible application of microscopic supercon-

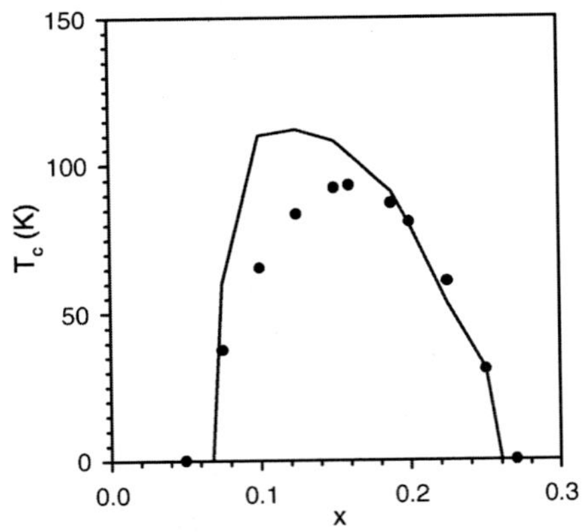

Figure 4.6. The relations of T_c and $T_{c,theo}$ vs. x. The solid line gives the relation between $T_{c,theo}$ and x. The $T_{c,theo}$ is determined by Eq. (4.5). The data are determined by $T_c = 93[1-82.6(x-0.16)^2]$ [120, 135].

ductivity of Cooper pairs without the long-range phase coherence. The further study on the microscopic superconductivity will lead to a series of novel significant findings. The word "superconductivity" in all literatures before this book means macroscopic superconductivity. Therefore, the idea "microscopic superconductivity" proposed in chapter 2 has significance in both theory and application.

Besides the discovery of the pseudogap in the high-T_c cuprates, there are many discoveries on the pseudogap phenomena in different non-superconducting materials, and many opinions on the mechanism of the pseudogap. Chapter 6 lists the following non-superconducting materials with the pseudogap. They are: Kondo semiconductors, such as CeNiSn, CeRhSb, and CeRhAs; magnetic manganese oxides with colossal magnetoresistance (CMR), such as $La_{1-x}Ca_xMnO_2$, $La_{2/3}Ba_{1/3}MnO_x$, $La_{1.2}Sr_{1.8}Mn_2O_7$, and $La_{0.82}Sr_{0.18}MnO_3$; and some quasi-one-dimensional inorganic materials, such as $(TaSe_4)I$ and $K_{0.3}MoO_3$. We suggest to make P-I-P junction and other weakly coupled systems, where P and the electrodes of the weakly coupled systems are the non-superconducting materials with the pseudogap. If there is the bias-induced external radiation of $\omega = 2|e|V/\hbar$, then we can say that the pseudogap correlates with Cooper pair without the long-range phase coherence.

4.2.4. Possible high Temperature Superconductivity in the Heavily Underdoped High-T_c Cuprates

Here, the heavily underdoped high-T_c cuprates (Bi2212 etc.) mean $x < 0.05$. Note that x indicates the number of $O_{p\sigma}$ holes in one unit cell of Cu^{++} lattice in the CuO_2 plane. From Fig. 4.8 we know that the heavily underdoped high-T_c cuprates (Bi2212 etc.) are in the pseudogap state.

Using the method given in chapter 5 we can obtain the dependence of the gap on temperature for any values of x. For example, for $x = 0.027$ the curve of gap $\Delta(T)$ vs. x is shown in Fig. 4.10. From Fig. 4.10 we know that the heavily underdoped high-T_c

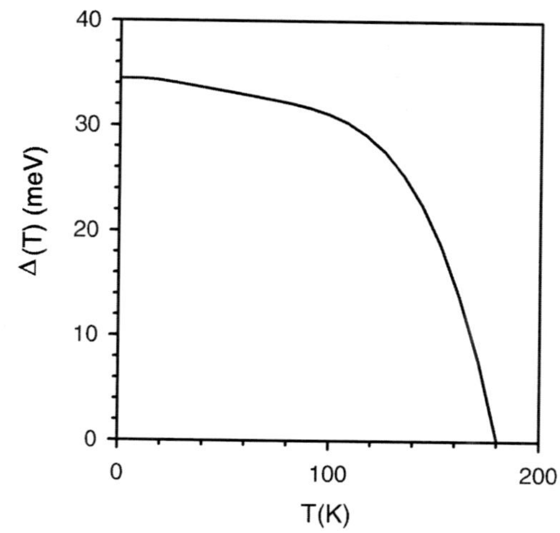

Figure 4.7. Theoretical curve of $\Delta(T)$ vs. T for Bi2212 and Y123 with T_c =84 K and $x = 0.126$.

cuprates (Bi2212 etc.) can have very large values of gap. From Eqs. (3.108), (3.123), and (3.124) we know that the large value of the gap leads to the large Josephson coupling energy, and thus to possible high-T_c. Let us try to find the superconductivity for the heavily underdoped high-T_c cuprate with $x = 0.027$. If we use the experimental value at $x > 0.05$, $\rho_n = 0.00131/x^2$ $(10^{-3}\Omega \cdot \text{cm})$ [136] for the R_n in Eq. (3.108), and substitute Eq. (3.108) into Eq. (3.124), then we obtain $T_{c,theo} = 0$, i.e., there is no superconductivity for the heavily underdoped high-T_c cuprates with $x = 0.027$. However, until now the ρ_n is an unknown number for the heavily underdoped high-T_c cuprates (Bi2212 and Y123). If we assume $\rho_n = 0.001/x^2$ $(10^{-3}\Omega \cdot \text{cm})$ which is only slightly less than $0.00131/x^2$, then 197 K $< T_{c,theo} <$ 245 K. If we take $\rho_n = 0.001/x^2$ $(10^{-3}\Omega \cdot \text{cm})$ for the heavily underdoped high-T_c cuprates (Bi2212 etc.) with any values of x, then there is a theoretical superconducting region which is shown in Fig. 4.11. However, until now no body knows the true value of ρ_n for the heavily underdoped high-T_c cuprates, and thus their superconductivity is only a conjecture. We suggest to adopt artificial methods to reduce the ρ_n of the heavily underdoped high-T_c cuprates (Bi2212 etc.) or to try some other heavily underdoped high-T_c cuprates, such as $HgBa_2Ca_2Cu_3O_8$, and $Tl_2Ba_2Ca_2Cu_3O_{10}$.

4.3. Inhomogeneities of Gap and Superconductivity

4.3.1. Introduction

Pan *et al.* found by scanning tunneling microscope (STM) that for their 84 K T_c $Bi_2Sr_2CaCu_2O_{8+\delta}$ (Bi2212) sample, the magnitudes of the gaps in the CuO_2 plane are different in different locations, which is called local gaps [137]. There is a distribution of magnitudes of the gap at 4.2 K on 600 $\AA$ × 600 $\AA$ area in the CuO_2 plane, with maximum value 65 meV, minimum value 25 meV, and the most probable value 40 meV [137]. That the distribution of the gap in the CuO_2 plane is inhomogeneous was also found in

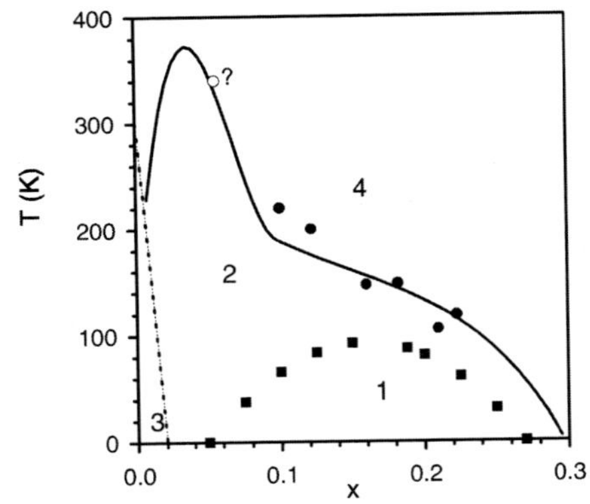

Figure 4.8. Phase diagram. The states in ranges 1, 2, 3, and 4 are superconducting state, pseudogap state, antiferromagnetic insulating state, and normal state, respectively. The data of T^* vs. x are taken from Ref. [94] for the solid circles, and from Ref. [20] for the circle, respectively. The solid squares indicate the experimental values of T_c vs. x [120, 135]. The solid line is the theoretical curve of T^* vs. x.

Refs. [138]– [143] [147]– [149]. The observed local gap and local superconductivity have the following nine features. First, the observed cumulative counts for different magnitudes of the local gaps are different [137, 138]. The maximum and minimum values of the cumulative counts are the local gaps with 30 and 50 meV, respectively [138]. They are 11 and 1, respectively. Second, the inhomogeneity of the gap in the CuO_2 plane is not stable at the same tip position in STM for long time measurement [138]. Third, when the point-contact mode was entered by producing a tip crash, dI/dV - V characteristic with a clear peak was replaced by a multi-peak structure [138]. Fourth, the different values of the local gaps in the CuO_2 plane were observed in Tl2212, Y123, and LSCO as well [139, 140, 149]. Fifth, the inhomogeneity of the gap in the CuO_2 plane exists in different hole doping level [141]. Sixth, the inhomogeneity of the gap in the CuO_2 plane exists at different temperatures [141]. Seventh, the inhomogeneity of the gap exists even above T_c, in the pseudogap range [141]. Eighth, a remnant of magnetic field expulsion is seen in form of a transient Meissner effect at several tens of degrees above T_c, judging from the optical conductivity [142, 143]. Using torque magnetic measurement on Bi2212, Wang *et al.* uncovered a field-enhanced diamagnetic signal above superconducting transition temperature T_c that increases with applied field to 32 tesla [144]. Using magnetic imaging (scanning SQUID microscopy) of $La_{2-x}Sr_xCuO_4$ thin films, Iguchi *et al.* observed inhomogeneous diamagnetic regions [145]. Xu *et al.* reported evidences for vortices in $La_{2-x}Sr_xCuO_4$ at temperatures as high as 150 K, which is significantly above T_c [146]. Ninth, the CuO_2 plane breaks into distinct domains of either "good" or "bad" superconductivity [147].

Analysis of the STM data in Ref. [137] led Pan *et al.* to suggest that the observed inhomogeneous electronic structure arises from the ionic potential associated with the off-stoichiometry oxygen dopants disordered in the BiO layer [137, 150]. It is obvious that the theory in Refs. [137, 150] can not explain that the values of the gap are not stable at the same tip position for long time measurement, and that the inhomogeneity appears in many kinds of the high-T_c cuprates without the BiO layer. Matsuda *et al.* and Lang *et al.* thought that

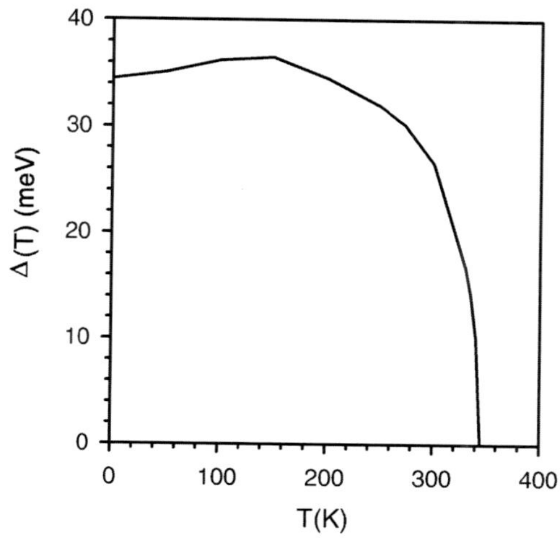

Figure 4.9. Theoretical curve of $\Delta(T)$ vs. T for Bi2212 and Y123 with $T_c = 0$ K and $x = 0.05$.

the gap inhomogeneity is a result of the in-plane charge distribution [141, 148]. However, there is no quantitative calculation based on the in-plane charge distribution. Joglekar *et al.* thought that the high-T_c cuprates are granular d-wave superconductors. They studied the dynamics of Josephson coupling between such granular d-wave superconductors, focused on the effect of nodal Cooper pairs and disorder impurities, and found that the nodal Cooper pairs give rise to a power-law Josephson coupling which leads to the stability of the superconducting phase in the CuO_2 plane [151]. Martin and Balasky thought that the inhomogeneity is induced by disordered doping, and constructed a semi-phenomenological model [152].

The basic aim of section 4.3 is to demonstrate that the inhomogeneities of the gap and the superconductivity come from the inhomogeneity of distribution of the antiferromagnetic short-range coherence lengths (AFSRCL) in the CuO_2 plane. In section 4.3.2 we make numerical calculations to give the inhomogeneities of distributions of the gap and the superconductivity in the CuO_2 plane. In section 4.3.3 we explain the nine features of the inhomogeneities. In section 4.3.4 we give some predictions. In section 4.3.5 we give a reconciliation of the sharp contradiction between Lang *et al.*'s observations [148] and basic concept of superconductivity. In section 4.3.6 we make conclusions and give a physical picture for the inhomogeneities of the gap and the superconductivity.

4.3.2. Numerical Results of the Inhomogeneity

Pan *et al.* found by STM that for their Bi2212 sample with $T_c = 84$ K, the distribution of magnitudes of the gaps at 4.2 K in 600 $\AA \times 600 \AA$ area of the CuO_2 plane is inhomogeneous, with maximum value 65 meV, minimum value 25 meV, and the most probable value 40 meV [137]. As we know, the antiferromagnetic short-range order is formed under thermal disturbance. Therefore, there is fluctuation around the statistically most probable value for AFSRCL's. TLSMI is an interaction mediated by the two nearest-neighbor Cu^{++} ions in

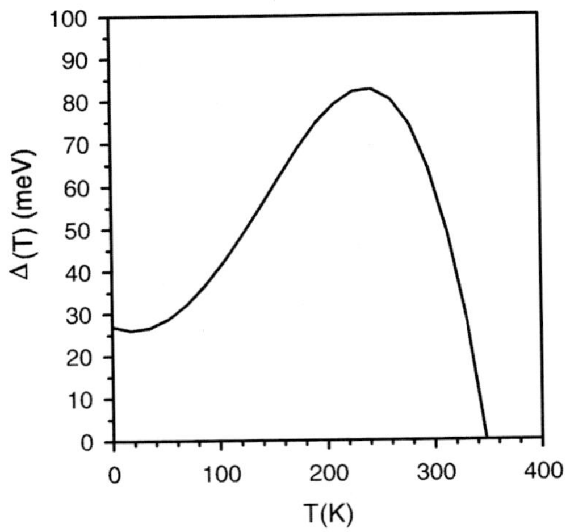

Figure 4.10. Theoretical curve of the gap $\Delta(T)$ vs. T for the heavily underdoped high-T_c cuprates (Bi2212 etc.) with $x = 0.027$.

the CuO_2 plane. Thus, TLSMI can be used to discuss the inhomogeneity of the high-T_c cuprates in Refs. [137]– [143] [147, 148]. From Eq. (3.54) in section 3.2.1, we know that TLSMI depends on N''/N'. From Fig. 3.10, we know that N''/N' is determined by AFSRCL ξ. Therefore, we can say that TLSMI depends on AFSRCL ξ as well. In our theory, the inhomogeneity comes from the fluctuation of AFSRCL ξ' at different locations in the CuO_2 plane around the most probable value ξ (=$a_{Cu}/\sqrt{x}$) [153]. Therefore, the inhomogeneity is inherent in the CuO_2 plane in our theory.

From Fig. 3.10, the different values of ξ' have different values of N''/N'. From Eqs. (3.54)and (3.57) the different values of N''/N' have different TLSMI, and thus different magnitudes of the gaps. From the empirical formula of T_c vs. x in Refs. [120, 135], we know that $x = 0.126$ if the sample in Ref. [137] is $T_c = 84$ K. The most probable value of ξ' is $\xi = a_{Cu}/\sqrt{0.126} = 2.817\ a_{Cu}$. Correspondingly, $N''/N' = 1.33$ from Fig. 3.10. The corresponding most probable value of the gap at Cu-O bond direction at 0 K, $\Delta(0K)$, is 34.8 meV from BCS gap equation Eq. (3.57). If the fluctuation values of ξ' at different locations are from 1.7 a_{Cu} to 7 a_{Cu}, then the corresponding values of N''/N' are from 1.1 to 1.66. Substituting these values of N''/N' into BCS gap equation yields that the fluctuation values of $\Delta(0K)$ are from 20.2 meV to 58.6 meV. The experimental values of $\Delta(0K)$ in Ref. [137] are: the most probable value is 40 meV, the minimum value is 25 meV, and the maximum value is 65 meV. Note that the experimental values of $2\Delta(0K)$ in Ref. [137] come from the peak-to-peak values of the derivative conductance. Although simple, this peak-to-peak method should give a larger gap value than the true one, considering the effect of thermal smearing and/or the quasiparticle damping [154]. Therefore, we can say that our theory can explain the distribution of the values of $\Delta(4.2K)$ observed in Ref. [137].

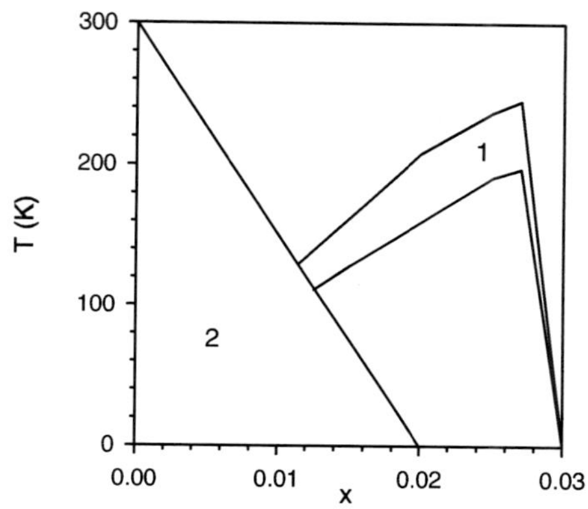

Figure 4.11. The region 1 is a region of possible superconducting state for the heavily underdoped high-T_c cuprates (Bi2212 etc.). The region 2 is a region of antiferromagnetic insulating state.

4.3.3. Explanations for the Nine Features of the Inhomogeneities

The explanation for the first feature is as follows. From solving the BCS gap equation we know that there is a one to one correspondence between the local gap and the value of ξ'. Because the value of ξ' has the most probable value ξ, the local gap also has the most probable value, and, correspondingly, the local gaps have the maximum observed cumulative count. The 30 meV in Ref. [138] is the local gap corresponding to the maximum observed cumulative count. The 50 meV in Ref. [138] is the local gap corresponding to the minimum observed cumulative count.

Second, the value of the local gap should not be stable at the same tip position in STM technology for long time measurement because the thermal perturbation sets the value of ξ' at any fixed location unstable.

The explanation for the third feature is as follows. As we know, the peak on dI/dV - V characteristic corresponds to a gap. When the point-contact mode is entered by producing a tip crash, the tunneling range contains many clusters with different values of ξ', and thus dI/dV - V characterizes a multi-peak structure instead of a clear peak.

Fourth, the inhomogeneity of the gap in the CuO_2 plane should be observed in many kinds of the high-T_c cuprates because the fluctuation of AFSRCL ξ' is dependent only on the thermal perturbation in the CuO_2 plane, and thus is independent of the kinds of the high-T_c cuprates.

Fifth, the inhomogeneity of the gap in the CuO_2 plane should exist in different $O_{p\sigma}$ hole doping level. The reasons are as follows. The fluctuation of AFSRCL ξ' is dependent only on the thermal perturbation in the CuO_2 plane. Therefore, the existence of the fluctuation of AFSRCL ξ' is independent of the hole doping level.

Sixth, the inhomogeneity of the gap in the CuO_2 plane should be observed at different temperatures. The reasons are as follows. The fluctuation of AFSRCL ξ' is dependent only on the thermal perturbation in the CuO_2 plane. Thermal perturbation exists at any temperature, even at 0 K, where there is still the zero-point oscillation of quantum mechanics.

Therefore, the existence of the fluctuation of AFSRCL ξ' is independent of the temperature.

The explanation for the seventh feature is as follows. According to the observation of Ref. [141], $x = 0.11$, and $T_c = 67$ K, and the inhomogeneity of the gap in the CuO_2 plane exists at 100 K. According to our numerical calculation, the temperature T^* of opening a gap at the Fermi surface in the CuO_2 plane is 190 K (See Fig. 4.5.). The gap at 67 K $< T <$ 190 K is called the pseudogap. The inhomogeneity of the pseudogap in the CuO_2 plane should be observed at temperature $T_c < T < T^*$ because the fluctuation of AFSRCL ξ' is dependent only on the thermal perturbation in the CuO_2 plane, and the existence of the fluctuation of AFSRCL ξ' is independent of the temperature.

The explanation for the eighth feature is as follows. The superconducting transition temperatures at some locations in the CuO_2 plane, T_c', are different. For example, our theoretical values of T_c' for the sample in Ref. [137], calculated by Eq. (3.124), are 34.1 K, 52.7 K, 84 K, 113.1 K, 126.3 K, 134.8 K, and 141.3 K, for the locations with $\xi' = 1.7\ a_{Cu}$ ($\Delta(0K) = 20.18$ meV), $\xi' = 2\ a_{Cu}$ ($\Delta(0K) = 24.85$ meV), $\xi' = 2.817\ a_{Cu}$ ($\Delta(0K) = 34.75$ meV), $\xi' = 4\ a_{Cu}$ ($\Delta(0K) = 46.16$ meV), $\xi' = 5\ a_{Cu}$ ($\Delta(0K) = 51.8$ meV), $\xi' = 6\ a_{Cu}$ ($\Delta(0K) = 55.47$ meV), and $\xi' = 7\ a_{Cu}$ ($\Delta(0K) = 58.57$ meV), respectively. The observed value of T_c in experiment for the bulk sample should be the most probable value 84 K, which corresponds to the most probable value $\xi = 2.817\ a_{Cu}$. To explain the eighth feature, we give an example. There are still some locations with $\xi' > 4\ a_{Cu}$ in the CuO_2 plane, for which $T_c' > 113.1$ K according to our calculations for the sample in Ref. [137]. Note that the observed value of T_c for the bulk sample in Ref. [137] is 84 K. Therefore, we can say that a remnant of magnetic field expulsion can be seen at several tens of degrees above T_c. The remnant of magnetic field expulsion is given by the locations with $\xi' > 4\ a_{Cu}$ in the CuO_2 plane. The existences of the diamagnetism observed in Refs. [144, 145] and the vortices observed in Ref. [146] at temperatures significantly above T_c come from that there are many locations on the CuO_2 plane whose values of local superconducting transition temperature T_c' are significantly above the most probable value of T_c.

The explanation for the ninth feature is as follows. Scanning tunneling spectroscopy on cleaved Bi2212 single crystals reveals inhomogeneities on length scales of $\approx 30\mathring{A}$. This scale indicates that there are nearly equal AFSRCL's in the area $\approx (30\mathring{A})^2$. In our theory, it is natural that the CuO_2 plane breaks into distinct domains of either "good" or "bad" superconductivity because the domains can have different AFSRCL, and therefore different superconducting transition temperature T_c'. For example, $T_c' = 34.1$ K ($\xi' = 1.7 a_{Cu}$) or 134.8 K ($\xi' = 6 a_{Cu}$), which are given in the paragraph for the eighth explanation. Since there are regions with "bad" superconductivity in the CuO_2 plane, the proximity effect, which implies leakage of Cooper pair wavefunction in a distance of order $\approx 20\ \mathring{A}$ (Cooper pair size) into the "bad" superconducting regions, can occur, and thus the "bad" superconducting regions have double gaps. Based on the same reason, the "good" superconducting regions also have double gaps.

4.3.4. Predictions

The dependences of the gap $\Delta(T)$ vs. T at different locations in the CuO_2 plane are much different. For example, the theoretical curves 1, 2, 3, 4, and 5 in Fig. 4.12 correspond to $\xi' = 1.7\ a_{Cu}$, $2\ a_{Cu}$, $2.817\ a_{Cu}$, $5\ a_{Cu}$, $7\ a_{Cu}$, and $x = 0.126$ for Bi2212 and Y123. All the five

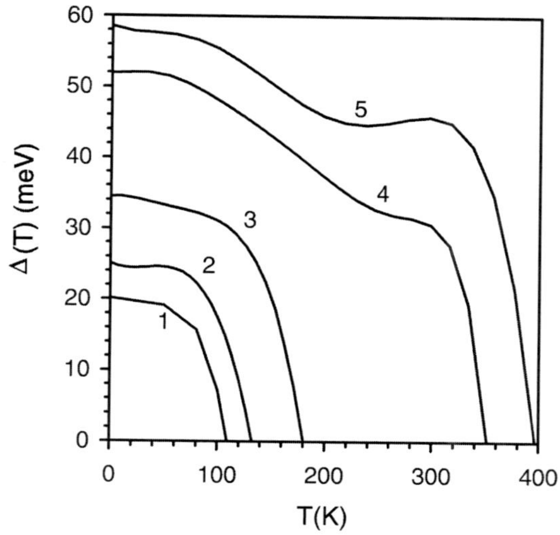

Figure 4.12. Theoretical curves of $\Delta(T)$ vs. T. The curves 1 - 5 correspond to AFSRCL $\xi' = 1.7, 2, 2.8, 5, 7\, a_{\text{Cu}}$, and $x = 0.126$.

theoretical curves in Fig. 4.12 have the pseudogaps, i.e., the gaps above the corresponding superconducting transition temperatures T_c'. Note that there is no discontinuity at T_c' for the five curves. The evaluation from the pseudogap at $T > T_c'$ to the gap at $T < T_c'$ is smooth. Both the gap and the pseudogap are gap of Cooper pairs. But the people call the gap at $T > T_c'$ the pseudogap due to a historical misunderstanding. The curves in Fig. 4.12 are difficult to be observed in experiment because of the unstable property of AFSRCL under thermal perturbation. However, from Fig. 4.12 we can see that even at room temperature 300 K there are gaps larger than 30 meV in the locations with AFSRCL larger than 5 a_{Cu}. These large gaps at room temperature should easily be observed by long time measurement at fixed location in the CuO_2 plane or by measurements at many locations in the CuO_2 plane.

4.3.5. Reconciliation of the Sharp Contradiction between Basic Concept of Superconductivity and Lang *et al.*'s Observations

A novel atom-scale technique, designed to locally distinguish superconducting from non-superconducting regions by using quasiparticle scattering resonances at Ni impurity atoms, has recently been proposed [148]. Lang *et al.* applied this technique to carry out spectral surveys on as-grown Ni-doped Bi2212 sample [148]. Lang *et al.* found that no Ni scattering resonances are observed in any region where $\Delta > 50$ meV, Ni scattering resonances are observed in regions where $\Delta < 35$ meV, and Ni scattering resonances go to zero gradually from 30 to 50 meV. Lang *et al.*'s conclusions are that purely superconducting regions when $\Delta < 35$ meV, a mixture of superconducting and normal regions when 35 meV $< \Delta < 50$ meV, and normal regions when $\Delta > 50$ meV. As is well known, these conclusions violate basic concept on superconductivity. The gap is the bound energy of Cooper pair. Therefore, according to the basic concept, the larger the gap is, the stronger the superconductivity should be. Our numerical calculations in the paragraph for the eighth explanation accord with this basic concept. Our theory can also reconcile the sharp contradiction between Lang *et al.*'s observations and the basic concept of superconductivity. The numerical calculations in the eighth explanation tell us that the longer the AFSRCL, the larger the gap. As is well known, the longer the AFSRCL, the stronger the antiferromagnetism. The Ni atom is magnetic. Therefore, the Ni atom is difficult to enter the regions with stronger antiferromagnetic correlation. Our conclusions are that Ni atoms can not enter the regions with $\Delta > 50$ meV, a few Ni atoms can enter the regions with 35meV $< \Delta < 50$ meV, and Ni atoms can easy enter the regions with $\Delta < 35$ meV.

4.3.6. Conclusions

In our theory the mobile $O_{p\sigma}$ hole distribution in the CuO_2 plane of the high-T_c cuprates is homogeneous, which is different from Refs. [141, 148]. We demonstrate that the thermal perturbation leads to the fluctuation of AFSRCL, and further leads to the fluctuation of pairing potential (TLSMI). The latter can cause the inhomogeneities of the gap and the superconductivity in the CuO_2 plane. The values of the gap and the critical temperature T_c in bulk sample are determined by the most probable value of AFSRCL. At $T = T_c$, a superconducting percolation channel is established by the locations with the most probable AFSRCL and the locations with AFSRCL larger than the most probable one. The proximity effect and the Josephson effect exist in the locations with lower values of T_c'. However, both effects are not important for the inhomogeneities. The gap and the superconductivity themselves are stable in the CuO_2 plane, and their stability does not need the help of the nodal Cooper pair, which is different from Ref. [151]. The main reference of section 4.3 is Ref. [155].

4.4. Relationship between Temperature and Gap

4.4.1. Types of Temperature-Dependence of Gap

The main reference of section 4.4 is Ref. [156]. Note that the conjecture on "fried" superconductor in Ref. [156] is wrong. For the bulk high-T_c cuprates, the values of J_K and J in Eqs. (3.5) and (3.54) will be given in chapter 5. Using these values, we can obtain the temperature dependence of gap by solving the BCS gap equation Eq. (3.57). The curves of gap vs. temperature for Bi2212 are shown in Fig. 4.13. The formulas to determine the J_K and J are as follows [83].

$$J_K = t_\circ^2 (1 - \frac{x}{4})^2 (B_6 + xA_7), \tag{4.6}$$

$$B_6 = \frac{1}{U_d - \varepsilon_p + \varepsilon_d - 2V_\circ} + \frac{1}{\varepsilon_p - \varepsilon_d + U_p}, \tag{4.7}$$

$$A_7 = \frac{1}{\varepsilon_p - \varepsilon_d - V_\circ} - \frac{1}{\varepsilon_p - \varepsilon_d} + \frac{1}{U_d - \varepsilon_p + \varepsilon_d - V_\circ} - \frac{1}{U_d - \varepsilon_p + \varepsilon_d - 2V_\circ}, \tag{4.8}$$

where $t_\circ$ is the hopping integral between O:p_σ orbital and Cu:$d_{x^2-y^2}$ orbital, U_p, U_d and $V_\circ$ are intra- and interatomic Coulomb repulsion on O:p_σ orbital, Cu:$d_{x^2-y^2}$ orbital and

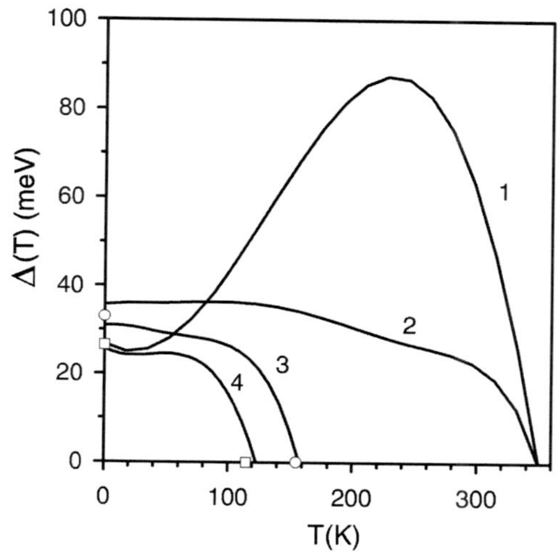

Figure 4.13. Theoretical curves of $\Delta(T)$ vs. T for Bi2212 and Y123. The curves 1, 2, 3, and 4 are for $x = 0.025$, 0.056, 0.16, and 0.208, respectively. The x is the number of $O_{p\sigma}$ holes in one unit cell of of the Cu lattice in the CuO_2 plane. The circles and squares are for $x = 0.16$ and 0.208, respectively [94].

between both orbitals, respectively.

$$J = \frac{2t_\circ^4}{\varepsilon_p - \varepsilon_d + V_\circ}\left[\frac{1}{U_d} + \frac{2}{2(\varepsilon_p - \varepsilon_d) + U_p}\right]. \tag{4.9}$$

However, in many measurements the high-T_c cuprates are not bulk samples, and thus the values of J_K and J are different more or less. To know which parameter changes value in Ekino *et al.*'s break junction, we introduce the break junction technique [16]. A very thin single crystal mounted on the glass-fiber substrate is cracked to form a junction at low temperature by applying bending force perpendicular to the CuO_2 plane. In this bending process the distance between Cu^{++} and O^{--} ions in the CuO_2 plane is extended more or less, and thus the hopping integral $t_\circ$ in Eqs. (4.6) and (4.9) becomes smaller. Therefore, the values of J_K and J in the break junction become smaller in comparison with the bulk samples. In Refs. [157]– [159] the S-I-S intrinsic Josephson tunneling junction of Bi2212 is employed. To make S-I-S, before the photolithograph process, a 25 nm thick Ag thin film and a 50 nm thick Au thin film, for example, were evaporated on the cleaved surface and then annealed at high temperature [157, 158]. The Ag atoms influence the surface of Bi2212, such as change the distance between Cu^{++} and O^{--} ions and the dielectric constant in the CuO_2 plane, which then causes the variations of $t_\circ$, $V_\circ$, J_K, and J. In Renner *et al.*'s experiments a special region (0.15 μm line) in the CuO_2 plane was detected by STM [160]. It is possible that in this selected region the dielectric constant has some variations. If the dielectric constant becomes smaller, then $V_\circ$ in Eqs. (4.7), (4.8), and (4.9) becomes larger. Eqs. (4.7), (4.8), and (4.9) tell us that J_K and J become larger and smaller, respectively, if $V_\circ$ becomes larger. In one word, the values of J_K and J have some variations in some cases in comparison with the values of the bulk samples of Bi2212 in Ref. [75] and Fig. 4.13.

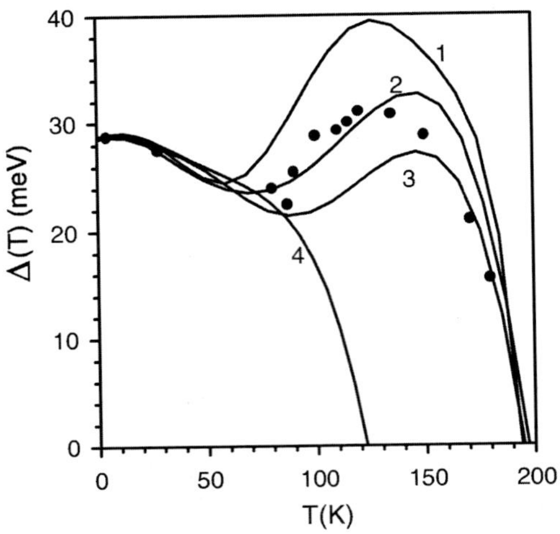

Figure 4.14. Theoretical curves of $\Delta(T)$ vs. T for Bi2212. The curves 1, 2, 3, and 4 are for $x = 0.188$, $J'_K = 0.65\ J_K$, $J' = 0.4J$, $0.47J$, $0.52J$, and $0.8J$, respectively. The data are from Ref. [16].

To find all the possible types of the temperature dependence of the gap for the different values of J_K and J, we take the following groups of values of x, J'_K, and J' for numerical calculations. Here J'_K and J' represent the changed values of J_K and J. (1) $x = 0.188$, $J'_K = 0.65J_K$, $J' = 0.4J$, $0.47J$, $0.52J$, and $0.8J$. The corresponding curves of $\Delta(T)$ vs. T are shown in Fig. 4.14. (2) $x = 0.188$, $J'_K = 0.54J_K$, $J' = 0.447J$. The corresponding curve is shown in Fig. 4.15. (3) $x = 0.208$, $J'_K = 0.45J_K$, and $J' = 0.7J$. $x = 0.127$, $J'_K = 0.35J_K$, and $J' = 0.63J$. The corresponding curves are shown in Fig. 4.16. (4) $x = 0.16$, $J'_K = 0.26J_K$, and $J' = 0.77J$. $x = 0.183$, $J'_K = 0.35J_K$, and $J' = 0.8J$. The corresponding curves are shown in Fig. 4.17. (5) $x = 0.124$, $J'_K = 1.44J_K$, and $J' = 0.58J$. The corresponding curve is shown in Fig. 4.18.

The possible types of the gap as a function of temperature in Figs. 4.14 - 4.18 are as follows. We express the temperature, at which the gaps in Figs. 4.14 - 4.18 are zero, as T^*_{change}. First, let us look at the relationships between T^*_{change} and T^* and between T^*_{change} and T_c. Note that the values of T^* are given in Fig. 4.5. Figs. 4.18 and 4.5 show $T^*_{\text{change}} =$ 402 K and $T^* = 189$ K for $x = 0.124$. Therefore, $T^*_{\text{change}} >> T^*$ in some cases. Figs. 4.14 and 4.5 show $T^*_{\text{change}} < T^*$ in some cases. Many curves show $T^*_{\text{change}} > T_c$. However, there is also a possibility that $T^*_{\text{change}} < T_c$. Fig. 4.17 shows $T^*_{\text{change}} = 92$ K for $x = 0.16$. The experimental value of T_c for $x = 0.16$ is 93 K [120, 135], and thus $T^*_{\text{change}} < T_c$. Second, let us look at the forms of the curves for the gap vs. temperature. There are two forms: non-BCS and BCS-like forms. The non-BCS form contains two forms: the high temperature peak, such as the curve in Fig. 4.18 and the curves 1 and 2 in Fig. 4.14, and the high temperature tail, such as the curve in Fig. 4.15. The value of the gap at high temperature peak is larger than the value of the gap at 0 K. The curves in Figs. 4.16 and 4.17 are BCS-like. We do not know the values of x for the samples in Refs. [16] [157]– [160] directly. However, we can use the given experimental values of T_c and the empirical formula $T_c = 93[1-82.6(x-0.16)^2]$ (K) [120,135] to know the experimental values of x. The values

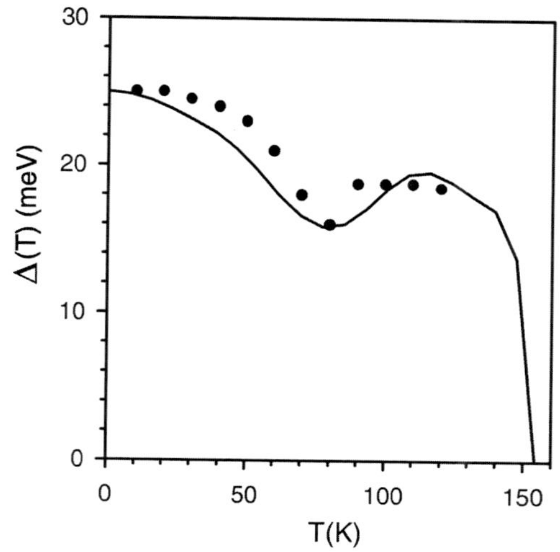

Figure 4.15. Theoretical curve of $\Delta(T)$ vs. T for Bi2212. The curve is for $x = 0.188$, $J_K' = 0.54J_K$, $J' = 0.447J$. The data are from Ref. [157].

of x are 0.124 [160], 0.188 [157], 0.127 and 0.208 [158], 0.16 and 0.183 [159], and 0.183 - 0.190 [16], respectively. From Figs. 4.14 - 4.18 we see clearly that all the experimental data of the gap vs. T can be fitted, in principle, by the possible theoretical curves of the gap vs. T. Because we do not know the exact values of J_K' and J' in Refs. [16] [157]– [160], we can not think that the data in Refs. [16] [157]– [160] have been explained by our theory exactly. However, we can say, at least, that the various data in Refs. [16] [157]– [160] can not deny yet the idea that the pseudogap at $T_c < T < T^*$ and the (superconducting) gap at $T < T_c$ are the same gap relating to the Cooper pairs.

To verify our theory, we make some predictions. The values of T_c for the samples in Refs. [16] [157]– [160] have not been observed. We predict that the values of T_c for these samples should have more or less variations in comparison with the experimental values of T_c of the corresponding bulk samples. The largest variation of the values of T_c should take place in the samples in Refs. [160] and [159]. For the sample of Ref. [160], the experimental value of T_c of the bulk is 83.5 K. The theoretical value of T_c, obtained by the theory in section 3.3, at the selected region in Ref. [160] is 129 K. For the samples in Ref. [159], the experimental values of T_c of the bulks are 93 K and 89 K. The theory in section 3.3 gives that the theoretical values of T_c are 74 K and 64 K for the samples in Ref. [159], respectively.

By the way, Suzuki *et al.* [157, 158] and Krasnov *et al.* [159] observed two peaks on their curves of dI/dV vs. V at the same time. Suzuki *et al.* thought that the peak at high voltage arises from a band-structure effect, such as a van Hove singularity [157], and from the pseudogap [158]. Krasnov *et al.* had the same point of view as [158]. Our theory agrees with the point of view in Ref. [157], i.e., the peak at high voltage does not relate to the gap of Cooper pairs. Refs. [158, 159] can not observe the pseudogap at temperature obviously larger than T_c due to the variations of J_K and J in their S-I-S junctions.

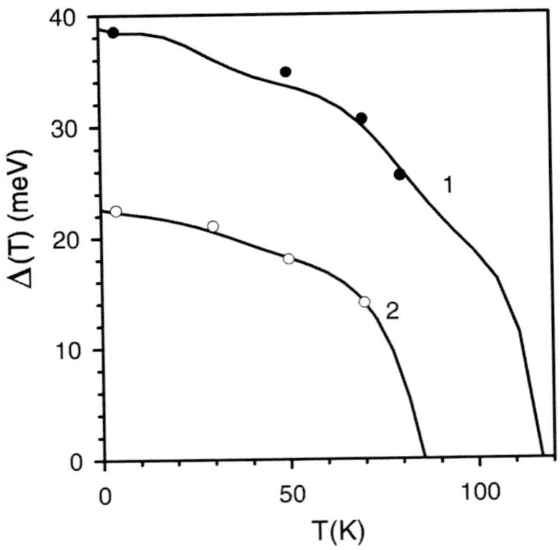

Figure 4.16. Theoretical curves of $\Delta(T)$ vs. T for Bi2212. The curves 1 and 2 are for $x = 0.127, J'_K = 0.35J_K, J' = 0.63J$, and $x = 0.208, J'_K = 0.45J_K, J' = 0.7J$, respectively. The data are from Ref. [158].

4.4.2. A Misunderstanding

Refs. [161]– [163] thought that the penetration depth and the maximum Josephson current seem to indicate the disappearance of the amplitude of wavefunction of Cooper pair at T_c, and thus if (superconducting) gap starts to evolve at T_c, then the pseudogap must be interpreted differently. In the analyses of Refs. [161, 162] the following formula was used [See Ref. [164] and Eq. (3.107) of this book.]:

$$I_c(T) = \frac{\pi\overline{\Delta(T)}}{2|e|R_n}\tanh\frac{\overline{\Delta(T)}}{2T}, \tag{4.10}$$

where $I_c(T)$ is the Josephson critical current at zero voltage, R_n the resistance, and e the electron charge. At zero temperature

$$I_c(0) = \frac{\pi\overline{\Delta(0)}}{2|e|R_n}. \tag{4.11}$$

Based on Eq. (4.10), Refs. [161]– [163] deduced $\Delta(T_c) = 0$ from experimental $I_c(T_c) = 0$, i.e., the amplitude of wavefunction of Cooper pair disappears at T_c, and thus the pseudogap comes from another mechanism. We think that the conclusions in Refs. [161]– [163] are wrong. The key point is that the explanations for Eqs. (4.10) and (4.11) in Refs. [161]– [163] are correct only for the conventional superconductors in which the pairing and the long-range phase coherence happen at nearly the same temperature $T = T_c \cong T^*$ (See sections 1.1.1 and 3.3.). In Mahan's derivation for Eqs. (4.10) and (4.11) it is assumed that, as soon as the pairs form in the system, they automatically have the long-range phase coherence, i.e., $T^* = T_c$ [164]. Actually, for the high-T_c cuprates, T_c is determined by the long-range phase coherence condition Eq. (3.124), while the zero-gap temperature T^* is determined by Eq. (3.69). Our theory (Fig. 4.8) shows that $T^* \neq T_c$ for the high-T_c cuprates.

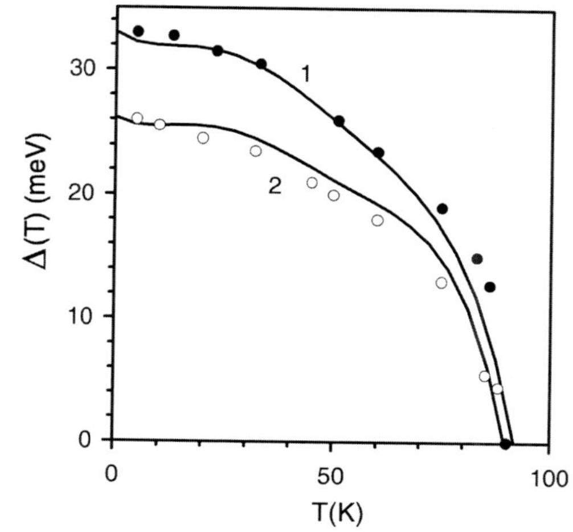

Figure 4.17. Theoretical curves of $\Delta(T)$ vs. T for Bi2212. The curves 1 and 2 are for $x = 0.16$, $J'_K = 0.26J_K$, $J' = 0.77J$ and $x = 0.183$, $J'_K = 0.35J_K$, $J' = 0.8J$, respectively. The data are from Ref. [159].

At T_c, the number density of supercurrent Cooper pairs is zero, and thus all Cooper pairs are in the states without the long-range phase coherence. In this case the current through the junction should be given by Eq. (2.290). In our theory, the $I_c(T_c) = 0$ observed in Refs. [161]– [163] comes from the arbitrariness of $\phi_{R,arbitrary}$ and $\phi_{L,arbitrary}$, and does not come from $\Delta(T_c) = 0$. Any conclusion for the high-T_c cuprates, based on Eqs. (4.10) and (4.11) as maximum supercurrent formulas, is not reliable. Refs. [165, 166] discussed the property of the high-T_c cuprates based on the simple formula $\lambda^2(0)/\lambda^2(T) \propto \Delta(T)\tanh[\Delta(T)/2T]$. Eq. (3.123) indicates that the number density of supercurrent Cooper pairs, and thus $\lambda(T)$, are a very complex function of the gap. Therefore, the conclusions in Refs. [165, 166] are not reliable.

4.5. Unified Explanation for Doping Dependences

There are experimentally observed doping dependences of the following six physical quantities: superconducting transition temperature T_c [120, 135], the pairing temperature T^* [20, 94], the gap at 0 K $\Delta(0)$ [94], $2\Delta(0)/T^*$ [94], $2\Delta(0)/T_c$ [94], and the number density of supercurrent Cooper pairs at 0 K $n_s(0)$ divided by the effective mass m^* [167]. There are also many papers which try to explain the dependence. Doniach and Inui showed that the doping dependence of T_c should be determined by the long-range phase coherence condition in the quantum XY-model [96]. Emery and Kivelson [168], and Chakraverty and Ramakrishnan [169] showed that the doping dependence of T_c should be determined by the long-range phase coherence condition in the classical XY-model. Szotek, Gyorffy and Temmerman explained the doping dependence of penetration depth, which is proportional to $n_s(0)/m^*$, using eight-band model [170]. Bernhard *et al.* thought that the reduction of $n_s(0)/m^*$ on the underdoped side can be understood in terms of a decrease of x, while some other mechanism is needed to account for the reduction of $n_s(0)/m^*$ in the overdoped

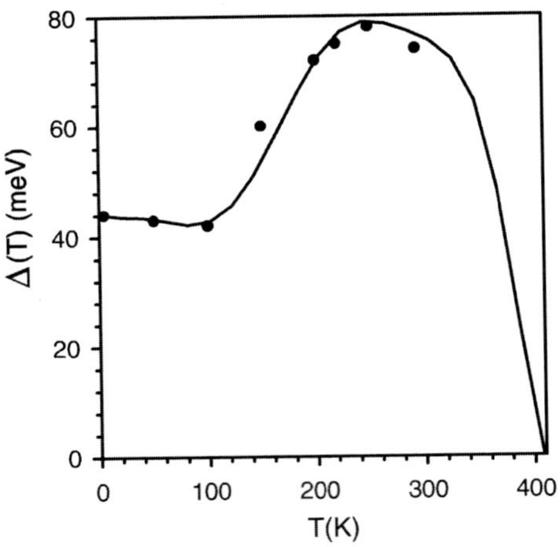

Figure 4.18. Theoretical curve of $\Delta(T)$ vs. T for Bi2212 with $x = 0.124$, $J'_K = 1.44J_K$, and $J' = 0.58J$. The data are from Ref. [160].

side [171]. Until now a unified explanation for all the six doping dependences is still lacking.

It is obviously that if a theory can well explain all the six doping dependences, then the theory is more convincing. Taking the two local spin-mediated interaction (TLSMI), proposed in Refs. [74, 75], and Eq. (3.53) of this book, as a pairing potential, and solving BCS gap equation Eq. (3.57) and the equation of T_c Eq. (3.124), this section explains all the six doping dependences.

The steps to obtain the six theoretical doping dependences are as follows. First, when we solve BCS equation Eq. (3.57), to obtain the pairing temperature T^* as a function of x, we consider only the zero-gap condition, which is different from the long-range phase coherence condition in Eq. (3.124). Second, we obtain the gap at 0 K $\Delta(0)$ as a function of x by taking $T = 0$ in BCS gap equation Eq. (3.57). $\Delta(0)$ is the value of the gap at 0 K and along the Cu-O bond direction. Third, it is easy to obtain $2\,\Delta(0)/T^*$ as a function of x. Fourth, using the empirical formula of T_c, $T_c = 93[1 - 82.6(x - 0.16)^2]$, for Bi2212 and Y123, we can obtain $2\Delta(0)/T_c$ as a function of x. Fifth, Eq. (3.125) can give the number density of supercurrent Cooper pairs at 0 K, $n_s(0)$, as a function of x. $m^* = 6m_\circ$ [172]. Sixth, Using Eq. (3.124), we can obtain $T_{c,theo}$ as function of x.

The theoretical curve and experimental data of T^* vs. x are given in Fig. 4.5. Fig. 4.19 gives the theoretical curve and experimental data of $\Delta(0)$ vs. x. Fig. 4.20 gives the theoretical curve and experimental data of $2\Delta(0)/T^*$ vs. x. Fig. 4.21 gives the theoretical curve and experimental data of $2\Delta(0)/T_c$ vs. x. Here, the values of $\Delta(0)$ vs. x for theory and experiment are taken from Fig. 4.19, and the values of T_c are taken from $T_c = 93[1 - 82.6(x - 0.16)^2]$. Fig. 4.22 gives the theoretical curve and experimental data of $n_s(0)/m^*$. In our theory, the maximum value of $n_s(0)/m^*$ is at $x = 0.19$, which coincides with the experiment in Ref. [167]. However, according to the eight-band theory the maximum value of $n_s(0)/m^*$ is at the underdoped region [170]. In our theory all the reductions of $n_s(0)/m^*$ on the underdoped and the overdoped sides come from the same equation Eq. (3.125), which is different from the conjecture in Ref. [171]. The theoretical curve and experimental

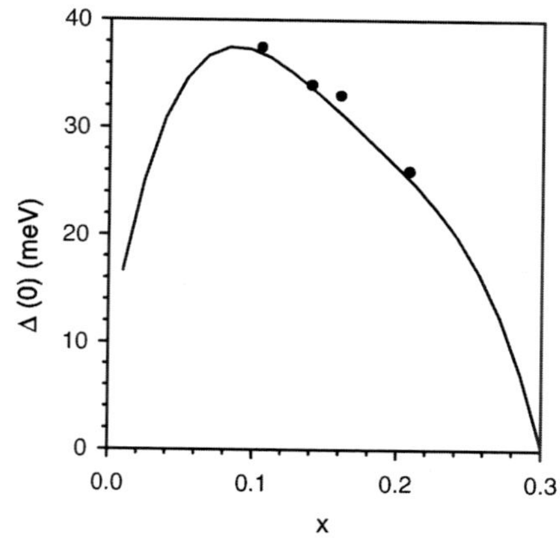

Figure 4.19. Theoretical curve of $\Delta(0)$ vs. x. The data are from Ref. [94].

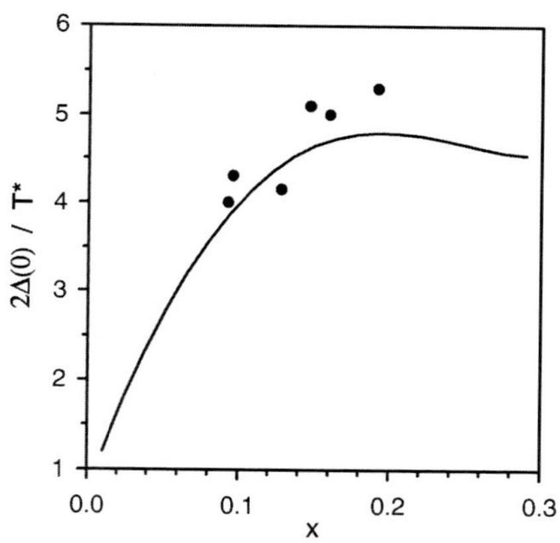

Figure 4.20. Theoretical curve of $2\Delta(0)/T^*$ vs. x. The data are from Ref. [94].

data of T_c vs. x are given in Fig. 4.6. According to Ref. [168], $T_{c,theo} \neq 0$ at $x = 0.05$. However, the experimental value of T_c at $x = 0.05$ is zero. The key problem in Ref. [168] is to assume $n_s(0) \propto x$ without any demonstration. Actually, our calculation shows that $n_s(0)$ is a very complex function of x as given by Fig. 4.22. The first attempt to explain the six doping dependences can be found in Ref. [173].

Zuev *et al.* reported measurements of the CuO_2 plane superfluid density n_s (magnetic penetration depth λ) of much underdoped films of Y123, with T_c's from 6 to 50 K, and found that T_c is not proportional to $n_s(0)$; instead, $T_c \propto n_s(0)^{1/2.3\pm0.4}$ [174]. This conflicts with a popular point of view that quasi-2D thermal phase fluctuations in the classical XY model determine the T_c. Liang *et al.* found the lower critical field $H_{c1}(0)$ for much underdoped Y123 with T_c between 8.9 and 22 K is proportional to $T_c^{1.64\pm0.04}$ [175]. Our theoretical curves and experimental data in Fig. 4.22 and Fig. 4.6 show that the proportionality between $n_s(0)$ and T_c should be violated.

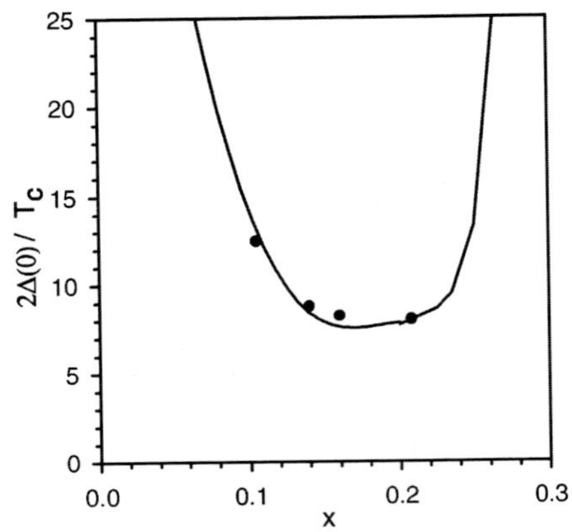

Figure 4.21. Theoretical curve and the data of $2\Delta(0)/T_c$ vs. x. The theoretical and experimental data of $\Delta(0)$ are taken from Fig. 4.19. T_c is determined by $T_c = 93[1-82.6(x-0.16)^2]$ [120, 135].

4.6. Isotope Effect

$\alpha_{T_c} = -[\ln T_c(^{18}O) - \ln T_c(^{16}O)]/[\ln M(^{18}O) - \ln M(^{16}O)]$ is the definition of the exponent of isotope effect. If $\alpha_{T_c} \neq 0$, then we say that there is the isotope effect. Williams *et al.* found the isotope effect in the high-T_c cuprates [176]. To explain the experiments in Ref. [176] in our theory, we have to mention the Fisher *et al.*'s work in Ref. [177]. Fisher *et al.* pointed out that, when one substitutes ^{18}O for ^{16}O, there are the following effects of zero-point motion on the effective Cu-O hopping integral t_o: (1) the change of t_o due to the zero-point vibration of the oxygen parallel or perpendicular to the Cu-O bond; (2) the change of t_o due to the change of the mean Cu-O bond angle; and (3) the change of t_o due to the equilibrium Cu-Cu distance in anharmonic coupling [177]. J and J_K are important for the pairing potential of Cooper pair, TLSMI, and dependent on t_o^4 and t_o^2, respectively. From Eqs. (3.57), (3.59), and (3.53), we know that the variation of J and J_K lead to the variation of the gap. From Eqs. (3.124), (3.125), and (3.108), we know that the variation of the gap leads to the variation of T_c. Therefore, from the theory in Ref. [177] and the theory in this book, the high-T_c cuprates possess the isotope effect. From Fisher *et al.*'s work in Ref. [177] and TLSMI theory, the pairing temperature T^* of the high-T_c cuprates also should have the isotope effect. The isotope effect of T^* was observed by neutron spectroscopic experiment of relaxation rate of crystal-field excitations in $La_{1.96-x}Sr_xHo_{0.04}CuO_4$ [178].

4.7. Admixtures of d-Wave Gap in Y123 Films

Smilde *et al.* found that there are admixtures of d-wave gap in untwinned Y123 superconducting films measured by angle-resolved electron tunneling [179]. The gap consists of the following three parts: 83% d-wave, 15% isotropic s-wave, and 2% anisotropic s-wave.

We explain the experimental results in Ref. [179] as follows. The electron-phonon coupling constants λ is 0.3 for Y123 and 0.2 for Bi2212 [180]. The phonon-mediated inter-

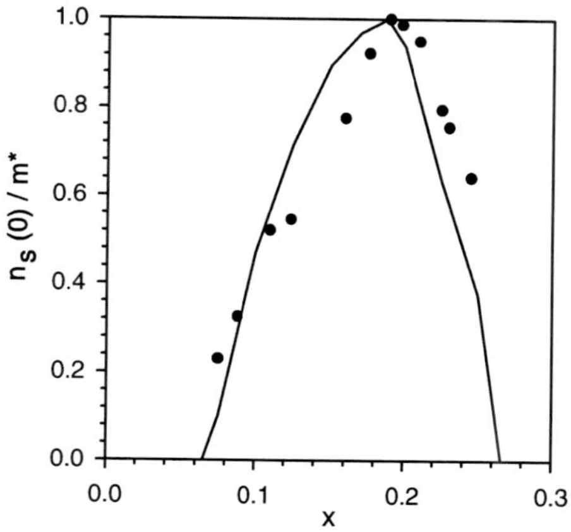

Figure 4.22. Theoretical curve of $n_s(0)/m^*$ vs. x. The data are taken from Ref. [167]. The maximums of the theoretical curve and the data are at $x = 0.19$.

action in Y123 is larger than that in Bi2212. Therefore, the phonon-mediated interaction in Y123 has contribution to superconductivity, and contributes an isotropic s-wave gap [See Eq. (3.65).]. The anisotropic s-wave gap comes from the second term in Eq. (3.65). Considering the nesting Fermi surface structure in Fig. 5.1 and the factor $\cos(k_x a)\cos(k_y a)$ in the second term of Eq. (3.65), the anisotropic s-wave has only a little contribution to the gap. Considering the nesting Fermi surface structure and the factor $\sin(k_x a)\sin(k_y a)$ in d-wave gap [See Eq. (3.66).], the anisotropic d-wave is primary in the gap.

4.8. $La_{2-x}Sr_xCuO_4$ (LSCO)

LSCO is a p-type cuprate [181]. Refs. [94, 182, 183] guessed that the observed gap of LSCO has the simple d-wave symmetry. The pseudogap was observed in LSCO [94]. The $2\Delta(0)/T_c$ is very large. The doping range of superconductivity for LSCO is at $0.05 < x < 0.27$ [94, 120, 135]. Here, the x represents the number of $O_{p\sigma}$ holes in unit cell of Cu^{++} lattice in the CuO_2 plane. The observed maximum value of T_c is about 40 K [94]. The values of the gap in the CuO_2 plane of nearly optimally doped LSCO are widely distributed from 5 to 12 meV, and the averaged gap is about 8 meV [149]. There are two specific heat jumps at T_c and T^* in LSCO [45].

The theory in chapter 3 can explain all the above observed properties of LSCO qualitatively or quantitatively. First, let us explain the lower value of T_c in comparison with Bi2212 and Y123. In LSCO there are the same CuO_2 plane and antiferromagnetic short-range order as in Bi2212 and Y123 [184]. Therefore, TLSMI exists between the $O_{p\sigma}$ holes in the CuO_2 plane. However, the observed maximum value of T_c for LSCO is much less than that for Bi2212 and Y123. The origin is as follows. Tab. 5.3 tells us that the hopping integral t_o of LSCO is less than that of Y123. This reduction leads to the reduction of J_K and J, and thus to the reduction of TLSMI. Correspondingly, the gap $\Delta(T)$, and thus the Josephson tunneling coupling energy E_J in section 3.3 are reduced. From Eq. (3.124) we know that the reduction of E_J leads to the reduction of T_c.

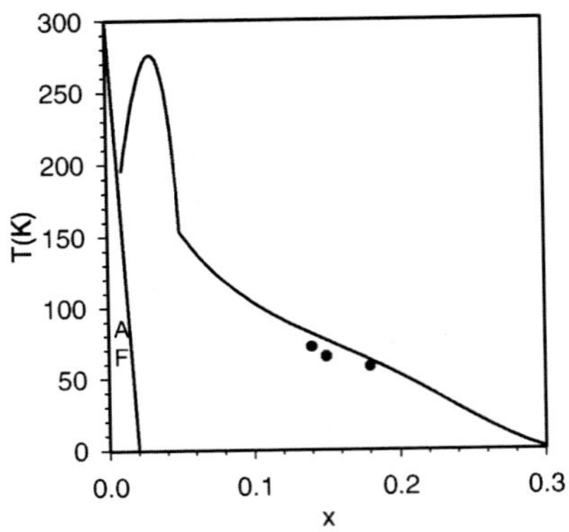

Figure 4.23. Theoretical curve of pairing temperature T^* vs. doping x for LSCO. Here, x represents the number of $O_{p\sigma}$ holes in one unit cell of Cu^{++} lattice in the CuO_2 plane. The data come from Ref. [94]. AF represents antiferromagnetic insulating region.

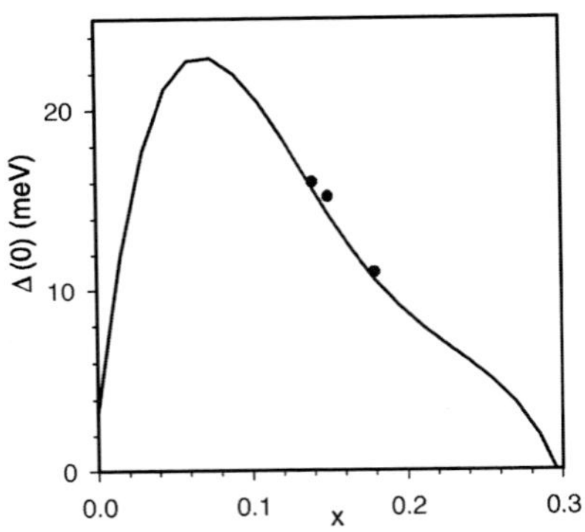

Figure 4.24. Theoretical curve of $\Delta(0)$ vs. x for LSCO. The data come from Ref. [94]. Here and in all the following diagrams, $\Delta(0)$ is the gap at $T = 0$ K and along the Cu-O bond direction.

Second, let us explain the origins of the pseudogap and the larger $2\Delta(0)/T_c$ in comparison with the conventional superconductivity. Under the zero-gap condition we obtain the pairing temperature T^*, at which the Cooper pairs without the long-range phase difference locking begin to appear. See Fig. 4.23. Only at temperature T_c which is lower than T^*, the Cooper pairs with the long-range phase difference locking begin to appear. Because $T^* > T_c$ or even $T^* \gg T_c$, $2\Delta(0)/T_c$ of LSCO is larger than that of the conventional superconductivity. For the latter, $T_c \approx T^*$. See Figs. 4.24, 4.25, and 4.26.

Third, let us explain that the superconductivity of LSCO exists only in a certain doping range. The higher the doping, the smaller the N''/N' in Eq. (5.2), which leads to the smaller TLSMI. The smaller the TLSMI, the smaller the gap, which leads to the smaller Josephson coupling energy E_J, and thus to the lower value of T_c or even to $T_c = 0$. On the other hand, the lower the doping, the larger the resistivity ρ_n because $\rho_n \propto x^{-2}$ (See Tab. 5.4.), which leads to smaller Josephson coupling energy, and thus to the lower value of T_c or even to

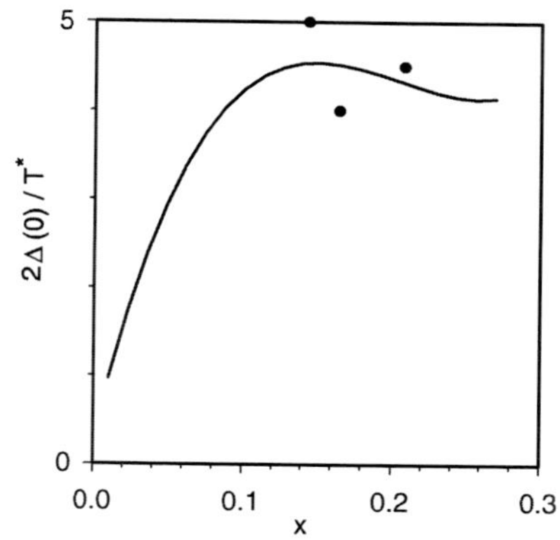

Figure 4.25. Theoretical curve for $2\Delta(0)/T^*$ vs. x for LSCO. The data come from Ref. [94].

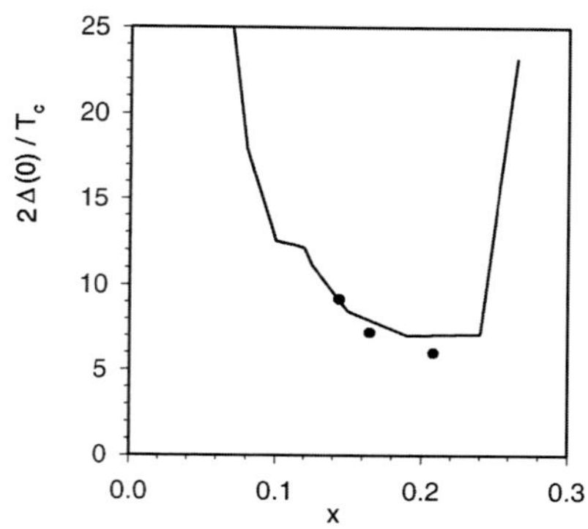

Figure 4.26. Theoretical curve for $2\Delta(0)/T_c$ vs. x for LSCO. The data come from Ref. [94].

$T_c = 0$. Fig. 4.27 gives theoretical and experimental doping dependence of T_c.

Fourth, the inhomogeneous distribution of the gap in the CuO_2 plan, observed in Ref. [149], can be explained qualitatively as follows. The superconductivity, gap, and pseudogap are related with the two local spin-mediated interaction, which has factors N'' and N'. According to the definitions in chapter 3 N'' and N' are dependent on the size of cluster with antiferromagnetic short-range order in the CuO_2 plane. Under thermal disturbance the size of the cluster has more or less variation on the different location in the CuO_2 plane. Therefore, the superconductivity, gap, and pseudogap have more or less variation on the different locations in the CuO_2 plane.

There are four predictions on LSCO. First prediction is on the the direction of maximum gap in the CuO_2 plane of LSCO. The gap in the CuO_2 plane of LSCO is the composite d-wave as well as Bi2212 and Y123. From the composite d-wave we can plot a curve of $\Delta(0)$ vs. ϕ. Here, $\phi = 0°$ is the Cu-O bond direction. From Figs. 4.28 and 4.29 we see that for much underdoped LSCO the direction of maximum gap of composite d-wave is not at the Cu-O bond direction. It is at the direction with about 15° away from the Cu-O bond direction. From Figs. 4.28 and 4.29 we see that for the popular simple d-wave the direction of maximum gap is always at the Cu-O bond direction.

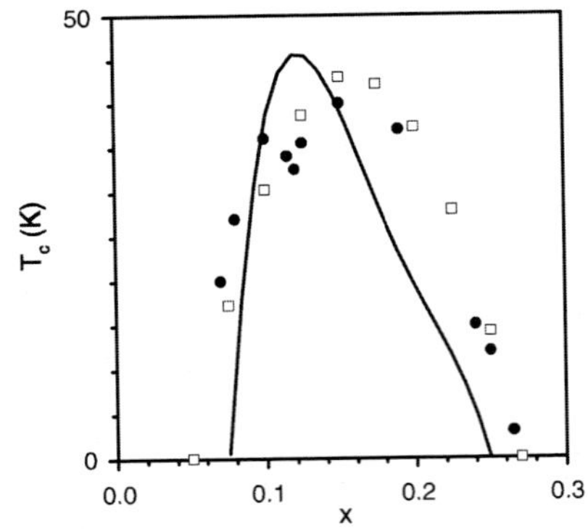

Figure 4.27. Theoretical curve of T_c vs. x for LSCO. The circles come from Ref. [94]. The squares come from Refs. [120, 135].

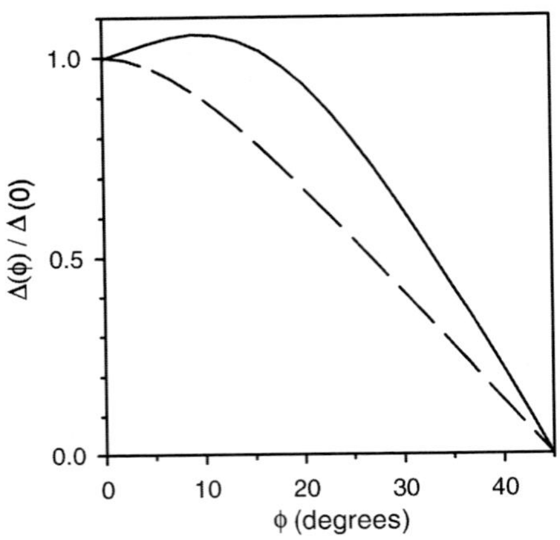

Figure 4.28. Anisotropy of the gap at 0 K for $x = 0.05$. The solid line is the composite d-wave curve. The dashed line is the simple d-wave curve. 0° corresponds to the Cu-O bond direction.

Second prediction is on the heavily underdoped LSCO. Here, heavily underdoped LSCO means $x < 0.05$. From Fig. 4.23 we see that the values of T^* of the heavily underdoped LSCO are very high. It is interesting to calculate the curves of pseudogap vs. temperature for the heavily underdoped LSCO. The theoretical curves of pseudogap vs. temperature for LSCO with $x = 0.035$ and 0.025 are shown in Fig. 4.30.

Third, $T_c = 0$ at $x = 0.27$. However, the gap is not zero. The gap at $x = 0.27$ is 3.5 meV.

Fourth, there are two kinds of Cooper pairs at $T < T_c$: supercurrent and non-supercurrent Cooper pairs. The reason is as follows. The gap and the Cooper pair without the long-range phase difference locking, i.e., non-supercurrent Cooper pair, begin to appear at T^*. $T^* > T_c$. At T_c, there are already many non-supercurrent Cooper pairs in LSCO. However, the number density of supercurrent Cooper pairs, i.e., Cooper pairs with the long-range phase difference locking, is still zero just at T_c. When temperature goes down from T_c, the supercurrent Cooper pairs appear more and more in LSCO, i.e., more and more

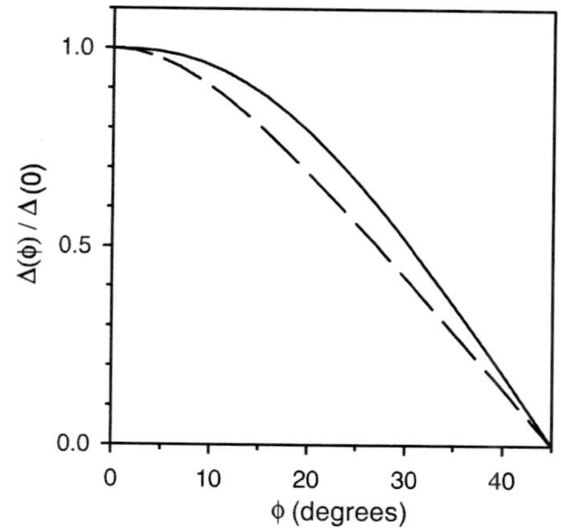

Figure 4.29. Anisotropy of the gap at 0 K for $x = 0.20$. The solid line is the composite d-wave curve. The dashed line is the simple d-wave curve. 0° corresponds to the Cu-O bond direction.

non-supercurrent Cooper pairs become supercurrent Cooper pairs.

The theory on LSCO can be found in Ref. [185] as well.

4.9. Fermi Arc Phenomena

4.9.1. Introduction

The Fermi surface is a set of points in momentum space describing gapless excitations. Norman *et al.* examine the temperature dependence of the Fermi surface in the CuO_2 plane of underdoped Bi2212 by angle-resolved photoemission spectroscopy, and find that, on cooling the sample, there is an initial breakup of the Fermi surface along the Cu-O bond direction (0°) at a temperature T^* into disconnected Fermi arcs, which then shrink with decreasing temperature before collapsing to the d-wave point nodes along 45° direction [186]. Fermi arc is a line of gapless excitations on Fermi surface in the CuO_2 plane. T^* is the temperature at which a pseudogap is opened. The similar variations of Fermi arc with temperature are also found in Refs. [187, 188]. Harris *et al.* find for their 78 K T_c underdoped Bi2212 that the higher the temperature, the larger the Fermi arc [187]. Harris *at al.* also find for their optimally doped 29 K T_c Bi2201, that there is a considerable Fermi arc near the d-wave node position at $T = 60$ K [188]. On the contrary, Mesot *et al.* find that for many under- and over-doped Bi2212 samples with $T_c > 75$ K there is no Fermi arc at 15 K [189]. However, Mesot *et al.*'s experiment does not mention that if there is Fermi arc at $T \approx T_c$ for under- and over-doped Bi2212 samples. Norman *et al.* also find that the behaviors of T^* vs. ϕ for under- and over-doped Bi2212 are different [186]. Here ϕ is an angle. $\phi = 0$ is along the Cu-O bond direction. For the underdoped Bi2212 the reduction of T^* with the increase of ϕ is fast. For the overdoped Bi2212 the values of T^* at $\phi = 0°$ and 22.5° are nearly the same. Now the Fermi arc phenomena still attract a great attention [190, 191].

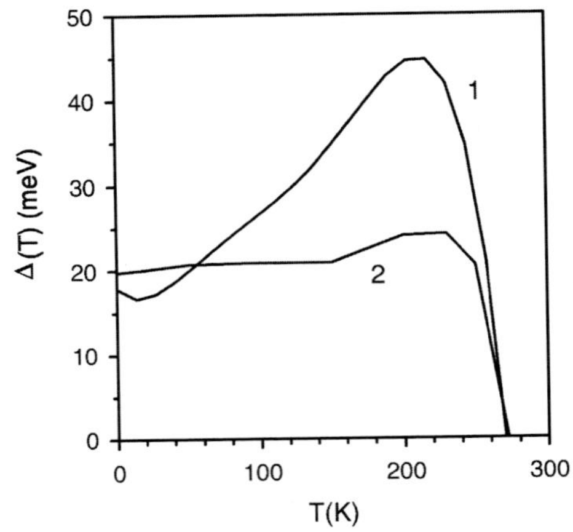

Figure 4.30. Theoretical curves of $\Delta(T)$ vs. T for heavily underdoped LSCO. The curves 1 and 2 correspond to $x = 0.025$ and 0.035, respectively.

As we know, many theories lead to the d-wave description for the gap in the high-T_c cuprates. According to the d-wave description the expression of gap is $\Delta(T,\mathbf{k}') = \Delta(T)[\cos(k'_x a_{Cu}) - \cos(k'_y a_{Cu})]$, where a_{Cu} is the nearest-neighbor distance between two Cu^{++} ions in the CuO_2 plane, and k'_x, k'_y are along the two orthogonal Cu-O bond directions. Based on this d-wave description, there is no Fermi arc, and there are only node points on the Fermi surface. To explain the observed Fermi arc, people try to use dirty d-wave. The dirty d-wave mechanism of the Fermi arc says that the impurity scattering broadens the d-wave node point into a region of finite width [187, 188]. It is noteworthy that Mesot *et al.* think based on their experiments that the dirty d-wave mechanism should be eliminated [189]. Until now there is no any other qualitative explanation and microscopic theory for the Fermi arc phenomena [190].

This section proposes a model for the anisotropy of Heisenberg and Kondo Hamiltonians, explains all the observed Fermi arc phenomena in Bi2212 and Bi2201, and predicts that Fermi arc can appear in both under- and over-doped high-T_c samples and at both temperatures, $T > T_c$ and $T \approx T_c$.

4.9.2. Anisotropy of d-pσ Model

It is obvious that the Fermi arc phenomena come from much complex anisotropy of the gap than the d-wave anisotropy. To consider the much anisotropy of the gap, we have to consider possible more complex anisotropy of pairing potential TLSMI. We will show that the d-pσ Hamiltonian of Eq. (3.5) in the CuO_2 plane, proposed by Fukuyama and Matsukawa [83] and many others, is of anisotropy. In Eq. (3.5) $J_K \propto t_\circ^2$ and $J \propto t_\circ^4$ (See section 4.4.1.). The definition of hopping integral $t_\circ$ is [192]

$$t_\circ = \langle a(1)|V(1)|b(1)\rangle, \tag{4.12}$$

where a(1) and b(1) are eigenstates of $H_{O^{--}}$ and $H_{Cu^{++}}$, respectively. $H_{O^{--}}$ and $H_{Cu^{++}}$ contain the kinetic energy of electron 1 plus its interaction with the nucleus and core electrons

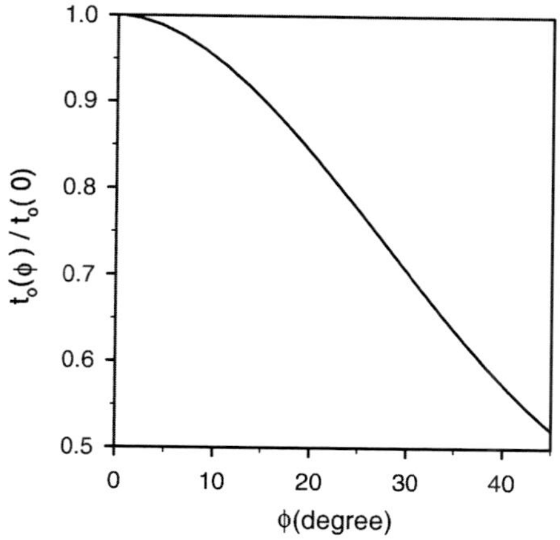

Figure 4.31. Anisotropy of the hopping integral $t_\circ(\phi)$. Here ϕ is an angle in the CuO_2 plane. $\phi = 0$ is along the Cu-O bond direction.

of O^{--} and Cu^{++}, respectively. $V(1)$ contains the interaction of electron 1 with the nucleus and core electrons of O^{--} ion. The hopping integral $t_\circ$ in Eq. (4.12) is an isotropic constant. The physical meaning of $t_\circ$ is the probability amplitude for hopping of an electron between nearest-neighbor O^{--} and Cu^{++} sites in the CuO_2 plane. In physics, this probability amplitude should be direction-dependent, i.e., $t_\circ \to t_\circ(\phi)$. However, until now there is no an exact theory on this anisotropy. This section gives a simple anisotropic model for the probability amplitude.

In our simple anisotropic model of $t_\circ$ we take free atom wavefunctions for $a(1)$ and $b(1)$ in Eq. (4.12). From Ref. [193] we know

$$a(1) = O_{2px} = N_2 \frac{2r}{a_\circ} e^{-\frac{2r}{2a_\circ}} \sin(\theta)\cos(\phi), \tag{4.13}$$

$$b(1) = Cu_{3d(x^2-y^2)} = N_3 \sqrt{\frac{3}{2}} \left(\frac{2r'}{a_\circ}\right)^2 e^{-\frac{2r'}{3a_\circ}} \sin^2(\theta')\cos(2\phi'), \tag{4.14}$$

$$N_2 = \frac{1}{4}\left(\frac{8}{2\pi a_\circ^3}\right)^{1/2}, \tag{4.15}$$

$$N_3 = \frac{1}{81}\left(\frac{8}{3\pi a_\circ^3}\right)^{1/2}, \tag{4.16}$$

$$V(1) = \frac{2e^2}{r}, \tag{4.17}$$

where ϕ is an angle in the CuO_2 plane, $\phi = 0$ if the direction is along the Cu-O bond, $a_\circ = 0.529$ Å, the z-component is vertical to the CuO_2 plane, the origin of coordinates r, θ, and ϕ is at the O^{--} site, and the origin of coordinates r', θ', and ϕ' is at the Cu^{++} site. Substituting Eqs. (4.13) - (4.17) into Eq. (4.12) and keeping the variable ϕ, we obtain a numerical relation between $t_\circ(\phi)$ and ϕ shown in Fig. 4.31. We can consider the anisotropy of J and J_K using Fig. 4.31. A much exact treatment for the anisotropy of d-pσ model is to

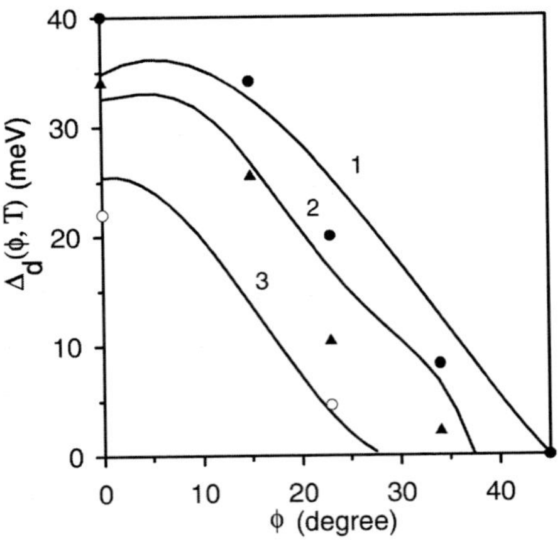

Figure 4.32. The theoretical curves of gap at different temperatures, $\Delta_d(\phi, T)$ vs. ϕ for Bi2212 sample with $x = 0.114$. The curves 1, 2, and 3 are at T = 13, 100, and 150 K, respectively. The solid circle, solid triangle, and circle are at 13, 100, and 150 K, respectively [187].

consider at the beginning of calculation that the jump of $O_{p\sigma}$ hole from Cu^{++} ion to O^{--} ion can have different directions or paths. However, until now an exact theory of anisotropy of t_o is still lacking [127].

The formula of TLSMI in section 3.1 can be rewritten as a form with more complex anisotropy.

$$U_{\mathbf{kk}'}(\phi) = -A(T,\phi)F_{\mathbf{kk}'}, \tag{4.18}$$

$$A(T,\phi) = \frac{J(\phi)J_K^2(\phi)(1+C)N''/N'w(J(\phi)/T)}{T^2 + 64J(\phi)J_K^2(\phi)\sum_{\mathbf{kp}} g(\mathbf{k},\mathbf{p})/\{1+32\pi^2 J_K^2(\phi)[N(E_F)]^2\overline{h(\mathbf{q})}\}}. \tag{4.19}$$

BCS equation of the gap function in Eq. (3.57) becomes

$$\Delta(T,\mathbf{k},\phi) = -\sum_{\mathbf{k}'} \frac{U_{\mathbf{kk}'}(\phi)\Delta(T,\mathbf{k}',\phi)}{2m(T,\mathbf{k}',\phi)} \tanh\frac{m(T,\mathbf{k}',\phi)}{2T}, \tag{4.20}$$

$$m(T,\mathbf{k},\phi) = \sqrt{[\varepsilon(\mathbf{k}) - E_F]^2 + |\Delta'(T,\mathbf{k},\phi)|^2}. \tag{4.21}$$

4.9.3. Numerical Results

The main solution of Eq. (4.20) below T^* is of composite d-wave form for Bi2212 and Y123, i.e.,

$$\Delta_d(T,\mathbf{k},\phi) = \Delta_3(T,\phi)F_2(\mathbf{k})F_5(\mathbf{k}), \tag{4.22}$$

$$F_2(\mathbf{k}) = \sin(k_x a)\sin(k_y a), \tag{4.23}$$

$$F_5(\mathbf{k}) = \cos^2(k_x a)\cos^2(k_y a). \tag{4.24}$$

Note that the value of $\mathbf{k}$ on Fermi surface is determined by the value of ϕ.

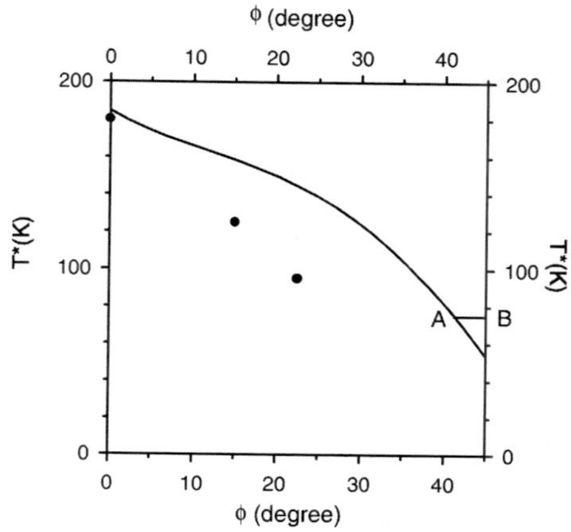

Figure 4.33. The theoretical curve of T^* vs. ϕ for Bi2212 sample with $x = 0.125$. For the line AB see section 4.9.4. The data are from Ref. [186].

The value of T_c in Ref. [187] is 78 K. Using the T_c empirical formula in Ref. [120] we know $x = 0.114$ for the sample in Ref. [187]. Fig. 4.32 clearly shows that the higher the temperature, the larger the Fermi arc for the 78 K T_c underdoped Bi2212. The reason is that at higher temperature the smaller gaps near the node point of composite d-wave disappear due to the anisotropy of $t_\circ(\phi)$ in Fig. 3.31.

Figs. 4.33 and 4.34 show that the behaviors of T^* vs. ϕ for underdoped ($x = 0.125$) and overdoped ($x = 0.188$) are different. The value of T^* at 28° is much lower than the value of T^* at 0° for $x = 0.125$, and nearly the same as for $x = 0.188$. So, Figs. 4.33 and 4.34 can well explain the observed relation between T^* and ϕ in Ref. [186]. The physical reasons of Figs. 4.33 and 4.34 are as follows. $J(\phi)$ and $J_K^2(\phi)$ of the pairing potential TLSML in Eq. (4.19) are decreased when ϕ is increased. However, $w(J(\phi)/T)$ of TLSMI in Eq. (4.19) is increased when $J(\phi)$ is decreased due to larger ϕ [See Eqs. (3.19), (7.41), and (7.66).]. We know from section 5.2 that for overdoped sample, i.e., for larger value of x, $J(\phi)$ is smaller than the corresponding value of underdoped sample. Due to larger $w(J(\phi)/T)$ at larger ϕ for the overdoped sample, there is one peak at larger ϕ in Fig. 4.34.

Our Figs. 4.33 and 4.34 show that at $T = 15$ K (much less than T_c) all samples including both under- and over-doped do not have Fermi arc. To make this point more clear, see Fig. 4.35. Fig. 4.35 shows that for Mesot *et al.*'s 83 K T_c underdoped Bi2212 sample there is no Fermi arc at 15 K, i.e., there is only a d-wave node point at 45° at 15 K [189].

4.9.4. Predictions

Our theory predicts that the Fermi arc can appear clearly at temperature slightly less than T_c. For example, let us interpret the meaning of the line AB in Fig. 4.33. At point A $T = T^* = 75$ K and $\phi = 41°$. At point B $T = 75$ K and $\phi = 45°$. As we know, T^* is a temperature at which a pseudogap just begins to be opened. Therefore, there is no any gap from A (= 41°) to B (= 45°), which means that Fermi arc between 41° and 45° appears at $T = 75$ K for the Norman *et al.*'s 85 K T_c ($x = 0.125$ from the T_c empirical formula) underdoped Bi2212

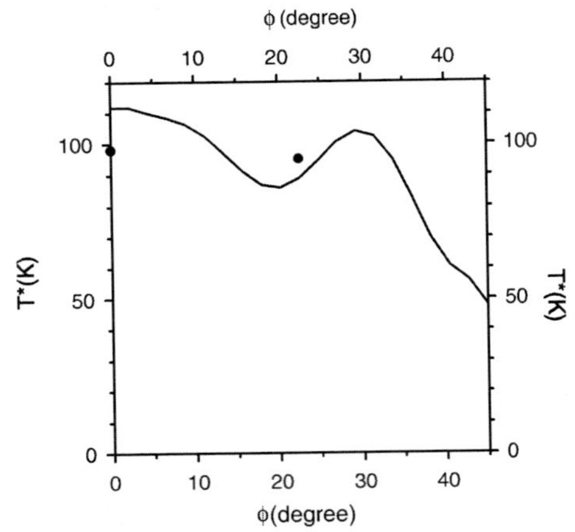

Figure 4.34. The theoretical curve of T^* vs. ϕ for Bi2212 sample with $x = 0.188$. The data are from Ref. [186].

sample. Figs. 4.33 and 4.44 can make general predictions for appearance of Fermi arc. Figs. 4.33 and 4.34 predict that Fermi arc can clearly appear at temperatures both above and slightly less than T_c in both under- and over-doped Bi2212. Using the lines like line AB, Figs. 4.33 and 4.34 can give the size of Fermi arc at any temperature.

The appearance of Fermi arc in Bi2201, reported in Ref. [188], is obvious in our theory because the Fermi arc is only dependent on the CuO_2 plane. Thus Fermi arc will appear in all the high-T_c cuprates due to nearly the same CuO_2 plane.

Until now there is no systematic experimental observation for the Fermi arc phenomena. Our predictions wait future experimental verifications. Our theory on Fermi arc can be found in Ref. [194] as well.

Note that the prediction in section 4.1 that the maximum value of the gap deviates from the Cu-O bond direction is still effective even we consider the anisotropy of J and J_K in Eq. (3.5) (See Figs. 4.32 and 4.35.).

4.10. Field Independence of the Pseudogap

The responses to the magnetic field (H) of pairing temperature T^* and superconducting transition temperature T_c in the underdoped high-T_c cuprate $YBa_2Cu_4O_8$ are reported up to 23.2 Tesla based on NMR measurements by Zheng *at al.* [195]. The T_c under zero field is 74 K. The T_c was found to be 55 K and the pairing temperature $T^* = 160$ K under the field H = 23.2 Tesla parallel to the c axis by measuring the frequency shift of NMR resonance circuit. Zheng *et al.* found that the magnetic field of 23.2 Tesla has no effect on the pairing temperature. Gorny *et al.* reported, for magnetic fields of 0, 8.8, and 14.8 Tesla, measurements of the temperature-dependent ^{63}Cu NMR spin lattice relaxation rate near and above T_c for near optimally doped Y123 with $T_c = 93$ K and $T^* = 110$ K [196]. Although the magnetic fields shift T_c down by as much as 8 K, the T^* does not shift down in temperature. The magnetic field independence of T^* also was observed in Ref. [197].

The explanations for the magnetic field independence of T^* and dependence of T_c are

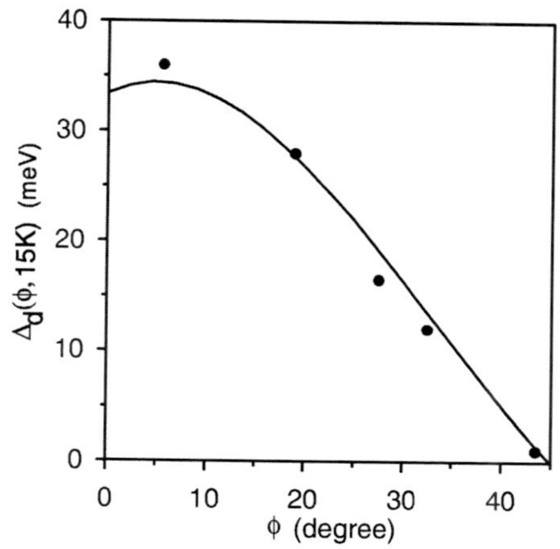

Figure 4.35. The theoretical curve of $\Delta_d(\phi, 15K)$ vs. ϕ for Bi2212 sample with $x = 0.125$. The data are from Ref. [189].

as follows. It is known that the magnetic field can destroy the long-range phase coherence between supercurrent Cooper pairs at $T < T_c$ [198]. Our theory in section 3.3 shows that there are Cooper pairs with the long-range phase coherence at $T < T_c$. The superconductivity is dependent on the Cooper pairs with the long-range phase coherence. Therefore, the theories in section 3.3 and Ref. [198] can explain the drop of T_c in magnetic field. The Cooper pairs at $T_c < T < T^*$ are without the long-range phase coherence. Therefore, T^* is not influenced by magnetic field.

4.11. Pseudogap in the Vortex Cores

Renner *et al.* reported the studies of vortex cores in under-doped ($T_c = 83.0$ K) and over-doped ($T_c = 74.3$ K) Bi2212 by local probe tunneling spectroscopy [199]. Renner *et al.* investigated these crystals at 4.2 k in a magnetic field applied perpendicular to the CuO_2 plane. The tunneling spectroscopy is carried out using an Ir tip mounted perpendicularly to the (001) surface of Bi2212 in a scanning tunneling microscopy. At the center of the core in a normal state, Renner *et al.* found a pseudogap at the Fermi level, which shows striking similarities with the normal state pseudogap measured above T_c.

The explanations for the observed pseudogap in the vortex cores are as follows. We know from Fig. 4.8 that the Cooper pairs without the long-range phase coherence begin to appear at pairing temperature T^*. We know from section 3.3 that the Cooper pairs with the long-range phase coherence begin to appear at superconducting transition temperature T_c. $T^* > T_c$. Therefore, at $T < T_c$ there are two kinds of Cooper pairs in the CuO_2 plane of Bi2212. The pseudogap in the vortex cores comes from the Cooper pairs without the long-range phase coherence at $T < T_c$.

Chapter 5

Methods of Numerical Calculations

5.1. Calculations for the RKKY−Like Interaction

Eqs. (3.53) - (3.55) give the renormalized TLSMI of the high-T_c cuprates, $U_{\mathbf{kk}'}$, is

$$U_{\mathbf{kk}'} = -A(T)F_{\mathbf{kk}'}, \tag{5.1}$$

$$A(T) = \frac{1}{2N_{Cu}} \frac{JJ_K^2(1+C)N''/N'w(J/T)}{T^2 + 64JJ_K^2 \sum_{\mathbf{kp}} g(\mathbf{k},\mathbf{p})/\{1+32\pi^2 J_K^2[N(E_F)]^2\overline{h(\mathbf{q})}\}}, \tag{5.2}$$

$$g(\mathbf{k},\mathbf{p}) = \left(\frac{1}{2N_{Cu}}\right)^2 \frac{f[\varepsilon(\mathbf{k}) - E_F]F_5(\mathbf{k})F_5(\mathbf{p})}{\varepsilon(\mathbf{p}) - \varepsilon(\mathbf{k})} e^{i(\mathbf{k}-\mathbf{p})\cdot(\mathbf{R}_{i'}-\mathbf{R}_i)}, \tag{5.3}$$

where $g(\mathbf{k},\mathbf{p})$ is a RKKY-related factor, and $\mathbf{R}_{i'}$ and $\mathbf{R}_i$ are the position vectors of the nearest-neighbor Cu^{++} ions in the CuO_2 plane. The RKKY-like interaction $RKKY'$ is

$$\begin{aligned} RKKY' &= \left(\frac{1}{2N_{Cu}}\right)^2 \sum_{\mathbf{kk}'} \frac{f(\mathbf{k}) - f(\mathbf{k}')}{\xi(\mathbf{k}) - \xi(\mathbf{k}')} e^{i(\mathbf{k}-\mathbf{k}')\cdot(\mathbf{R}_{i'}-\mathbf{R}_i)} F_5(\mathbf{k})F_5(\mathbf{k}') \\ &= -\sum_{\mathbf{kk}'} 2g(\mathbf{k},\mathbf{k}'). \end{aligned} \tag{5.4}$$

Taking the average of $F_5(\mathbf{k})F_5(\mathbf{k}')$ on the Fermi surface and transforming the summations over $\mathbf{k}$ and $\mathbf{k}'$ into the integrations over $\mathbf{k}$ and $\mathbf{k}'$, we have

$$RKKY' = \left(\frac{1}{2N_{Cu}}\right)^2 \int \frac{d\mathbf{k}d\mathbf{k}'}{(2\pi)^4/S^2} \frac{2f(\mathbf{k})}{\xi(\mathbf{k}) - \xi(\mathbf{k}')} e^{i(\mathbf{k}-\mathbf{k}')\cdot(\mathbf{R}_{i'}-\mathbf{R}_i)} \overline{F_5(\mathbf{k})F_5(\mathbf{k}')}, \tag{5.5}$$

where S is the area occupied by the $2N_{Cu}$ Cu^{++} ions, $S = a^2 2N_{Cu}$, a is the nearest-neighbor distance between two O^{--} ions. $2N_{Cu} = N_O$, N_O is the total number of O^{--} ions in the CuO_2 plane.

The Fermi surface is shown in Fig. 5.1. The expression of the Fermi surface is

$$2.9127\cos(k_x a)\cos(k_y a) = E_F, \tag{5.6}$$

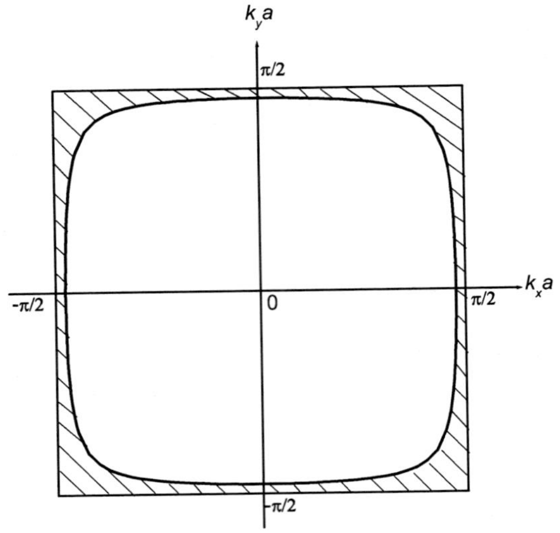

Figure 5.1. The schematic plot of the Fermi surface of $O_{p\sigma}$ holes in the Brillouin zone of the O^{--} ion lattice in the CuO_2 plane. The shading area denotes the region occupied by the $O_{p\sigma}$ holes at T = 0 K.

which comes from section 4.1 and Ref. [118]. $E_F > 0$. The unit of E_F is eV. The bottom of the $O_{p\sigma}$ holes is a square surrounded by $k_x a = \pi/2$, $k_x a = -\pi/2$, $k_y a = \pi/2$, and $k_y a = -\pi/2$. The Fermi surface shown in Fig. 5.1 possesses nesting structure. It is obvious that this nesting Fermi surface benefits the composite d-wave in Eq. (3.66), and does not benefit the second s-wave in Eq. (3.65). Ref. [118] gives that for Bi2212 $x = 0.192$, and $E_F = 0.294$ eV. Using rigid band approximation for the dispersion relation Eq. (5.6), we have Tab. 5.1.

The method to obtain the value of x from E_F is given by the program EF-x.nb (Mathematica 4.0). The EF-x.nb consists of the following four sentences.

(1) EF=0.2142;

(2) A=2.9127;

Table 5.1. The numerical relations between E_F and x. E_F is the Fermi energy of $O_{p\sigma}$ holes. x is the number of $O_{p\sigma}$ holes in one unit cell of Cu^{++} lattice in the CuO_2 plane.

x	0.01	0.015	0.02	0.025	0.027	0.03	0.035
$E_F(eV)$	0.009	0.014	0.0196	0.0255	0.0273	0.0315	0.0373
x	0.045	0.05	0.056	0.075	0.1	0.125	0.15
$E_F(eV)$	0.0502	0.0568	0.0651	0.0924	0.1302	0.1722	0.2142
x	0.16	0.17	0.1828	0.188	0.192	0.2	0.208
$E_F(eV)$	0.2365	0.252	0.2772	0.2856	0.295	0.3098	0.3266
x	0.225	0.25	0.27	0.29			
$E_F(eV)$	0.3612	0.4127	0.4587	0.4998			

(3) NIntegrate[1,x,0,ArcCos[EF/(A)], y,0,ArcCos[EF/(A)/Cos[x]]];
(4) (Pi*Pi-%*4)/(Pi*Pi)//N

We make the following explanations for the program EF-x.nb. The area of Brillouin zone of the O^{--} lattice in the CuO_2 plane is $(2\pi)^2$. Let N_O denote the total number of O^{--} ions in the CuO_2 plane. $2N_O$ $O_{p\sigma}$ holes can be put into $(2\pi)^2$. The area of the square in Fig. 5.1 is only π^2. Therefore, there are $N_O/2$ $O_{p\sigma}$ holes in the square. The shading area in Fig. 5.1 is the total number of $O_{p\sigma}$ holes of N_O O^{--} ions. Every Cu^{++} ion has x $O_{p\sigma}$ holes in the CuO_2 plane. The total number of the $O_{p\sigma}$ holes of N_O O^{--} ions is $xN_O/2$. Therefore, (shading area)/(square area)=x. The 4th sentence of the program gives that the value of x is about 0.15 corresponding to $E_F = 0.2142$.

To make analytical derivation, we take effective mass approximation for the $\xi(\mathbf{k})$ and $\xi(\mathbf{k}')$ in Eq. (5.5).

$$\begin{aligned}\xi(\mathbf{k}) &= \varepsilon(\mathbf{k}) - E_F \\ &= 2.9127\cos(k_x a)\cos(k_y a) - E_F \\ &\cong 2.9127 - \frac{\hbar^2 k^2}{2m^{***}} - E_F. \end{aligned} \tag{5.7}$$

$$\xi(\mathbf{k}') \cong 2.9127 - \frac{\hbar^2 k'^2}{2m^{***}} - E_F. \tag{5.8}$$

Let us estimate the m^{***} for different values of E_F. For example, $E_F = 0.21$ eV.

$$\begin{aligned}2.9127\cos(k_x a)\cos(k_y a) &= 0.21 \\ &\cong 2.9127 - \frac{\hbar^2 k^2}{2(m^{***}/m_\circ)m_\circ}, \end{aligned} \tag{5.9}$$

where $m_\circ$ is the mass of a free electron. If we assume $k_x a = k_y a$, then we have $k^2 = k_x^2 + k_y^2 = 0.468 \times 10^{16}/\text{cm}^2$ and $m^{***}/m_\circ = 0.652$ from Eq. (5.9). Using the same method, we obtain $m^{***}/m_\circ = 0.756$ when $E_F = 0.0525$ eV, and $m^{***}/m_\circ = 0.8838$ when $E_F = 0$ eV. Therefore, $m^{***}/m_\circ \cong 1$ for any value of E_F.

Taking the effective mass approximation and assuming $T = 0$ K, Eq. (5.5) becomes

$$RKKY' = \frac{2}{(2\pi)^4}\int d\mathbf{k}a^2 \int d\mathbf{k}'a^2 \frac{1}{-\frac{\hbar^2 k^2}{2m^{***}} + \frac{\hbar^2 k'^2}{2m^{***}}} e^{i(\mathbf{k}-\mathbf{k}')\cdot(\mathbf{R}_{i'}-\mathbf{R}_i)} \overline{F_5(\mathbf{k})F_5(\mathbf{k}')}. \tag{5.10}$$

The region of integration over $\mathbf{k}$ is approximately shown in Fig. 5.2. Making variable transformation from $\mathbf{k}'$ to $\mathbf{k}$+$\mathbf{p}$ and introducing $\gamma = \hbar^2/(2m^{***})$, Eq. (5.10) becomes

$$\begin{aligned} RKKY' = \frac{2a^4}{(2\pi)^4\gamma}\overline{F_5(\mathbf{k})F_5(\mathbf{k}')}\int_0^\infty dpp \int_0^{2\pi} d\phi e^{ip\cos\phi|\mathbf{R}_{i'}-\mathbf{R}_i|} \\ \frac{1}{p^2}\int_{\frac{\sqrt{2}\pi}{2a}}^{k_F} dkk \int_0^{2\pi} d\theta \frac{1}{1+\frac{2k}{p}\cos\theta}. \end{aligned} \tag{5.11}$$

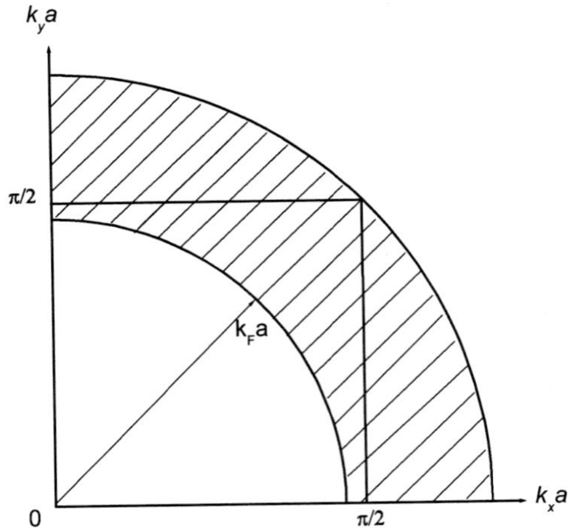

Figure 5.2. The shading area denotes the region of integration over **k** in one quadrant. k_F is the Fermi wave number at Cu-O bond direction.

Using the following two integration formulas,

$$\int_0^{2\pi} d\theta \frac{1}{1+\varepsilon\cos\theta} = \begin{cases} \frac{2\pi}{\sqrt{1-\varepsilon^2}}, & \varepsilon^2 < 1 \\ 0, & \varepsilon^2 > 1 \end{cases} \tag{5.12}$$

and

$$\int_0^{2\pi} d\phi e^{ip\cos\phi|\mathbf{R}_{i'}-\mathbf{R}_i|} = 2\pi J_\circ(p|\mathbf{R}_{i'}-\mathbf{R}_i|) = 2\pi J_\circ(pa_{Cu}), \tag{5.13}$$

Eq. (5.11) becomes

$$RKKY' = \frac{2a^4}{(2\pi)^4\gamma}\overline{F_5(\mathbf{k})F_5(\mathbf{k}+\mathbf{p})}\int_{\frac{\sqrt{2}\pi}{2a}}^{k_F} dkk\int_{2k}^{\infty} dp\frac{1}{p}2\pi J_\circ(pa_{Cu})\frac{2\pi}{\sqrt{1-\left(\frac{2k}{p}\right)^2}}. \tag{5.14}$$

Making variable transformation from p to $z=\sqrt{p^2-4k^2}$, Eq. (5.14) becomes

$$RKKY' = \frac{2a^4}{(2\pi)^2\gamma}\overline{F_5(\mathbf{k})F_5(\mathbf{k}+\mathbf{p})}\int_{\frac{\sqrt{2}\pi}{2a}}^{k_F} dkk\int_0^{\infty} dz\frac{J_\circ(a_{Cu}\sqrt{z^2+4k^2})}{\sqrt{z^2+4k^2}}. \tag{5.15}$$

Making variable transformation from z to $x=z/(2k)$ and using the formula in Ref. [200], Eq. (5.15) becomes

$$RKKY' = -\frac{a^4}{4\pi\gamma}\overline{F_5(\mathbf{k})F_5(\mathbf{k}+\mathbf{p})}\int_{\frac{\sqrt{2}\pi}{2a}}^{k_F} dkkJ_\circ(ka_{Cu})Y_\circ(ka_{Cu})$$

$$= -\frac{a^4}{4\pi\gamma}\overline{F_5(\mathbf{k})F_5(\mathbf{k}+\mathbf{p})}\left\{\int_0^{k_F} dkkJ_\circ(ka_{Cu})Y_\circ(ka_{Cu})\right.$$

$$\left.-\int_0^{\frac{\sqrt{2}\pi}{2a}} dkkJ_\circ(ka_{Cu})Y_\circ(ka_{Cu})\right\}. \qquad (5.16)$$

Using the formula in Ref. [201], Eq. (5.16) becomes

$$RKKY' = -\frac{a^4}{4\pi\gamma}\overline{F_5(\mathbf{k})F_5(\mathbf{k}+\mathbf{p})}\left\{2\left(\frac{k_F}{2}\right)^2[Y_\circ(k_Fa_{Cu})J_\circ(k_Fa_{Cu})\right.$$

$$\left.+Y_1(k_Fa_{Cu})J_1(k_Fa_{Cu})] - \left(\frac{\sqrt{2}\pi}{2a}\right)^2\frac{1}{2}[Y_\circ(\pi)J_\circ(\pi)+Y_1(\pi)J_1(\pi)]\right\}. \qquad (5.17)$$

Because $[Y_\circ(\pi)J_\circ(\pi)+Y_1(\pi)J_1(\pi)] = 0.00105$ is very small in comparison with the first term, we neglect the second term in Eq. (5.17). Considering that the true Fermi surface in Fig. 5.1 is not the circle in Fig. 5.2 and the maximum gap is not at the Cu-O bond direction, we should give an adjustable factor for the k_F in Eq. (5.17). Our numerical calculations indicate that $k_F \longrightarrow 0.86k_F$ is better. The a in Eq. (5.17) is equal to $3.9/\sqrt{2}\times 10^{-8}$ cm. At last, we have

$$RKKY' = -0.6\overline{F_5(\mathbf{k})F_5(\mathbf{k}+\mathbf{p})}\left(\frac{0.86k_F}{10^8}\right)^2\left(\frac{m^{***}}{m_\circ}\right)$$

$$[Y_\circ(0.86k_Fa_{Cu})J_\circ(0.86k_Fa_{Cu})+Y_1(0.86k_Fa_{Cu})J_1(0.86k_Fa_{Cu})], \qquad (5.18)$$

where the unit of Eq. (5.18) is $1/eV$. Eq. (5.18) is an expression of $RKKY'$ at $T = 0$ K. Now we consider the influence of temperature on $RKKY'$. Refs. [202, 203] showed that for a three dimension (3-D) system the temperature dependence of RKKY interaction is

$$RKKY \propto e^{-\frac{\pi^2}{6}\left(\frac{k_BT}{E_F}\right)^2(k_Fr)^2}, \qquad (5.19)$$

where r is the distance between two local spins. Ref. [201] showed that the spatial variation of 2-D RKKY interaction is slower than the 3-D RKKY interaction. Therefore, for the RKKY-like interaction in the CuO_2 plane of the high-T_c cuprates we take approximately

$$RKKY' = -0.6\overline{F_5(\mathbf{k})F_5(\mathbf{k}+\mathbf{p})}\left(\frac{0.86k_F}{10^8}\right)^2\left(\frac{m^{***}}{m_\circ}\right)[Y_\circ(0.86k_Fa_{Cu})$$

$$J_\circ(0.86k_Fa_{Cu})+Y_1(0.86k_Fa_{Cu})J_1(0.86k_Fa_{Cu})]e^{-\frac{\pi^2}{6}\left(\frac{k_BT}{E_F}\right)^2(0.86k_Fa_{Cu})}, \qquad (5.20)$$

in which the $(k_Fr)^2$ in Eq. (5.19) becomes the $(0.86k_Fa_{Cu})$ in Eq. (5.20).

The Eq. (5.20) is for the case of hole conduction in the CuO_2 plane. For the case of electron conduction in the CuO_2 plane the dispersion relation Eq. (5.6) becomes

$$-2.9127\cos(k_xa)\text{Cos}(k_ya) = -E_F, \qquad (5.21)$$

Table 5.2. The numerical relations between N''/N' and x.

x	0.01	0.015	0.02	0.025	0.027	0.03	0.035
N''/N'	1.76	1.702	1.66	1.63	1.618	1.602	1.586
x	0.045	0.05	0.056	0.075	0.1	0.125	0.15
N''/N'	1.545	1.516	1.505	1.4375	1.3775	1.33	1.29
x	0.17	0.188	0.2	0.225	0.25	0.27	
N''/N'	1.25	1.2125	1.175	1.075	1	0.91(0.865)	

where $E_F > 0$. Thus the $RKKY'$ for the case of electron conduction is obtained from Eq. (5.10).

$$RKKY'_e = \frac{2}{(2\pi)^4}\int d\mathbf{k}a^2 \int d\mathbf{k}'a^2 \frac{1}{\frac{\hbar^2k^2}{2m^{***}} - \frac{\hbar^2k'^2}{2m^{***}}} e^{i(\mathbf{k}-\mathbf{k}')\cdot(\mathbf{R}_{i'}-\mathbf{R}_i)}\overline{F_5(\mathbf{k})F_5(\mathbf{k}')}$$

$$= -RKKY'. \tag{5.22}$$

Considering $[Y_\circ(z)J_\circ(z) + Y_1(z)J_1(z)] < 0$ for electron in the first band $z < \pi/2$ and for electron in the second band $\pi < z < 3\pi/2$, we can say that there is no change of sign in comparison with the case of hole conduction in the first band $\pi/2 < z < \pi$.

5.2. The Values of Parameters

The values of parameters appearing in the expressions of TLSMI in Eqs. (5.1) - (5.3) can be determined by experiments and first principle calculations, at least, approximately.

The empirical formula of T_c vs. x for Bi2212 and Y123 is $T_c = 93[1 - 82.6(x-0.16)^2]$ (K), where x is the $O_{p\sigma}$ hole number in one unit cell of the Cu lattice in the CuO_2 plane [120, 135]. If a reference does not give the value of x, then we always use the above T_c formula to get the corresponding value of x from the observed T_c.

We use ξ to express the antiferromagnetic short-range coherence length (AFSRCL). $\xi = a_{Cu}/\sqrt{x}$ if $x < 0.15$ [153]. $\xi < a_{Cu}/\sqrt{x}$ if $x > 0.15$ [204]. Using the values of ξ and Fig. 3.10, we can obtain the values of N''/N' for any value of x. The relations between N''/N' and x used in chapter 4 are shown in Tab. 5.2.

To look for the expression of J_K in Eq. (3.5), we write the exact form of Kondo Hamiltonian [83].

$$H_K = \frac{t_\circ^2}{U_d}\sum_i\sum_{\alpha ss'} \widehat{\mathbf{S}}_i\cdot \vec{\sigma}_{ss'}\, p^+_{\alpha s}p_{\alpha s'}\left(A_1 + A_2\sum_{\beta\neq\alpha} n_\beta\right)$$

$$+\frac{t_\circ^2}{U_d}\sum_i\sum_{\alpha\neq\beta ss'} \widehat{\mathbf{S}}_i\cdot \vec{\sigma}_{ss'}\, p^+_{\beta s}p_{\alpha s'}(1-n_{\beta -s})(1-n_{\alpha -s'})(B_6 + A_7\sum_{\gamma\neq\alpha,\beta} n_\gamma), \tag{5.23}$$

$$A_1 = U_d\left[\frac{1}{\varepsilon_p - \varepsilon_d + U_p} + \frac{1}{U_d - (\varepsilon_p - \varepsilon_d) - 2V_\circ}\right], \tag{5.24}$$

$$A_2 = U_d \left[\frac{1}{\varepsilon_p - \varepsilon_d - V_\circ + U_p} - \frac{1}{\varepsilon_p - \varepsilon_d + U_p} + \frac{1}{U_d - (\varepsilon_p - \varepsilon_d) - V_\circ} - \frac{1}{U_d - (\varepsilon_p - \varepsilon_d) - 2V_\circ} \right], \tag{5.25}$$

$$B_6 = U_d \left[\frac{1}{\varepsilon_p - \varepsilon_d} + \frac{1}{U_d - (\varepsilon_p - \varepsilon_d) - 2V_\circ} \right], \tag{5.26}$$

$$A_7 = U_d \left[\frac{1}{\varepsilon_p - \varepsilon_d - V_\circ} - \frac{1}{\varepsilon_p - \varepsilon_d} + \frac{1}{U_d - (\varepsilon_p - \varepsilon_d) - V_\circ} - \frac{1}{U_d - (\varepsilon_p - \varepsilon_d) - 2V_\circ} \right], \tag{5.27}$$

where the definitions of $t_\circ$, ε_p, ε_d, U_d, U_p, and $V_\circ$ are given in Eq. (3.1), $n_\beta (= n_{\beta\uparrow} + n_{\beta\downarrow})$ indicates the number of $O_{p\sigma}$ holes at site β. To obtain a better expression of J_K, we rewrite Eq. (5.23) as

$$\begin{aligned} H_K = & \frac{t_\circ^2}{U_d} \sum_i \sum_{\alpha\beta s s'} \widehat{\mathbf{S}}_i \cdot \vec{\sigma}_{ss'} \, p^+_{\beta s} p_{\alpha s'} (1 - n_{\beta - s})(1 - n_{\alpha - s'})(B_6 + A_7 \sum_{\gamma \neq \alpha, \beta} n_\gamma) \\ & - \frac{t_\circ^2}{U_d} \sum_i \sum_{\alpha s s'} \widehat{\mathbf{S}}_i \cdot \vec{\sigma}_{ss'} \, p^+_{\alpha s} p_{\alpha s'} (1 - n_{\alpha - s})(1 - n_{\alpha - s'})(B_6 + A_7 \sum_{\gamma \neq \alpha} n_\gamma) \\ & + \frac{t_\circ^2}{U_d} \sum_i \sum_{\alpha s s'} \widehat{\mathbf{S}}_i \cdot \vec{\sigma}_{ss'} \, p^+_{\alpha s} p_{\alpha s'} \left(A_1 + A_2 \sum_{\beta \neq \alpha} n_\beta \right) \\ \equiv & \, I + II + III. \end{aligned} \tag{5.28}$$

To estimate the magnitudes of the terms I, II, and III, we should know the values of $t_\circ$, ε_p, ε_d, U_d, U_p, and $V_\circ$. The first principle numerical calculations have not yet given unified values for them. Han *et al.* gave a table listing these values, which is quoted here as Tab. 5.3 (with a few changes) [205].

To make the numerical calculations in chapter 4, we use the average values of $n_{\beta-s}$, $n_{\alpha-s'}$, $\sum_{\gamma\neq\alpha,\beta} n_\gamma$, and $\sum_{\beta\neq\alpha} n_\beta$ in Eq. (5.28).

$$\overline{n_{\beta-s}} = \overline{n_{\alpha-s'}} = \frac{x}{4} \tag{5.29}$$

because $\overline{n_{\beta-s}}$ represents the average number of $O_{p\sigma}$ holes with spin $-s$ ($\uparrow$ or $\downarrow$) at O^{--} site β, one Cu^{++} ion corresponds two O^{--} ions, and x represents the average number of $O_{p\sigma}$ holes with spins $\uparrow$ and $\downarrow$ at any Cu^{++} site.

$$\overline{\sum_{\gamma\neq\alpha,\beta} n_\gamma} = 2 \times \frac{x}{2} \tag{5.30}$$

because $\overline{n_\gamma} = \overline{n_{\gamma\uparrow}} + \overline{n_{\gamma\downarrow}} = x/2$, α, β, and γ are the sites of O^{--} ions around i-th Cu^{++} ion, and thus α, β, and γ can take 1, 2, 3, and 4 in the CuO_2 plane.

$$\overline{\sum_{\beta\neq\alpha} n_\beta} = 3 \times \frac{x}{2}. \tag{5.31}$$

Table 5.3. The values of parameters in strong correlation Hamiltonian given by Eq. (3.1) (unit eV).

Author	*Sample*	U_d	U_p	$V_\circ$	$\varepsilon_p - \varepsilon_d$	$t_\circ$
Shen	La_2CuO_4	5-6	0	0	0.3	2.3-2.4
	$YBa_2Cu_3O_7$	6	0	0	0.5	2.5
Fujimori	La_2CuO_4	5	0	0	0.4	1.9
	$YBa_2Cu_3O_7$	5	0	0	0.5	2.3
Mila	La_2CuO_4	10	0	0-1	4	-1.3
Hubbard	*3d*	10	3	3	–	–
Schruter	–	9	6	2.5	1.5	1.5
Emeri – Hirsch	–	5-6	2-3	1-2	1	1.3-1.5
Han	La_2CuO_4	7	4.1	0.58	1.6	1.3
	$YBa_2Cu_3O_7$	7	–	0.55	2.0	1.6
Matsukawa– Fukuyama	$YBa_2Cu_3O_7$	8	4	1	2	1

In the following we show that in comparison with the term I in Eq. (5.28) the terms II and III can be neglected. To estimate the magnitudes of I, II, and III, we take $\varepsilon_p - \varepsilon_d = 2$ eV, $U_d = 8$ eV, $U_p = 4$ eV, $V_\circ = 1$ eV, and $t_\circ = 1.1$ eV. Thus, $A_1 = 0.4167U_d$, $A_2 = -0.01667U_d$, $B_6 = 0.75U_d$, and $A_7 = 0.45U_d$. If $x = 0.27$, then

$$\frac{II + III}{I} \cong 0.1. \tag{5.32}$$

If $x = 0$, then

$$\frac{II + III}{I} \cong 0.1. \tag{5.33}$$

Therefore, for any value of x we can neglect the terms II and III approximately. Neglecting the terms II and III in Eq. (5.28), we have

$$H_K = J_K \sum_{i\alpha\beta ss'} \widehat{\mathbf{S}}_i \cdot \vec{\sigma}_{ss'} \, p^+_{\beta s} p_{\alpha s'}, \tag{5.34}$$

$$J_K = 1.1^2 \left(0.75 + 0.45 \times 2 \times \frac{x}{2}\right)\left(1 - \frac{x}{4}\right)^2. \tag{5.35}$$

The unit of J_K is eV.

The observed value of J in Eq. (3.5) is dependent on x. $J(x = 0) = 1333$ K [93]. $J(x = 0) = 1500$ - 2500 K [206]. $J(x = 0) = 1392$ K [207]. We take $J(x = 0) = 1583$ K. The J is a linearly decreasing function of x [93]. J can be described by $J = (1583 - 2973x)$ (K) for $x < 0.27$ [93].

The electron-phonon coupling constant $\lambda = N(E_F)V_{ep} \cong 0.25$ at $x = 0.15$ [208]. The Coulomb pseudo-potential $\mu^* = N(E_F)V_c^* = 0.15$ at $x = 0.15$ [209].

Table 5.4. The experimental values of $\rho_n(T \approx T_c)$.

Sample	$T_c(K)$	x	$\rho_n(T \approx T_c)(10^{-3}\Omega.cm)$	$b(=\rho_n x^2)$
$YBa_2Cu_3O_{7-\delta}$	90	0.14	0.040-0.053	0.000784-0.001039
$YBa_2Cu_3O_{7-\delta}$	90	0.14	0.043	0.00084
$YBa_2Cu_3O_{7-\delta}$	90	0.14	0.1	0.00196
$YBa_2Cu_3O_{7-\delta}$	90	0.14	0.2	0.00392
$YBa_2Cu_3O_{7-\delta}$	90	0.14	0.06429-0.07857	0.00126-0.00154
$YBa_2Cu_3O_{7-\delta}$	80	0.119	0.075	0.001062
$YBa_2Cu_3O_{7-\delta}$	60	0.0945	0.1	0.000892
$YBa_2Cu_3O_{7-\delta}$	60	0.0945	0.1154	0.001029
$YBa_2Cu_3O_{7-\delta}$	50	0.0852	0.2125	0.00154
$Bi_2Sr_2CaCu_2O_{8+\delta}$	87	0.132	0.01625-0.13	0.000283-0.00226

The resistance of normal state at $T \approx T_c$, R_n, appearing in Eq. (3.108), only can be determined very approximately. After analysing the data in Ref. [136], we find that the better dependence of ρ_n on x is $\rho_n = b/x^2$. Here, ρ_n represents the resistivity. We quote the table in Ref. [136] as Tab. 5.4 (with a few changes). The values of b quite scatter. We take $b = 0.00131$ ($10^{-3}\Omega$·cm) in our numerical calculations.

The resistance $R_n = \rho_n d/S$, where d is the distance between two nearest-neighbor squares in the simplified granular model (See Fig. 3.21.). $d \cong d'$, and S is the area of junction in the granular model of the high-T_c cuprates (See section 3.3.). d' is the thickness of one layer of the CuO_2 plane. $S = d'a_\circ$. $a_\circ$ is the average distance between two centers of two nearest-neighbor Cooper pairs at T_c. Eq. (3.100) gives

$$a_\circ = \sqrt{\frac{a_{Cu}^2}{x/2}}. \tag{5.36}$$

The Eq. (3.100) is not exact because not all $O_{p\sigma}$ holes become Cooper pairs at T_c. The exact expression of $a_\circ$ should be

$$a_\circ = \sqrt{\frac{a_{Cu}^2}{|\Psi(T_c)|^2 x/2}}, \tag{5.37}$$

where $|\Psi(T_c)|^2$ is the ratio of the number of $O_{p\sigma}$ holes forming Cooper pairs to the total $O_{p\sigma}$ hole number. $|\Psi(T)|^2 = \overline{Q(T)}$ for London local superconductors (See section 2.10 and Ref. [210].). From Eqs. (2.410) and (2.405) we know $|\Psi(T)|$ is

$$|\Psi(T)| = \sqrt{\pi T[\overline{\Delta(T)}]^2 \sum_{n=-\infty}^{+\infty} \frac{1}{\{[(2n+1)\pi T]^2 + [\overline{\Delta(T)}]^2\}^{\frac{3}{2}}}}. \tag{5.38}$$

The curves of $|\Psi(T)|^2$ vs. T are shown in Fig. 5.3. If $x = 0.16$ for Bi2212 and Y123, then $T_c = 93$ K. From Fig. 5.3 we know $|\Psi(93K)|^2 = 0.7 \approx 1$. Therefore, for convenience we prefer to take Eq. (5.36) rather than Eq. (5.37).

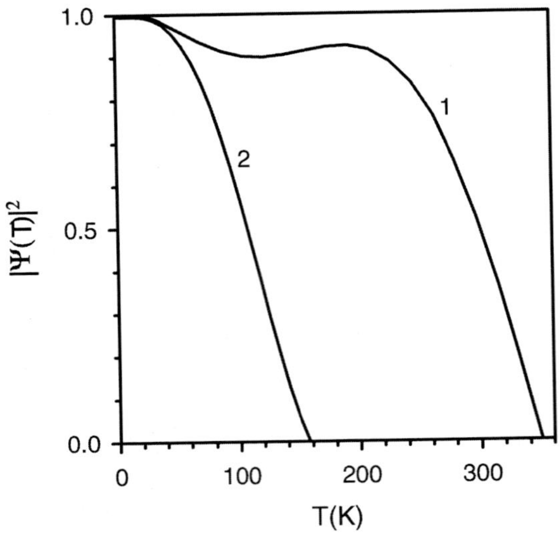

Figure 5.3. The curves of $|\Psi(T)|^2$ vs. T. The curves 1 and 2 are for $x = 0.025$ and 0.16, respectively.

The capacitance C in Eq. (3.123) is

$$C = \varepsilon_\infty \frac{S}{d} \tag{5.39}$$

for the granular model of the high-T_c cuprates. $S = d'a_\circ$. $d \cong d'$. ε_∞ is the high-frequency dielectric constant. The experimental values of ε_∞ are between 2 and 5 [211]. In our numerical calculations we take $\varepsilon_\infty = 2.71$.

5.3. Program to Calculate T_c and Gap

The gap equation of the composite d-wave is given by Eqs. (3.69) - (3.71). We quote as follows.

$$\Delta_3(T)F_2(\mathbf{k})F_5(\mathbf{k}) = -\sum_{\mathbf{k}'} \frac{U_{\mathbf{kk}'d}\Delta_3(T)F_2(\mathbf{k}')F_5(\mathbf{k}')}{2m(T,\mathbf{k}')} \tanh\frac{m(T,\mathbf{k}')}{2T}, \tag{5.40}$$

$$m(T,\mathbf{k}) = \sqrt{[\varepsilon(\mathbf{k}) - E_F]^2 + |\Delta_3(T)F_2(\mathbf{k})F_5(\mathbf{k})|^2}, \tag{5.41}$$

$$U_{\mathbf{kk}'d} = -A(T)F_2(\mathbf{k})F_2(\mathbf{k}')F_5(\mathbf{k})F_5(\mathbf{k}'), \tag{5.42}$$

where $\Delta_3(T)F_2(\mathbf{k})F_5(\mathbf{k})$ is the composite d-wave gap. $F_2(\mathbf{k})$ and $F_5(\mathbf{k})$ are given in Eqs. (3.32) and (3.35).

$$F_2(\mathbf{k}) = \sin(k_x a)\sin(k_y a), \tag{5.43}$$

$$F_5(\mathbf{k}) = \cos^2(k_x a/2)\cos^2(k_y a/2). \tag{5.44}$$

$A(T)$ is given by Eqs. (5.2), (5.3), and (5.4).

$$A(T) = \frac{1}{2N_{Cu}} \frac{JJ_K^2(1+C)N''/N'w(J/T)}{T^2 - \frac{32JJ_K^2 RKKY'}{1+32\pi^2 J_K^2[N(E_F)]^2 \overline{h(\mathbf{q})}}}, \tag{5.45}$$

where $RKKY'$ is given by Eq. (5.20). The solution of Eq. (5.40) is the composite d-wave gap. The solution of Eq. (5.40) under zero-gap condition is the pairing temperature T^*. Substituting the obtained composite d-wave gap into Eq. (3.124), we obtain $T_{c,theo}$.

Eq. (5.40) is a non-linear integral equation. In the following we interpret a program in our numerical methods. The program is written by Mathematica 4.0. The name of our program is GENERAL.nb. At first, we give this program. Then, we interpret all those sentences in the program one by one.

GENERAL.nb

(1) t = 91
(2) n = 632
(3) xx = .188;
(4) N″ = 1.2125;
(5) EF′ = .272;
(6) mm = 0.989;
(7) delta = .86;
(8) ferco = 1.05;
(9) CC = 1.05;
(10) kFaCu = 1.4142^{2}*ArcCos[(EF'/(2.774))$^{.5}$]*delta;
(11) kFxaO = ArcCos[EF'/(2.774)];
(12) JK = 1.1^{2}*(1 - xx/4)^{2}*(.75 + .45 xx)*11600;
(13) J = (1439.1 - xx 2702.7)*1.1;
(14) a' = BesselJ[0, kFaCu]*BesselY[0, kFaCu]+BesselJ[1, kFaCu]*BesselY[1, kFaCu];
(15) z = ArcCos[EF'/((2.774)*Cos[x])];
(16) NIntegrate[(Cos[x/2]*Cos[z/2])^{2}, {x, 0, kFxaO}];
(17) NIntegrate[1, {x, 0, kFxaO}];
(18) %%/%;
(19) NIntegrate[1, {x, 0, ArcCos[(EF' + .01)/(2.774)]}, {y, 0, ArcCos[(EF' + .01)/(2.774)/Cos[x]]}];
(20) NIntegrate[1, {x, 0, ArcCos[(EF' + -.01)/(2.774)]}, {y, 0, ArcCos[(EF' - .01)/(2.774)/Cos[x]]}];
(21) 4 (% - %%)/(.02)/4/3.1416^{2};
(22) c' = (%/ferco)^{2}*(JK/11600)^{2}*32*Pi^{2}*(%%%%)^{2};
(23) NIntegrate[(Cos[x/2]*Cos[z/2])^{4}, {x, 0, kFxaO}];
(24) NIntegrate[1, {x, 0, kFxaO}];
(25) %%/%;
(26) NIntegrate[(Cos[x/2])^{2}, {x, -kFxaO, kFxaO}];
(27) %/(2*kFxaO);
(28) f' = %^{2}*%%%;
(29) b' = .6*(kFaCu/3.9)^{2}*f'*a'/(1 + c');
(30) w = Exp[-3.1416^{2}/6*(t/EF'/ferco/11600)^{2}*(kFaCu)];
(31) a = (1/2/Pi)^{2}*J*JK^{2}*N"*CC/(t^{2} + mm*32*J*JK^{2}/11600*w*b');
(32) b = J/t;
(33) c = Exp[b/4] + Exp[-b/4];
(34) e = 32 + 2*(2 + c)^{2};
(35) f = 32 + c*(2 + c)^{2};

(36) g = 1 + (e/f)^4;
(37) j = (Sin[x]*Sin[y])^2;
(38) k = (Cos[x/2]*Cos[y/2])^4;
(39) l = (2.774*Cos[x]*Cos[y] - EF')*11600*ferco;
(40) m = Sqrt[l^2 + n^2*j*k];
(41) o = (Tanh[m/2/t]);
(42) p = 4*a*g*j*k*o*n/2/m;
(43) q = 1.57079;
(44) q1 = ArcCos[EF'/(2.774)];
(45) NIntegrate[p, {x, 0, q1}, {y, z, q}, PrecisionGoal → 3];
(46) NIntegrate[p, {x, q1, q}, {y, 0, q}, PrecisionGoal → 3];
(47) n' = % + %%
(48) aa = ArcCos[(EF'/(2.774))$^{.5}$];
(49) n*(Sin[aa])^2*(Cos[aa/2])^4/11.6
(50) NIntegrate[Sin[x]*Sin[z]*(Cos[x/2]*Cos[z/2])^2, {x, 0, kFxaO}];
(51) NIntegrate[1, {x, 0, kFxaO}];
(52) ave = %%/%;
(53) coulom = 3.1019*Sqrt[xx]*10$^{(-12)}$/(1.38*10$^{(-16)}$);
(54) jocou = Pi*4109/4*n*ave*1.4142*3.9*10$^{(-8)}$/Sqrt[xx]/ (0.00131/xx^2)/
10$^{(-3)}$*Tanh[n*ave/2/t];
(55) alpha = .86;
(56)ratio=alpha*2*4*jocou/
(alpha*4*jocou*(1.075*(t/(4*jocou)) +0.9247*(t/(4*jocou)) ^{2})+coulom)
(57) ns0=(alpha * 4 * jocou * (2 − 1.075 * (t/(4 * jocou))
-0.9247*(t/(4*jocou))^2)-coulom)$^{(2/3)}$/364.8
(58) 91
(59) 632
(60) 632.042
(61) 21.1835
(62) 0.995+1.88983 × 10$^{(-43)}$i
(63) -0.018+0.031849i

The * in GENERAL.nb represents multiplication.

The 1st sentence is a given temperature. The t represents temperature (K).

The n in the 2nd sentence represents $\Delta_3(T)$ (K). The composite d-wave gap is

$$\Delta'_d(T,\mathbf{k}) = \Delta_3(T)F_2(\mathbf{k})F_5(\mathbf{k}). \tag{5.46}$$

However, at this step n is only a trial value.

The xx in the 3rd sentence represents x, i.e., the $O_{p\sigma}$ hole number in one unit cell of Cu^{++} lattice in the CuO_2 plane.

The N″ in the 4th sentence represents N″/N' in Eq. (5.45).

The E'_F in the 5th sentence represents $E'_F = E_F$/ferco. E_F is given by Tab. 5.1. The ferco is given by the 8th sentence.

The mm in the 6th sentence represents $m^{***}/m_\circ$ in Eq. (5.20). According to the estimation in Eq. (5.9), $m^{***}/m_\circ \cong 1$. Here we take mm=0.989 for all values of x approximately.

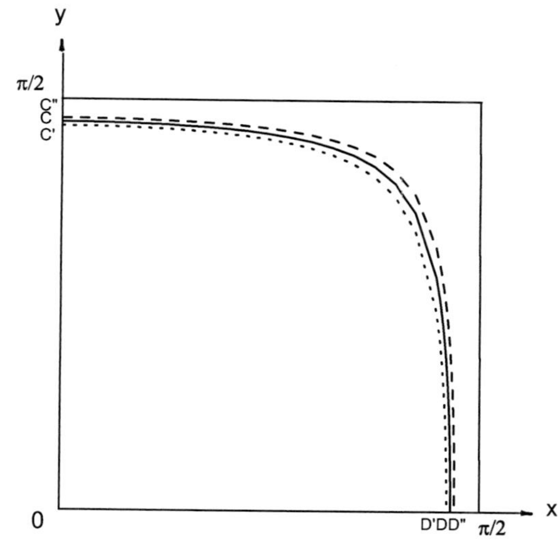

Figure 5.4. The curve CD is given by 2.774cos(x)cos(y) $= E'_F$. $E'_F = 0.272$ eV. $x = k_x a$. $y = k_y a$.

The delta in the 7th sentence represents the factor 0.86 in Eq. (5.18).

The ferco in the 8th sentence represents an adjustable factor for the Fermi energy and dispersion relation. In all our calculations in the chapter 4 we take the true fermi energy of $O_{p\sigma}$ holes as $E_F = E'_F \times$ferco. The dispersion relation Eq. (5.6) becomes 2.9127/1.05cos(x)cos(y)=E'_F. x=$k_x a$. y=$k_y a$. 2.9127/1.05=2.774.

The CC in the 9th sentence represents (1+C) in Eq. (5.45). We take CC = 1.05 for our numerical calculations. Actually, it is better to take CC = 1.

The kFaCu in the 10th sentence represents $k_F a_{Cu}0.86$ in Eq. (5.18). Both the dispersion relations, 2.774cos(x)cos(y)=E'_F and 2.9127cos(x)cos(y)=E_F, give the same k_F. $k_F a = \sqrt{2}$Arccos($\sqrt{E'_F/2.774}$). The a is the nearest-neighbor distance between two O^{--} ions in the CuO_2 plane. Therefore, $k_F a_{Cu} = \sqrt{2}^{2}$Arccos($\sqrt{E'_F/2.774}$).

The kFaO in the 11th sentence represents $\overline{OD}$ in Fig. 5.4.

The JK in the 12th sentence represents the J_K in Eq. (5.35).

The J in the 13th sentence represents the J in section 5.2.

The a' in the 14th sentence appears in Eq. (5.20).

The z in the 15th sentence represents a function of x for the dispersion relation in Fig. 5.4. z (or y) represents the curve CD in Fig. 5.4.

The 16th, 17th, and 18th sentences represent the average of [cos(x/2) cos(y/2)]^2 on the Fermi surface, which is needed by I_r in Eq. (3.52) and Eq. (5.2).

The 19th, 20th, and 21th sentences give the density of states of a single spin per O^{--} ion, which is needed by I_r in Eq. (3.52) and Eq. (5.2).

The 22th sentence represents I_r in Eq. (3.52). I_r appears in Eq. (5.2). The reason of a factor ferco is as follows. The curves C"D" and C'D' in Fig. 5.4 possess dispersion relations [2.774cos(x)cos(y)=$E_F - 0.01$]ferco and [2.774cos(x)cos(y)=$E_F + 0.01$]ferco, respectively. Therefore, the energy difference between C'D' and C"D" is 0.02ferco rather than 0.02 in the 21th sentence.

The 23th, 24th, and 25th sentences represent the average of [cos(x/2) cos(y/2)]^4 on the Fermi surface.

The 26th and 27th sentences represent the average of [cos(x/2)]^2 on the Fermi surface. It can be shown by direct calculations that

$$\overline{[\cos(x/2)\cos(y/2)]^4} \times \overline{[\cos(x/2)]^2}^2 \cong$$

$$\overline{[\cos(x/2)\cos(y/2)\cos((x+p_x a)/2)\cos((y+p_y a)/2)]^2}. \quad (5.47)$$

The expression of the right side in Eq. (5.47) is equal to $\overline{F_5(\mathbf{k})F_5(\mathbf{k}+\mathbf{p})}$ in Eq. (5.20).

The f' in the 28th sentence represents $\overline{F_5(\mathbf{k})F_5(\mathbf{k}+\mathbf{p})}$.

The a in the 31th sentence represents the $A(T)$ of Eq. (5.45) roughly. Considering that the gap equation has $\Sigma_{\mathbf{k}}\cdots$, the $1/(2N_{Cu})$ in Eq. (5.45) becomes $[1/(2\pi)]^2$, i.e.,

$$\frac{1}{2N_{Cu}}\sum_{\mathbf{k}}\cdots = \frac{1}{(2\pi)^2}\int dk_x a dk_y a\cdots. \quad (5.48)$$

In comparison with $A(T)$, there is no w(J/T) in a.

The g in the 36th sentence represents w(J/T).

The j in the 37th sentence represents a part of the composite d-wave.

The k in the 38th sentence represents another part of the composite d-wave.

The l in the 39th sentence represents the true dispersion relation. The unit is K.

The p in 42th sentence represents the integral expression of the gap equation Eq. (5.40). Because we perform integration only in the first quadrant, there is an extra factor 4 in the expression of p.

The 45th and 46th sentences represent the gap equation Eq. (5.40). The integration region is shown by the shading area in Fig. 5.5. Actually, at $T > 0$ K there are region occupied by $O_{p\sigma}$ holes and empty region unoccupied by $O_{p\sigma}$ holes above and below the Fermi surface, respectively. In our numerical calculations we approximately take the region occupied by $O_{p\sigma}$ holes above the Fermi surface as the empty region unoccupied $O_{p\sigma}$ holes below the Fermi surface, and thus the integration region is given by the shading area in Fig. 5.5.

The n' in the 47th sentence corresponds to the value of n in 2nd sentence. n' is calculated from the gap equation. If n' $\cong$ n, then the trial value of the composite d-wave gap in the 2nd sentence, n, becomes the true value.

The 49th sentence gives the value of the gap at Cu-O bond direction. The unit is meV.

The 52th sentence gives the average of directional factors in the composite d-wave gap on the Fermi surface.

The coulom in the 53th sentence represents the $2e^2/C$ in Eq. (3.123).

The jocou in the 54th sentence represents $E_J(T)$ in Eq. (3.108).

The alpha in the 55th sentence represents α in Eq. (3.125).

The 56th sentence is the $T_{c,theo}$ equation of Eq. (3.124). If the ratio is nearly equal to 1, then the t in the 1st sentence gives the value of $T_{c,theo}$.

The ns0 in the 57th sentence represents the relative value of the number density of supercurrent Cooper pairs at 0 K, $n_s(0)$. We define $n_s(0) = 1$ if $x = 0.188$.

The 91 in the 58th sentence is the input temperature (K) in the 1st sentence.

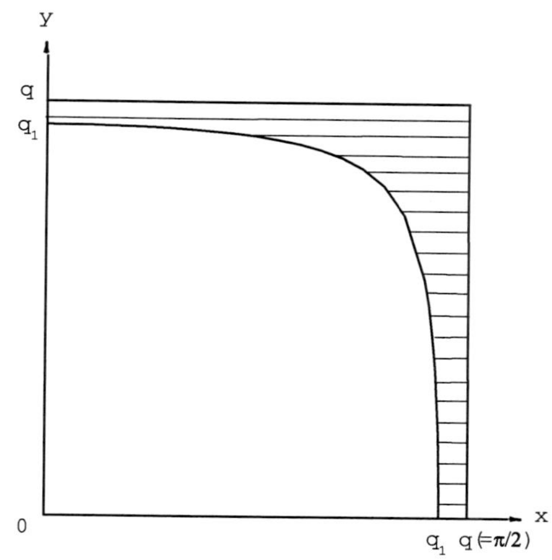

Figure 5.5. The shading area denotes one quadrant of the integration region of the gap equation Eq. (5.40).

The 632 in the 59th sentence is the n (K) in the 2nd sentence.

The 632.042 in the 60th sentence is given by the 47th sentence. Because 632 $\cong$ 632.942 we know that the composite d-wave gap is 632sin(x)sin(y)[cos(x/2) cos(y/2)]^2 (K) at 91 K.

The 21.1835 in the 61th sentence is given by the 49th sentence. From the 61th sentence we know that the magnitude of the gap at Cu-O bond direction and 91 K is 21.1835 meV.

The 0.995 in the 62th sentence is given by the 56th sentence. 0.995 $\approx$ 1 means that at 91 K the Josephson coupling energy is nearly equal to the sum of the zero-point fluctuation energy and the thermal disturbance energy. Therefore, $T_{c,theo} = 91$ K.

The -0.018 in the 63th sentence is given by 57th sentence. $-0.018 \approx 0$ indicates that at 91 K there are no supercurrent Cooper pairs.

From the above numerical calculations we know that for Bi2212 and Y123 with $x = 0.188$, the theoretical value of T_c, $T_{c,theo}$, is 91 K. The corresponding experimental value from $T_c = 93[1 - 82.6(x - 0.16)^2]$ is 87 K.

5.4. Examples of Application of General.nb

Using the GENERAL.nb, we can obtain the $\Delta_3(0K)$ and $n_s(0K)$. For example, we study Bi2212 and Y123 with $x = 0.188$. From Tab. 5.1 we find $E'_F = E_F/1.05 = 0.272$ (eV). From Tab. 5.2 we find $N''/N' = 1.2125$. To obtain $\Delta_3(0K)$ we take t = 1 K in the 1st sentence. To get n' = n we should make many trials for finding the better value of n in the 2nd sentence. After many trials we obtain n = 800 (K), and the n' in the 47th sentence is 800.114 (K). From n' $\approx$ n, we know that n = 800 is the solution of the gap equation. The 61th sentence gives that the magnitude of the gap at 1 K and Cu-O bond direction is 26.8147 meV. The 63th sentence gives $n_s(0K) \approx n_s(1K) = 1.0$.

Using the GENERAL.nb, we can also obtain the temperature dependence of the composite d-wave gap. For example, we make the theoretical numerical calculations of the curve 2 in Fig. 4.14. The curve 2 indicates the temperature dependence of the composite d-wave gap at Cu-O bond direction, $\Delta(T)$. The curve 2 corresponds to Bi2212 with $x = 0.188$. From Tab. 5.1 we obtain $E'_F = 0.272$ eV. From Tab. 5.2 we obtain $N''/N' = 1.2125$. For the curve 2 $J_K \rightarrow 0.65J_K$ and $J \rightarrow 0.47J$. The value of ferco in the 8th sentence is still an unknown constant. To fit the experimental value of $\Delta(T)$ at 0 K, $\Delta(0) = 28.5$ meV, we take t = 1 in the 1st sentence and n = 856 in the 2nd sentence. If n = 856, then the 49th sentence tells us that $\Delta(1K) = 28.69$ meV. When the values of J_K and J change, the value of F_F also changes. We adjust the value of ferco in the 8th sentence. After many trials, we take ferco = 0.808 because n' = 855.397 ≈ 856 = n at this value. At t = 50 (K), n = 760 (K), n' = 759.9 (K) ≈ n. From the 49th sentence $\Delta(50K) = 25.47$ (meV). At t = 80 (K), n = 718 (K), n' = 719.29 (K) ≈ n. $\Delta(80K) = 24.06$ (meV). At t = 90 (K), n = 772 (K), n' = 772.384 (K) ≈ n. $\Delta(90K) = 25.876$ (meV). At t = 110 (K), n = 912 (K), n' = 913.6 (K) ≈ n. $\Delta(110K) = 30.57$ (meV). At t = 130 (K), n = 1012 (K), n' = 1013.43 (K) ≈ n. $\Delta(130K) = 33.92$ (meV). At t = 150 (K), n = 1007 (K), n' = 1006.19 (K) ≈ n. $\Delta(150K) = 33.75$ (meV). At t = 170 (K), n = 838 (K), n' = 839.94 (K) ≈ n. $\Delta(170K) = 28.08$ (meV). At t = 197 (K), n = 0.001 (K), n' = 0.000100074 (K) ≈ n. $\Delta(197K) = 0$ (meV). The 197 K is the pairing temperature T^* corresponding to zero-gap. Note that in comparison with the theoretical curve 2 in Fig. 4.14 the above theoretical values have a few changes because we use an improved program, GENERAL.nb, which is better than that used in chapter 4.

Using the GENERAL.nb to calculate $T_{c,theo}$ is more complex than to calculate $\Delta(T)$. The steps to calculate $T_{c,theo}$ are three. First, choose a trial temperature in 1st sentence. Second, choose an appropriate value of n in the 2nd sentence to get n' = n. Third, if the ratio in the 56th sentence is not nearly equal to one, then we should choose another trial temperature in the 1st sentence. Repeat the three steps until the ratio in the 56th sentence is nearly equal to one. In the latter case the value of t in the 1st sentence represents $T_{c,theo}$.

5.5. $La_{2-x}Sr_xCuO_4$ (LSCO)

From Tab. 5.3 we see that the value of $t_\circ$ of LSCO is less than the value of Bi2212 and Y123. We have taken $t_\circ = 1.1$ eV for Bi2212 and Y123. Therefore, we take $t_\circ = 1.0$ eV for LSCO. Correspondingly, the 12th and 13th sentence of GENERAL.nb have extra factors $(1.0/1.1)^2$ and $(1.0/1.1)^4$, respectively.

The observed resistivity in LSCO at $T \approx T_c$ (normal state) is five times of the observed resistivity in Bi2212 and Y123 at $T \approx T_c$ (normal state) [212]. We have taken $\rho_n = 0.00131/x^2$ ($10^{-3}\Omega.cm$) for Bi2212 and Y123. Therefore, $\rho_n = 0.00131/x^2 \times 5$ ($10^{-3}\Omega.cm$) for LSCO. Correspondingly, in the 54th sentence of GENERAL.nb the factor has an extra factor 5.

In section 5.2 we define that d is the distance between two nearest-neighbor squares in the simplified granular model (See Fig. 3.21.). We have taken $d = d'$ for Bi2212 and Y123. The size of a Cooper pair in the CuO_2 plane of LSCO is larger than that in Bi2212 and Y123. From Figs. 3.20 and 3.21 we know that the value of d in LSCO, d_e, is less than the value of d in Bi2212 and Y123, d_y. we assume $d_e/d_y = 0.353$. Correspondingly, the 53th sentence of GENERAL.nb has an extra factor 0.353, and the 0.00131 in the 54th sentence has an extra factor 0.353.

Making the above changes in GENERAL.nb, we obtain a modified GENERAL.nb for LSCO. Using this modified GENERAL.nb, we can obtain all the numerical results in section 4.8.

5.6. Limit of Supercurrent in the High-T_c Cuprates

The GL coherence length of the high-T_c cuprates is very small. So, the high-T_c cuprates belong to type-II superconductors different from type-I superconductors in an important fundamental way. For small magnetic fields and, hence, also for small transport currents, type-II superconductors reside in the Meissner state. In this state they behave like type-I superconductors, i.e., they expel the magnetic field and current into a thin surface layer.

The difference first appears when the magnetic field at the surface exceeds the value H_{c1}. Then the type-II superconductor must enter the Shubnikov phase, i.e., flux lines must penetrate into the superconductor. In the Shubnikov phase the ideal type-II superconductor has a finite electrical resistance even at very small transport currents due to the motions of flux lines. To prevent the motion of flux line, one can introduce spatial inhomogeneity into the CuO_2 plane of the high-T_c cuprates. Such inhomogeneities can be localized defects, such as departures from stoichiometry at the atomic scale, or more extended defects, such as dislocations, grain boundaries, inclusions of second phases, twin planes, etc. The energy of a flux line will be small if it passes through such an inhomogeneity. The inhomogeneity tends to "pin" flux lines in particularly favorable positions relative to ideal material. It is the pinning which allows the system to sustain the high critical supercurrent density. The type-II superconductor with the inhomogeneity is called hard superconductor.

There is a problem. What is the theoretical upper limit of the critical supercurrent density, $J_{c,max}$? Let us use the granular model of the high-T_c cuprates to calculate $J_{c,max}$. According to the granular model in section 3.3, the high-T_c cuprate consists of a square lattice of Josephson tunneling junctions. Eq. (3.107) tells us that the critical supercurrent through one Josephson tunneling junction is

$$I_c = \frac{\pi\overline{\Delta(T)}}{2|e|R_n}\tanh\left[\frac{\overline{\Delta(T)}}{2T}\right]. \tag{5.49}$$

The area of cross section of this junction is $a_\circ d'$. The expression of $a_\circ$ is given by Eq. (5.36). d' is the thickness of one layer of the CuO_2 plane. $d' = 2.6$ Å is the diameter of O^{--} ion. From Eq. (5.49) we can obtain the critical supercurrent density in the CuO_2 plane. Considering that the lattice constant in the direction vertical to the CuO_2 plane is $c' = 11.9$ Å for Y123 [213], and there are only two CuO_2 planes in one unit cell,

$$J_{c,.max} = \frac{I_c}{a_\circ d'}\frac{2d'}{c'}. \tag{5.50}$$

Let us give a numerical example. We consider Y123 with $T_c = 90$ K. From the empirical formula $T_c = 93[1-82.6(x-0.17)^2]$, we obtain the $O_{p\sigma}$ hole number in one unit cell of the Cu^{++} lattice in the CuO_2 plane is $x = 0.14$. Section 5.2 gives $\rho_n = 0.00131/x^2$. $R_n = \rho_n d/S$. $S = a_\circ d'$. $d \approx d'$. Fig. 4.7 gives that the value of gap at Cu-O bond direction is $\Delta(T) = 32$

meV at 77 K. We take $\overline{\Delta(T)} = \Delta(T)/2$. In this case

$$J_{c,max} = \frac{\pi\Delta(T)}{2|e|\rho_n c'} \tanh\frac{\Delta(T)}{4T}. \tag{5.51}$$

Substituting the above numerical values for Y123 with $T_c = 90$ K into Eq. (5.51) gives $J_{c,max} = 6.3 \times 10^9$ A/cm^2 at 77 K. The highest critical supercurrent density was observed in superconducting bridge constriction of Y123 at 77 K, and it is 1 .3 $\times 10^9$ A/cm^2 [214].

5.7. Doped Magnetic Materials

Assume that the system is 3-D cubic lattice. The TLSMI is given by Eq. (3.79), in which $w(J/T)$ should be given by Eqs. (7.50) and (7.51). Take that the number of free holes of doped magnetic materials in one unit cell is 0.15. Assume that the *RKKY*-like interaction, E_F, N''/N', the form of Fermi surface, and the gap function in Eq. (3.79) are the same as that of Bi2212. Considering that the nesting Fermi surface benefits TLSMI under d-wave condition for Bi2212, we give an additional factor 1/15.8 for TLSMI in the 31st sentence of the GENERAL.nb. We consider high local spin with $S = 5/2$ instead of $S = 1/2$. When we make the corresponding modifications in GENERAL.nb, we obtain $T_c > 450$ K.

Under the same conditions as the above for $S = 5/2$, if $S = 3/2$, then the doped magnetic materials will be in normal state at any temperature. If we do not consider the additional factor 1/15.8, then the doped magnetic materials will be in pseudogap state at $0 < T < 750$ K, and in superconducting state at $T > 750$ K.

Chapter 6

Other Systems

6.1. $Nd_{2-x}Ce_xCuO_4$ (NCCO)

NCCO is a n-type superconductor [181]. The doping range of superconductivity for NCCO is at $0.12 < x < 0.20$ [215]. Applying the phase-sensitive capability of tunneling spectroscopy, Kashiwaya *et al.* did not observe the zero-bias conductance peak, and thus concluded that the d-wave symmetry of the gap should be rejected [216]. The main experimental evidence for the s-wave symmetry of the gap is the studies of the in-plane penetration depth $\lambda_{ab}(T)$ [217,218]. Considering the paramagnetism arising from Nd^{3+} ions, Cooper altered the limiting temperature dependence of $\lambda_{ab}(T)$ - $\lambda_{ab}(0)$ from exponential behavior to T^1 or T^2 behavior associated with d-wave symmetry [219] (Also see Ref. [220].). However, considering the temperature-dependent susceptibility of Nd^{3+} moments and correcting the measured temperature dependence of $\lambda_{ab}(T)$, Alff *et al.* still insisted that the exponential dependence indicate isotropic s-wave symmetry [221]. The quasiparticle tunneling conductance spectra of NCCO closely resemble those of d-wave LSCO and Y123 [216,221]. Tsuei and Kirtley presented a very powerful and conclusive evidence that NCCO has d-wave gap by using tricrystal phase-sensitive experiment [222]. Luo *et al.* suggested that the pairing in electron-doped cuprates has d-wave symmetry in the whole doping range [223]. There is no detectable imaginary component of the gap in NCCO, i.e., the symmetry of the gap is pure d-wave [222, 224]. It is noteworthy that the values of ratio $2\Delta(0K)/T_c$, coming from different measurements, are quite different. The values of $2\Delta(0K)_{depth}/T_c$ from penetration depth experiments are 3.9 - 4.5 for $T_c = 21$, 21.5, and 22 K [218], and 2.9 for $T_c = 24$ K [221]. The value of $2\Delta(0K)_{peak}/T_c$, measured by peak-to-peak value in the conductance spectra, is about 6 for $T_c = 24$ K [221]. Refs. [222, 225] pointed out that the zero-bias conductance peak can be suppressed by disorder. Recently, there has been a surge of interest in electron-doped cuprates [226]– [230].

Our theory in chapter 3 can make the following discussions for the above experimental results. In NCCO there are the same CuO_2 plane and antiferromagnetic short-range order as in Bi2212, Y123, and LSCO. Therefore, TLSMI exists between the electrons in the CuO_2 plane of NCCO. If the electrons are at the bottom of lower Hubbard band, the symmetry of the gap should be the composite s-wave. The reason is as follows. The composite s-wave contains a factor $\cos(k_x a)\cos(k_y a)$. This factor has larger contribution to the gap equation due to the small values of $k_x a$ and $k_y a$ near the bottom of the lower Hubbard band. On the contrary, the composite d-wave has only a small contribution to the gap equation due to the factor $\sin(k_x a)\sin(k_y a)$ and small $k_x a$ and $k_y a$. The composite s-wave contains two terms [See Eq. (3.65).]. One term comes from TLSMI. Another one comes from electron-phonon interaction. The superconductivity caused by both TLSMI and electron-phonon interaction might have higher T_c than only one interaction. If the electrons are at the bottom of upper Hubbard band, then the symmetry of the gap should be the composite d-wave. The reason is as follows. The composite d-wave contains a factor $\sin(k_x a)\sin(k_y a)$. This factor has larger contribution to the gap equation due to that $k_x a$ and $k_y a$ are nearly equal to $\pi/2$ at the bottom of upper Hubbard band. We support the experimental evidence in Ref. [222, 223], i.e., that the symmetry of the gap of NCCO is d-wave. Therefore, our theory infers that the electrons in NCCO are at the bottom of upper Hubbard band, and the gap is the composite d-wave. The composite d-wave is anisotropic (See section 4.1.). We explain $\Delta(0K)_{depth} < \Delta(0K)_{peak}$ as follows. There are maximum value and zero-point value in the composite d-wave gap. The value of $\Delta(0K)_{peak}$, measured by peak-to-peak value in the conductance spectra, is the maximum value of the gap. Eq. (1.10) tells us that $\lambda_{ab}^2(T) \propto 1/n_s(T)$. Eq. (3.123) tells us that $n_s(T)$ only relates to the average value of the gap. Therefore, the value of $\Delta(0K)_{depth}$ is the average of the composite d-wave gap. From this point we can understand that $\Delta(0K)_{depth} < \Delta(0K)_{peak}$. The reason that there should be a doping range of superconductivity for NCCO is the same as that for LSCO in section 4.8.

6.2. Sr_2RuO_4 (SRO)

The crystal structure of SRO is similar to LSCO. The most essential difference in the electronic structure near the Fermi surface E_F between the ruthenate and the cuprate is that the bands which cross E_F are of Ru4d(yz, zx, xy)-O2pπ antibonding character in SRO, which correspond to three quasi-2D Fermi surfaces, α, β and γ sheets, respectively [231], while those in the cuprates are of Cu3d(x^2-y^2)-O2pσ antibonding nature. This is due to the difference of the number of d electrons in the outermost shell, a Ru atom having four 4d electrons while a Cu atom having nine 3d electrons. The γ sheet, i.e., the Ru4d(xy)-O2pπ band, is mainly in the plane of RuO_2 [232]. The α sheet is hole-like, and the β and γ sheets are electron-like [231,233,234]. The Riseman *et al.*'s small-angle neutron-scattering measurements of magnetic flux lines in Ref. [235] suggested that the pairing occurs primarily on the electrons of the γ sheet. However, Riseman *et al.* pointed out in Ref. [236] that the nature of the pairing in SRO might be more complicated than they suggested in Ref. [235]. The ^{17}O NMR experiments show that the strong ferromagnetic spin enhancements are only in the Ru4d(xy) orbital [237]. Since the cubic perovskite $SrRuO_3$ is ferromagnetic with Curie temperature 150 K and local spin S = 1 due to the strong Hund's coupling, a popular assumption is that SRO has local spin S = 1 [237, 238]. Based on the scenario of ferromagnetic spin-fluctuation with S = 1, Rice and Sigrist proposed that we should look for a two-dimensional triplet pairing state [238]. The triplet superconductivity in SRO was identified by ^{17}O Knight shift [239] and Andereef reflection [240]. While recent phase-sensitive experiments using flux modulation in Josephson junction consisting of SRO and a conventional superconductor have been interpreted as the conclusive evidence for a chiral spin-triplet pairing, Zutic and Mazin propose an alternative interpretation [241]. Zutic and Mazin show that an overlooked chiral spin-triplet pairing is also compatible with the observed phase shifts in Josephson junctions and propose further experiments which would distinguish it from its spin-triplet counterpart [241]. The time-reversal symmetry breaking superconductivity in SRO was identified by muon spin-relaxation measurements [242]. Considering that a non-unitary state is very difficult to be stabilized unlike the unitary state [243, 244], the superconductivity in SRO should be the Γ_5^- pairing state in Ref. [238]. That both resistivities ρ_{ab} along the RuO_2 plane and ρ_c along the c axis are proportional to T^2 below 25 K indicate that a crossover to a canonical Fermi-liquid state takes place below 25 K prior to the superconducting transition [245]. The highest transition temperature is $T_c \cong 1.5$ K [246]. Laube *et al.* reported that for their sample T_c is 1.02 K, the zero-gap temperature T^* is a little larger than 1.02 K, i.e., there is the pseudogap at $1.02 < T < 1.26$ K [240]. The specific heat [247] and nuclear quadrapole resonance [248] measurements revealed that approximately 0.6 of the normal density of states remain in the superconducting phase in SRO sample. The latter experimental result was that the residual density of states is less than 20% [249]. The temperature dependence of the penetration depth [250] and specific heat measurements [251, 252] indicated that there is a node of the gap or a very small gap in SRO.

To explain the residual density of states observed in Refs. [247]– [249] Agterberg *et al.* presented a model for the superconducting transition in SRO in which the superconductivities in the band with Ru d_{xy} orbital character and in the bands with Ru $\{d_{xz}, d_{yz}\}$ orbital characters are weakly coupled. This model attributes the observed residual density of states to thermal excitations across a secondary gap that is smaller than the primary gap driving the superconductivity transition. Bonalde *et al.* pointed out that according to the Agterberg *et al.*'s model the secondary gap would lead to a crossover between the two exponential temperature behaviors at very low temperature or show up as a second superconducting transition [250], and also pointed out that specific heat [253] and NMR [254] experiments on high quality samples show no evidence of a crossover and a second transition down to 0.1 K. Kuwabara and Ogata proposed a mechanism leading to the spin-triplet superconductivity based on the antiferromagnetic spin-fluctuation [255].

Our theory in chapter 3 can make the following discussions for the properties of SRO. First, the superconductivity and gap in SRO can be only in the γ sheet. The reason is as follows. To have TLSMI$\neq 0$ needs, at least, $J \neq 0$ and $J_K \neq 0$. $J \propto t_\circ^4$ and $J_K \propto t_\circ^2$ (See [83] and chapter 5.). The $t_\circ$ is the hopping integral between Ru orbital and O orbital. The ^{17}O NMR experiments show that the strong ferromagnetic spin enhancements are only in the Ru4d(xy) orbitals [237]. Therefore, only the hopping integral $t_\circ$ between Ru4d(xy) and Opπ orbitals has contribution to TLSMI, which means that only the electrons in the γ sheet can have TLSMI. At present there is not any experiment to indicate that any strong magnetic correlation exists in the Ru4d(xz, yz) orbitals. Therefore, we have no reason to think that there is TLSMI for the α and β sheets. Second, the symmetry of the gap should be p-wave, i.e., spin-triplet pairing. For the ferromagnetic long- and short-range order $J < 0$, TLSMI in Eqs. (3.29) and (3.53) is negative for the two electrons with parallel spins, $\uparrow\uparrow$ and $\downarrow\downarrow$. In this case the space part of the wavefunction of Cooper pair should be odd parity, which means that the gap caused by TLSMI in ferromagnetic long- or short-range region should be, for example, p-wave. Third, the gap in the RuO_2 plane is anisotropic. The two p-waves in Eqs. (3.67) and (3.68) have the same contribution to the superconductivity and gap. In principle, the gap in our theory can be $\Delta'_{p1}(T,\mathbf{k}) + \Delta'_{p2}(T,\mathbf{k})$ or $\Delta'_{p1}(T,\mathbf{k}) + i\Delta'_{p2}(T,\mathbf{k})$. Considering that the latter is more stable than the former [91,92], we take the latter, which means that it is a time-reversal symmetry breaking superconductivity as observed in Ref. [244]. Due to the symmetry of the two p-waves in Eqs. (3.67) and (3.68) $\Delta_4(T) = \Delta_5(T)$. The magnitude of the p-wave gap in SRO is $\Delta_p(T, \mathbf{k}) = \Delta_4\sqrt{F_3^2(\mathbf{k}) + F_4^2(\mathbf{k})F_5(\mathbf{k})}$. Although there is no node of the gap, the magnitude of the gap along the direction 45° away from the Ru-O bond direction is very small. The anisotropy of the p-wave gap can explain the observed residual density of states. Third, the existence of pseudogap for SRO at $1.02 < T < 1.26$ K shows that the size of Cooper pair should be short. If the size of Cooper pair is too large, then it is very difficult to distinguish pairing temperature T^* from superconducting transition temperature T_c. Fourth, our theory guesses that the origin of lower value of T_c comes from the smaller value of the hopping integral $t_\circ$ between Ru orbital and O orbital in the RuO_2 plane. Rough numerical calculations in Ref. [256] indicate $t_\circ(\mathrm{SRO})/t_\circ(\mathrm{LSCO}) = 0.7$.

6.3. Gap and Superconductivity of Chain in Y123

The superconductivity in the high-T_c cuprates is generally thought to occur in the CuO_2 plane, while the rest of material layers serve only structural or charge-balance functions. Y123 is a high-T_c cuprate with a special feature interleaved between its CuO_2 planes: the Cu-O chains. These one-dimensional structures give rise to partially filled energy bands, allowing the possibility of metallic one-dimensional electronic state. Far-infrared spectroscopy detected a supercurrent component in the Cu-O chains [257]. Scanning tunneling spectroscopy detected gap in the Cu-O chains [258]. Gagnon *et al.* measured the thermal conductivity κ of Y123 along the a and b axes of detwinned single crystals [259]. Both κ_a and κ_b show the usual peak below $T_c = 93.8$ K, but a clear additional contribution to the b-axis thermal conduction can be observed at well below T_c. The difference $\kappa_{chain} = \kappa_a - \kappa_b$ exhibits a peak similar to the peak in the plane conductivity κ_a but with an onset temperature of 55 K [259]. This result suggests that the Cu-O chains give rise to a sudden growth in superfluid density [259]. Applying the microwave complex conductivity in the superconducting state of Y123, Sridhar *et al.* revealed enhanced pair conductivity below ~ 65 K [260]. Sridhar *et al.* pointed out that the Fermi surface of Y123 is believed to consist of multiple sheets associated with two plane bands and a chain band, and this raises the possibility of superconductivity with multiple pairing energies [260].

There is a model on the origins of the gap and superconductivity in the Cu-O chains [258]. Because the Cu-O chains reside in less than a nm from the CuO_2 plane, it is conceivable that the chains superconduct due to the proximity effect with the CuO_2 planes [258]. However, Refs. [259, 260] show that the superconductivities in the Cu-O chains and the CuO_2 planes are at different temperatures, which indicates that the CuO_2 planes and the Cu-O chains undergo their own superconducting transition.

For the gap and the superconductivity in the Cu-O chains our theory can make the following discussions. Section 3.2.5 gave a systematic theory on the gap, caused by TLSMI, in the Cu-O chain. We assume that the carriers in the Cu-O chain are the $O_{p\sigma}$ holes as well as in the CuO_2 plane. The $O_{p\sigma}$ holes should be near the top of the lower Hubbard band, which is shown in Fig. 6.1. From the BCS gap equation Eq. (3.57) and TLSMI in Eqs.

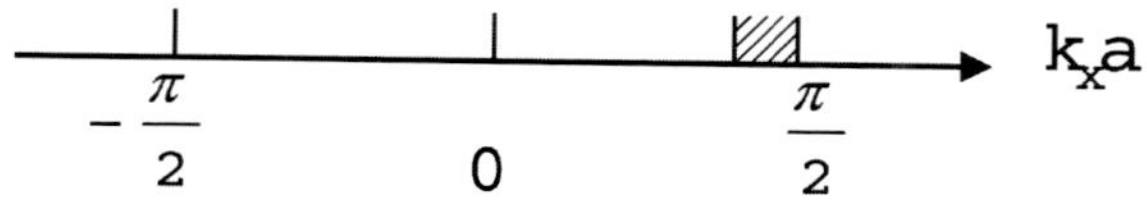

Figure 6.1. The shading area denotes the $O_{p\sigma}$ holes at the top of lower Hubbard band. The a is the lattice constant of the Cu-O chain.

(3.92) - (3.99) we can obtain that the possible forms of gaps in the Cu-O chain are

$$\Delta'_{\text{Cu-O},s}(T,\mathbf{k}) = \Delta_{\text{Cu-O},1}(T) + \Delta_{\text{Cu-O},2}(T)\cos(k_x a)F_{\text{Cu-O},3}(\mathbf{k}), \tag{6.1}$$

$$\Delta'_{\text{Cu-O},p}(T,\mathbf{k}) = \Delta_{\text{Cu-O},3}(T)\sin(k_x a)F_{\text{Cu-O},3}(\mathbf{k}) \tag{6.2}$$

There are only s- and p-wave solutions in the Cu-O chain, which is different from the case of the CuO_2 plane. For the $O_{p\sigma}$ holes the p-wave has larger contribution to the gap equation, and the second component of the s-wave gap in Eq. (6.1) has only a less contribution to the gap equation. At present there are no data on the electron-phonon interaction constant λ and Coulomb pseudo-potential μ^* in Cu-O chain. Therefore, we do not know the contribution of the first component of the s-wave gap in Eq. (6.1) to the gap equation.

6.4. Heavy-Fermion Superconductivity

There are following main experiments on the heavy-fermion systems. Steglich gave an experimental rule between T_c and antiferromagnetic transition temperature T_N, i.e., $T_N \approx 10T_c$, which is called "ten rule" [261]. For UPt_3, $CeCu_2Si_2$, UBe_{13}, $CeRu_2Si_2$, UPd_2Al_3 the T_c and T_N are 0.5, 0.7, 0.8, 1.5, 2 K and 5, 8, 8.8, 17, 14.5 K, respectively. Donovan *et al.* reported on the observation of a pseudogap in UPt_3. The pseudogap progressively disappears above $T = T_N = 5$ K [262]. The ^{195}Pt Knight shift study demonstrated that UPt_3 is the odd-parity superconductor with parallel spin pairing [263]. For UPt_3 there are two transition temperatures, separated by about 0.06 K [264, 265]. In addition, evidence for unconventional behavior of UPt_3 was obtained from specific heat [266], thermal conductivity [267], penetration depth [268], and magnetization [264]. This evidence exhibited power law at low temperature instead of the exponential behavior. The tunneling spectroscopy experiment suggests that the superconductivity in UPd_2Al_3 possesses d-wave gap [269]. Kimura *et al.* report the pressure-induced superconductivity in the noncentrosymmetric heavy-fermion $CeRhSi_3$ [270]. The superconductivity emerges above about 12 kbar even though the antiferromagnetic ordering persists [269, 270].

At present we can not explain the above experiments quantitatively due to lack of the values of related parameters. The qualitative explanations are as follows. Long time ago Edvards introduced a duality model to describe the itinerant f and d electrons, which assumed that the itinerant f and d electron system can be separated into two parts, one is mobile charges, and the other one is local spins [271, 272]. The Hamiltonian for the itinerant f and d electrons given by Edwards contains free part, Kondo Hamiltonian and Heisenberg Hamiltonian, which look like the same as Eq. (3.5) [271, 272]. Therefore, we can think that there is TLSMI below T_N in the heavy-fermion system. Considering that the phonon-mediated superconductivity can not be p- or d-wave, we assume that the superconductivity in the heavy-fermion system is caused by TLSMI. In Edwards model $T_N \propto (AJ+BJ_K^2)$ [271]. The larger the T_N, the larger the TLSMI, and so the larger the T_c. The TLSMI can, in principle, cause the superconductivity with s-, d-, or p-wave gap. The symmetry of the gap in a concrete heavy-fermion system is mainly determined by its lattice structure and the form of Fermi surface. According to the second feature of the TLSMI superconductivity in section 3.2.2, if the superconductivity of antiferromagnetic UPt_3 is p-wave, then the spin function of the spin-triplet state should be $(\uparrow\downarrow + \downarrow\uparrow)/\sqrt{2}$. The two similar transition temperatures in UPt_3 might come from two similar p-waves. The possibility possessing two p-waves for a system can be seen in Eqs. (3.67) and (3.68). The pressure-induced superconductivity in $CeRhSi_3$ [270] can be understood by our theory. The pairing potential of Cooper pairs in Eq. (3.53) is nearly proportional to JJ_K^2, which is proportional to $t_\circ^8$. $t_\circ$ is the hopping integral of the carriers between adjacent related orbitals. At high pressure the $t_\circ$ becomes larger, and thus the pairing potential becomes larger as well. Just this larger pairing potential leads to the superconductivity in $CeRhSi_3$.

6.5. Organic Superconductors

The organic superconductor $(TMTSF)_2X$ (X=PF_6, ClO_4, AsF_6, etc.) is quasi-one-dimensional conductor [273]. The organic superconductor $\kappa(\beta)-(BEDT-TTF)_2X$ ($X=I_3$, AuI_2, $Cu[N(CN)_2]Br$, $Cu(SCN)_2$, etc.) is very anisotropic quasi-two-dimensional conductor [274]. There are the following basic experimental facts. The pseudogap is revealed in $(TMTSF)_2PF_6$ and $(TMTSF)_2ClO_4$ with $T^*/T_c = 10$ [275]. They also have large low temperature gap, $\Delta_{experiment}(0K)/\Delta_{BCS}(0K) = 12$. Here $2\Delta_{BCS}(OK) = 3.5\ T_c$. A possibility of p-wave pairing in $(TMTSF)_2X$ was discussed in connection with the suppression by impurities [276], exceeding the Clogston-Chandrasekhar parametric limit [277], the absence of the Hebel-Slichter peak [278], and the revival of superconductivity in high magnetic fields [273]. The pseudogap is revealed in $(BEDT-TTF)_2X$ [273]. The non-BCS relation between the gap and temperature is observed in $\beta-(BEDT-TTF)_2AuI_2$ with $\Delta_{experiment}(0K)/\Delta_{BCS}(0K) = 8$ [279].

Our theory can explain the above experiments qualitatively and make some predictions. Toyota *et al.* thought that the organic superconductors are highly correlated Fermi liquid as well as the high-T_c cuprates [280]. All organic superconductors possess antiferromagnetism [261]. For example, Lee *et al.* observed coexistence of superconductivity and antiferromagnetism by nuclear magnetic resonance and electrical transport in $(TMTSF)_2PF_6$ [281]. For antiferromagnets $J \neq 0$ and $J_K \neq 0$, and thus TLSMI$\neq 0$. The pseudogap, the large low temperature gap, and the non-BCS relation between the gap and temperature all are the typical phenomena of the Cooper pairs caused by TLSMI. The 1-D solutions in section 3.2.5 tell us that both s- and p-wave are possible. If the wave number along the conduction direction k_x is about 0 or $\pi/(2a_x)$, then the superconductivity caused by TLSMI in $(TMTSF)_2X$ is s- or p-wave, respectively.

6.6. The Ladder Materials

Various types of Cu oxides with layer or chain structures have been synthesized and re-examined to get deeper insight into the anomalous metallic state of the high-T_c cuprates, which has not been fully understood yet. Among various one-dimensional analogs, a system is unique in which two strongly coupled Cu-O chains (legs) form ladders. Two-leg spin-1/2 ladders are found in vanadyl pyrophosphate $(VO)_2P_2O_7$ and in some cuprates, such as $SrCu_2O_3$. The $Sr_{14}Cu_{24}O_{41}$ system which was an already known two-leg ladder cuprate has attracted much attention owing to the discovery of superconductivity ($T_c \sim 12$ K) under pressure [282] and the pseudogap in hole-doped $Sr_{14-x}Ca_xCu_{24}O_{41}$ [283, 284]. Osafune *et al.* inferred that the observed pseudogap signals a formation of hole pairs in the ladders [283].

Our theory makes the following conjectures. Dagotto and Rice thought that the Cu_2O_3 two-leg ladder plane is an antiferromagnetic spin-1/2 lattice [285]. There might be TLSMI between two holes in the ladder, and therefore the superconductivity and gap might be caused by TLSMI. Hence the appearance of the pseudogap can be understood by our theory. The low value of T_c might come from that there is no Fermi surface nesting. We have experiences from the numerical calculations for the high-T_c cuprates that the nesting structure of the Fermi surface is crucial for appearance of hole superconductivity and gap caused by TLSMI.

6.7. The Infinite-Layer Structure

The infinite-layer parent structure of the copper oxide superconductors is the simplest structure containing the CuO_2 planes that are apparently essential to high-T_c superconductivity. Azuma *et al.* reported the isolation of an alkaline-earth-deficient infinite-layer phase, $(Ca_{1-x}Sr_x)_{1-y}CuO_2$ ($y \cong 0.1$), with T_c up to 110 K, and suggested that it is p-type superconductivity, with the carriers arising from calcium and strontium vacancies [286]. Another infinite-layer material is $Sr_{1-x}R_xCuO_2$ in which R is a rare-earth element. It was thought to be a n-type superconductor with $T_c = 43$ K [286]. Infinite-layer structure $LaBaCuO_2BO_3$ was synthesized by Zou *et al.* [287]. It transits to a diamagnetic state at 32 K [287].

Our theory explains qualitatively the above experiments as follows. The electronic structure of $CaCuO_2$ from LSDA and LSDA+U calculations is nearly the same as that of the other high-T_c cuprates [288]. Therefore, the discussions on the superconductivity and gap are nearly the same as Bi2212 and Y123 in chapter 4, LSCO in section 4.8, and NCCO in section 6.1. It is obvious from our theory that the pseudogap should be observed in the infinite-layer structure of the copper oxide superconductors, and the pseudogap and T^* are independent of the thickness of structure. However, a related experimental report has not yet been seen.

6.8. $Fe_{1-x}S$

Considering that TLSMI depends only on the Hamiltonian in Eq. (3.5) rather than on the CuO_2 plane, we make a probable explanation for the superconductivity at 110 K in $Fe_{1-x}S$

with $0.07 \leq x \leq 0.125$ [289]. The FeS is an antiferromagnetic insulator with $T_N = 613K$. The $Fe_{1-x}S$ is a hole conductor with a few $S_{p\sigma}$ holes which are similar to the $O_{p\sigma}$ holes in the high-T_c cuprates. The hopping integral $t_\circ$ from site S to site Fe is finite. Chapter 5 gives that J and J_K in Eq. (3.5) are proportional to $t_\circ^4$ and $t_\circ^2$, respectively. Therefore, J and J_K are not equal to zero. The Hamiltonian terms of $Fe_{1-x}S$ should be nearly the same as Eq. (3.5), which can also be seen from the duality model for the itinerant f or d electron [271, 272]. Therefore, there is the pair potential TLSMI in $Fe_{1-x}S$. The superconductivity of $Fe_{1-x}S$ is caused probably by TLSMI as well as the high-T_c cuprates. The technology to prepare $Fe_{1-x}S$ is difficult. In comparison with $Fe_{1-x}S$, the FeS_{1+x} might be easy to be prepared. In our theory both $Fe_{1-x}S$ and FeS_{1+x} have the similar superconductivity.

6.9. $SmRh_4B_4$

For $SmRh_4B_4$ $T_c = 1.815$ K and $T_N = 0.87$ K [290]. The Josephson critical current changes from 20 μA to 60 μA when the temperature is lower than T_N, which means that a large gap appears below T_N [290].

The above phenomenon can be understood by our theory. Because T_c is much higher than T_N, the superconductivity above T_N comes from the phonon-mediated interaction. When the temperature goes down to lower than T_N, there is the antiferromagnetism, and so there is TLSMI in the system. From Eq. (3.65) we know that both the s-wave component in TLSMI and the phonon-mediated interaction can together cause the superconductivity and gap. Therefore, the gap at $T < T_N$ becomes much larger than the gap at $T > T_N$. Because the s-wave component in TLSMI is anisotropic, the gap at $T < T_N$ should be anisotropic as well.

6.10. MnSi

MnSi is weak ferromagnetic metal. Blagoev *et al.* proposed a microscopic theory, based on the interaction mediated by spin fluctuations among the fermions, and predicted a s-wave superconductivity in the ferromagnetic phase, and a p-wave superconductivity in the paramagnetic phase [291].

We note that, so far, experiments have failed to find any kind of superconductivity in the paramagnetic phase of this compound [291]. Therefore, the interaction mediated by phonon can not cause the superconductivity in this system. From the Edwards duality model for f and d electron [271, 272], there is TLSMI in these systems. If the superconductivity in the ferromagnetic state is caused by TLSMI, then it should be p-wave, which is different from the prediction of the s-wave in Ref. [291]. The detailed analysis for the p-wave is the same as that in section 3.2.2.

6.11. Many Antiferromagnetic Materials

All the discussions in this section are purely theoretical. From the numerical calculations in chapter 5 and the discussions in chapter 3 we have experiences that the following four conditions are needed to get high-T_c: (a) large TLSMI needs large J_K and high local spin;

(b) the large contribution of the composite d-wave gap to the gap equation needs a nesting Fermi surface of hole carriers; (c) to satisfy the long-range phase coherence condition needs large ε_∞; (d) to have a larger Josephson coupling energy needs lower resistance at $T > T_c$. From the duality model of the f and d electron $T_N \propto (AJ + BJ_K^2)$ [271,272]. So, the better materials might be those with high T_N. T_N are 1040, 953, 738, 723, 600, and 230 K, for αVS, $\alpha - Fe_2O_3$, $LaFeO_3$, CrSb, NiO, and CuO, respectively [292]. For comparison we notice $T_N = 326$, and 500 K for La_2CuO_4 and $YBa_2Cu_3O_6$, respectively [293]. They are all antiferromagnetic insulators. To have a nesting Fermi surface we should introduce a few holes. If some material among the above materials has large T_N and ε_∞, high local spin, and nesting Fermi surface at the same time, then this material is of much possibility to become a high-T_c superconductor. Of course, due to the large numerical factor in the denominator of the TLSMI expression in Eq. (3.79) it is not easy to have high-T_c caused by TLSMI in 3 dimensional antiferromagnetic materials of low local spin.

It is interesting to note that the single crystal of CuO has an activation energy of 0.11 eV, and gives negative resistance, whereas the single crystal CuO annealed in oxygen atmosphere shows an anonymous shoulder feature on resistance-temperature curve around 150 K [294]. Our theory can give this shoulder a plausible explanation. The sample used by Zheng *et al.* is a underdoped CuO_{1+x}. The CuO is antiferromagnetic. Therefore, the TLSMI exists in CuO_{1+x}. In a few small regions of CuO_{1+x} the superconductivity appears due to the TLSMI, and these small superconducting regions lead to the shoulder. We predict that if we increase the hole doping, then CuO_{1+x} might become a high temperature superconductor with $T_c = 150$ K.

By the way for the ferromagnetic materials we can make the same discussions as for the antiferromagnetic materials. The only difference is that the gap in ferromagnetic materials should be the p-wave (See section 3.2.2.).

We give some examples for the possible superconductivity of some doped magnetic materials. If the magnetic ion has high local spin $S = 5/2$, then $T_c = 450$ K. If $S = 3/2$, then some doped magnetic materials are in normal state at any temperature. If we do not consider the additional factor 1/15.8 for TLSMI in 31st sentence of GENERAL.nb in case of $S = 3/2$, then some magnetic materials are in the pseudogap state at $0 < T < 750$ K, and begin to have superconductivity at $T > 750$ K. The phenomenon that superconductivity just appears at high temperature comes from the complex temperature dependence of TLSMI. See section 5.7 for details. It can be seen from these examples that high local spin benefits superconductivity.

6.12. Doping C_{60}

The superconductivity appears at relatively high temperature in alkali metal-doped C_{60} fullerenes. The highest T_c was attended in Tl_2RbC_{60}. For it T_c=48 K, and $2\Delta(4.2K)/T_c$=5.3 [295]. A decrease of 0.65 K in T_c in sample with 75% substitution of ^{13}C for ^{12}C yields an isotopic exponent of α=0.37, which is interpreted as evidence for phonon-mediated pairing [296, 297]. The another important experiment is $2\Delta(4.2K)/T_c$=5.3 for Rb_3C_{60} with T_c=28 K [298]. There are three features in the observed data of dI/dV vs. V, which should be noted [298]. First, the data go up gradually from V=0 to the value of V $>$ 0 corresponding to the first main peak, which is like the high-T_c cuprates. Second, the tunneling spectroscopy of a point contact junction has multi-peak structure corresponding to multi-gap. Third, occasionally, the authors did not observe a gap in the surface by using the point contact junction.

Our theory thinks that the superconductivity in these systems is caused by TLSMI. The reasons are as follows. It is argued that the π-band width is only 500 K [299]. So, even a weak on-site Coulomb repulsion can lead to relatively strong correlation. Actually, the Kondo resistance looks like to be observed in normal state, which means, at least, the existence of the local spins [300]. It is interesting to note that CsC_{60} has antiferromagnetic transition at 30 K [301]. Although the antiferromagnetic transition has not yet been observed in Tl_2RbC_{60} and Rb_3C_{60}, we infer that there are the antiferromagnetic short-range order, and thus TLSMI in the superconducting doping C_{60}. If the superconductivity is caused by the combination of the phonon-mediated interaction and TLSMI, then many experimental facts can be explained. For example, the above three observed features of the data of dI/dV vs. V can be explained as follows. The first feature comes from the property of the composite s-wave. The composite s-wave in Eq. (3.65) is also not a constant gap. The second and the third features come from the sixth feature of the gap caused by TLSMI in section 3.2.2. At present we can not obtain $2\Delta(0K)/T_c = 5.3$ by numerical calculations because we do not know the values of related parameters. From the experiments [94, 95] and our experiences in the numerical calculations for the high-T_c cuprates in chapters 4 and 5, we know that $2\Delta(0K)/T^* \cong 4.5$, which means $T^* > T_c$ if $2\Delta(0K)/T_c = 5.3$. We predict that the pseudogap should be observed at $T^* > T > T_c$ in Rb_3C_{60} and Tl_2RbC_{60}. That the superconductivity caused by TLSMI can have the isotope effect as well as the phonon-mediated superconductivity can be found in section 4.6.

6.13. A15 Compounds

The superconducting mechanism of strong-coupling old-fashioned "high-T_c" materials, such as Nb_3Sn and Nb_3Ge, is still an open problem. They have many unusual properties in both their normal and superconducting states. Very careful tunnel effect measurements for Nb_3Sn were made, which show an anomalous behavior of $\alpha^2(\omega)F(\omega)$ [302]. It turns out that the values of interest calculated from $\alpha^2(\omega)F(\omega)$ and Eliashberg equation are $\lambda = 0.76$ and $\mu^* = -0.1$. The negative value of Coulomb pseudo-potential μ^* may imply that some new pairing mechanism operates in Nb_3Sn [303]. Moreover, the multi-gap structure was observed in Nb_3Sn, which is $0.2 < 2\Delta(0K)/T_c < 4.8$ [304]. It can be deduced that the Nb_3Ge possesses non-phonon mechanism as well [305].

There are itinerant d electrons coming from atoms Nb in the A15 compounds. From the duality model of the d electron in Refs. [271, 272] we know that J and J_K are not zero. If we assume that there is the antiferromagnetic or ferromagnetic short-range order in A15 compounds, then there is TLSMI in them. The composite s-wave part of TLSMI can have contribution to the superconductivity and the gap together with the phonon-mediated interaction. We infer that the combination of the both interactions causes the higher T_c and the multi-gap structure in A15 compounds.

6.14. Cuprate Superlattice

Soon after the discovery of the high-T_c superconductivity in the high-T_c cuprates, researchers in a number of laboratories around the globe embarked upon deposition of thin films of the high-T_c cuprates, driven by the prospects of the superconducting electronics industry to emerge. The art of growth of the thin films progressed rapidly. In the subsequent years, a number of the high-T_c cuprate superlattices were synthesized; they encompassed several high-T_c compounds and a variety of "spacer" layers including insulators, semiconductors, metals, lower-T_c superconductors, and even ferromagnets [306]. The interesting experiments in this structure are as follows. Bozovic *et al.* made two single-phase films, n(Bi2212)/m(Bi2201), and three superlattices, n:m = 1:1, 1:2, and 1:5. The three superlattices showed that the onsets in resistance were all above 90 K. Thus, there was no significant decrease of T_c as the number of Bi2201 layers increased between the single-slab Bi2212 layers [307]. Bozovic *et al.* showed the resistant transition in one-unit-cell (31 Å) thick Bi2212 film, sandwiched between two Bi2201 layers; the latter were 10 unit cell (about 124 Å) thick [308]. The transition midpoint is at about 95 K. Li *et al.* [309] and Norton *et al.* [310] deposited 3($BaCuO_2$)/1($CaCuO_2$) and 2($BaCuO_2$)/1($SrCuO_2$) superlattices, respectively. The onsets in resistance are 70 - 80 K and 65 K, respectively. Notice that $SrCuO_2$ and $CaCuO_2$ are insulators, while $BaCuO_2$ is not even stable. Many works in this field are concentrated on the effect of the layer thickness on the large transition width in the curve of resistivity vs. temperature, and more than 10 mechanisms for the widening had been suggested [306].

We analyze the above experiments as follows. The experiment results in Refs. [307] and [308] leave very little room for doubt about the essentially quasi-2-D nature of the superconductivity, and thus do not favor the interlayer tunneling model in Ref. [311]. The quasi-2-D nature of the superconductivity in Refs. [307] and [308] is very obvious in our theory. From the point of view of our theory, the large transition width might come from the change of the antiferromagnetic short-range coherence length in some places of the CuO_2 plane. Although Bozovic and Eckstein claimed that one of the aims to make the high-T_c cuprate superlattice is to do new and important experiment [306], it is a pity that nobody measures the pseudogap and the pairing temperature T^*. Our theory thinks that it is obvious that the pseudogap and T^* should be observed because the superconductivity and the gap is caused by TLSMI. We feel that the $T^* > T_c$ has been observed in Ref. [307] although the authors were not conscious of this point. According to our theory for the high- T_c cuprates, T^* is independent of the layer thickness.

6.15. Borocarbides

The borocarbides, such as $LuNi_2B_2C$ and YNi_2B_2C, are a family of superconductors, with T_c ranging from 1 to 20 K [312]. A number of the borocarbides exhibit magnetic phases associated with the rare-earth moments. For example, $T_N = 15.5$ K for $TbNi_2B_2C$ [312]. The Tm, Er, Ho, and Dy nickel borocarbides show coexisting antiferromagnetism and superconductivity [313]. The NMR data shows no evidence of a Hebel-Slichter peak below T_c, unlike most s-wave superconductors [314]. The de Hass-van Alphen effect data also suggest an anisotropic gap [315]. Microwave absorption is significant at low temperature, again suggesting a finite density of states in the gap [316]. For YNi_2B_2C $2\Delta(0K) = 5.2\ T_c$ [317]. The NMR relaxation rate data in $Y(Ni_{.95}Co_{.05})B_2C$ show a simple activation behavior $1/T_1 \sim \exp(-2\Delta/T)$ with a value of $2\Delta(0K)$=2.9 $T_c \ll 5.2\ T_c$ [314].

A plausible unified explanation for all the experimental data can be given by our theory as follows. There is at least an antiferromagnetic short-range order, and thus TLSMI, in the borocarbides. The gap can have the form given by Eq. (3.65), where the isotropic gap $\Delta_1(T)$ comes from the phonon-mediated interaction, and the anisotropic gap $\Delta_2(T)F_1(\mathbf{k})F_5(\mathbf{k})$ comes from TLSMI.

6.16. C15 Laves Phases

The superconductor HfV_2 with T_c = 9.2 K possesses a Laves C15 phase structure. It has been found to have no NMR Hebel-Slichter peak in $1/T_1$ below T_c, unlike the isotropic s-wave superconductors [318]. The results of superconducting state could imply that there is a node point in the gap function [319].

At present, our theory can not give a definite explanation for the non-s-wave superconductivity in HfV_2. The Hf and V are transition metal. The d-electrons in them have Hamiltonian like Eq. (3.5) [271,272]. If there is magnetic short-range order, then there are TLSMI, and thus non-s-wave or anisotropic s-wave superconductivity.

The d-electron compound $ZrZn_2$ was investigated by Pfleiderer *at al.* [320]. Under hydrostatic pressure the superconductivity and ferromagnetism disappear at the same pressure, so the ferromagnetic state appeared to be a prerequisite for superconductivity.

At present we can not explain the superconductivity in $ZrZn_2$ quantitatively due to lack of the values of related parameters. The qualitative explanation is as follows. Long time ago Edvards introduced a duality model to describe the itinerant f and d electron, which assumed that the itinerant f and d electron system can be separated into two parts, one is mobile charges, and the other one is local spins [271,272]. The model Hamiltonian of the itinerant d electrons, given by Edvards, contains free, Kondo, and Heisenberg Hamiltonians, which look like the same as Eq. (3.5). Therefore, there is TLSMI in $ZrZn_2$. That the superconductivity is caused by TLSMI can explain that the ferromagnetism appears to be a prerequisite for superconductivity. According to section 3.2.2, it is p-wave superconductivity.

6.17. Kondo Semiconductors

Kondo semiconductors, such as CeNiSn, CeRhSb, and CeRhAs, are a class of strongly correlated f-electron materials whose low-energy excitation exhibits a pseudogap at low temperature [321, 322]. The Kondo semiconductors can be distinguished from the conventional semiconductors by the fact that the pseudogap disappears above a characteristic temperature T^*, and the pseudogap is very sensitive to any impurity [321]. These facts support the proposition that the pseudogap originates from many-body effects. A number of models have been proposed to account for the unusual pseudogap properties [323].

Here we suggest a possible model for the pseudogap. According to the Edwards duality model for the 4f-electrons of Ce [271, 272], the Hamiltonian of Ce-based Kondo semiconductors is nearly the same as Eq. (3.5). Therefore, if there is magnetic short-range order, then there is TLSMI. The appearance of the pseudogap is natural in our theory. The Ce-based Kondo semiconductors with the pseudogap are in the heavily underdoped region where $T_c = 0$. If the pseudogap is really caused by TLSMI, then after appropriate doping of electrons or holes the materials should become superconductors. We suggest to prepare weakly coupled systems, such as P-I-P junction, to detect the external radiation whose frequency ω is $2|e|V/\hbar$. Here, V is the dc bias applied on the junction, and P represents the Ce-based Kondo semiconductor. If we find the external radiation, then we can say that the pseudogap is related with the Cooper pairs without the long-range phase coherence (See sections 1.1.6 and 2.7.4.).

6.18. Colossal Magnetoresistive Oxides

The phenomenon that an extremely large negative magnetoresistance appears near the Curie temperature of manganese oxides is called colossal magnetoresistance (CMR). The colossal magnetoresistance was observed in many magnetic manganese oxides, such as $La_{1-x}Ca_xMnO_2$ [324], $La_{2/3}Ba_{1/3}MnO_x$ [325], $La_{1.2}Sr_{1.8}Mn_2O_7$, and $La_{0.82}Sr_{0.18}MnO_3$ [326]. Dessau *et al.* performed direct electronic structure measurement of the colossal magnetoresistive oxides using high energy resolution spectroscopy [326, 327]. A key result of these measurements is the discovery of a pseudogap at the Fermi surface below as well as above the magnetic order temperature. There are many possible conjectures on the origin of the pseudogap [326, 327].

We note that the resistance goes down when the temperature goes down from the magnetic order temperature [324, 325]. We guess that the pseudogap belongs to the Cooper pair without the long-range phase coherence. We infer that there is the long-range phase coherence between the Cooper pairs in some local places, which leads to the superconductivity in these local places, and thus leads to the decrease of the whole resistance. We suggest to make weakly coupled systems, such as P-I-P junction, to detect the external radiation whose frequency ω is $2|e|V/\hbar$. Here, V is the dc bias applied on the junction, and P represents the colossal magnetoresistive oxide. If we find the external radiation, then we can say that the pseudogap is related with the Cooper pairs without the long-range phase coherence (See sections 1.1.6 and 2.7.4.). It is very interesting to note that Mannella *et al.* reported experimental evidence that a very similar pseudogap state with a nodal-antinodal dichotomous character exists in the ferromagnetic metallic groundstate of the colossal magnetoresistive bilayer manganite $La_{1.2}Sr_{1.8}Mn_2O_7$ [190]. Mannella *et al.*'s findings might imply that the pseudogap in $La_{1.2}Sr_{1.8}Mn_2O_7$ comes from Cooper pairs without the long-range phase coherence as well as the pseudogap in the high-T_c cuprates.

6.19. $(TaSe_4)I$ and $K_{0.3}MoO_3$

$(TaSe_4)I$ and $K_{0.3}MoO_3$ are quasi-one-dimensional inorganic systems. The pseudogap was also observed in these systems [328, 329]. Joynt thought that the origin of the pseudogap in poorly conducting solids is among the most fundamental problems of present-day condensed matter physics, and proposed a model for it [330].

According to the Edwards duality model for the d electrons of Ta and Mo [271, 272], the Hamiltonian of $(TaSe_4)I$ and $K_{0.3}MoO_3$ are the same as Eq. (3.5). Therefore, if there is short-range magnetic order, then there is TLSMI. The analysis for the origin of the pseudogap in our theory is the same as for Ce-based Kondo semiconductors in section 6.17. We suggest to make weakly coupled systems, such as P-I-P junction, to detect the external radiation whose frequency ω is $2|e|V/\hbar$. Here, V is the dc bias applied on the junction, and P represents $(TaSe_4)I$ or $K_{0.3}MoO_3$. If we find the external radiation, then we can say that the pseudogap is related with the Cooper pairs without the long-range phase coherence (See section 1.1.6 and 2.7.4.).

6.20. UGe_2

UGe_2 is known to be qualitatively similar to the classical d-electron ferromagnet. The superconductivity observed below 1 K, in a limited pressure range, seems to arise from the same electrons that produce band magnetism, and is of a spin-triplet pairing [331, 332].

The superconductivity in UGe_2 has only been detected in ferromagnetic phase, and does not been observed at pressure above the critical pressure for the suppression of ferromagnetism [332]. The superconductivity in UGe_2 can be qualitatively explained as the same as that for $ZrZn_2$ in section 6.16.

6.21. Superfluidity of ^{3}He

^{3}He has a high zero-point energy which ensures that in the absence of applied pressure it remains liquid at all temperatures. ^{3}He atom is a fermion. By the time the first experimental results appeared, it was already thought likely on theoretical grounds that ^{3}He would undergo a phase transition to a BCS-like Cooper pairing state, the Cooper pair being p-wave and spin triplet. That is, the symmetry of the Cooper pair spatial wavefunction would be that of an atomic p orbital, with odd parity, and the spin wavefunction would be symmetric. Anderson and Morel had proposed an "equal spin pairing" model in which the spin part of wavefunction of Cooper pair is $|\uparrow\uparrow>$ or $|\downarrow\downarrow>$ [333]. Balian and Werthamer pointed out that a full treatment of symmetric spin, or triplet, pairing should be include the combination $|\uparrow\uparrow> + |\downarrow\downarrow>$ [334]. Furthermore, they were able to exhibit that the state (BW state) which with the simplest interparticle interaction had a lower energy than the Anderson and Morel (AM) state. NMR data strongly suggested that there are two phases, one as an AM state (the A phase) and the other as a BW state (the B phase). The phase diagram of liquid ^{3}He can be obtained from Ref. [335]. For example, the transition temperature of superfluidity is $T_c = 2.58$ mK at 33.41 bar for liquid ^{3}He-A.

Legget pointed out that despite the strong interatomic potential, many of the properties of liquid ^{3}He at low temperatures resemble those of a weakly interacting Fermi gas [335, 336]. According to Landau Fermi liquid theory when $T < T_L = 100$ mK, the ^{3}He atom can be described by quasiparticle, rather than real particle [336]. The actual quantitative calculations of the Cooper pairing potential from first principle, that is from a knowledge of the interatomic forces, is an extremely formidable task, and the resulting estimates of T_c have been scattered among many orders of magnitude [336]. The mechanism of superfluidity in liquid ^{3}He is still open.

Ref. [335] pointed out that the liquid ^{3}He is of near ferromagnetism. There is strongly ferromagnetic fluctuation in nearly ferromagnetic materials [337]. The strongly ferromagnetic fluctuation can cause the attractive interaction, TLSMI in chapter 3, between two ^{3}He atoms with $\mathbf{k}$ and $-\mathbf{k}$ in spin triplet states. We guess that TLSMI causes the superfluidity in liquid ^{3}He.

6.22. Superfluidity of ^{4}He

Liquid ^{4}He in superfluidity state (He II), for example, liquid ^{4}He at $T < 2.17$ K and zero pressure, behaves as it were a mixture of superfluid and normal fluids freely intermingling without any viscous interaction. The superfluid component has weight ρ_s, zero entropy, flows without viscosity, and participates in macroscopic quantum effects such as quantized vortices [338]. The normal fluid has weight $\rho_n = 1 - \rho_s$, and does not exhibit these properties. It is widely accepted that two fluid behavior in He II is linked to the presence of Bose-Einstein condensation, i.e., macroscopic large number of particles occupy a single state [339]. The existing microscopic theory of He II is in a much less satisfactory state than that of superconductor [338, 339].

The form of many-body wavefunction of ^{4}He in superfluidity state still is a focus of explanation for the superfluidity of ^{4}He [339]. We will propose a form of wavefunction. According to the physical picture in sections 2.8 and 3.3 the high-T_c superconductivity in cuprates comes from the Bose-Einstein condensation of Cooper pairs (bosons) and the long-range phase coherence. The size of Cooper pair is only about 25 $\AA$. At first at pairing temperature T^* there is the Bose-Einstein condensation of Cooper pairs without the long-range phase coherence. Then at superconducting transition temperature T_c ($< T^*$), where the Josephson coupling energy overcomes both the zero-point energy and thermal disturbance energy, there is the long-range phase coherence. The above physical picture can be used to liquid ^{4}He. ^{4}He atom is a boson. At first at T_B there is the Bose-Einstein condensation of ^{4}He atoms. Then at transition temperature of superfluidity T_c there is the long-range phase coherence between ^{4}He atoms in superfluid component. Because there is only one specific heat jump at T_c [338], it can be concluded that T_B is nearly equal to T_c. Referring to Eq. (2.322), we can write the wavefunction of superfluid component. Assume the He II consists of M ^{4}He atoms, and N ($< M$) ^{4}He atoms belong to the superfluid component. Then the wavefunction of the superfluid component is

$$\psi_{oN}(\mathbf{r}_1, \cdots, \mathbf{r}_N) = \phi_1(\mathbf{r}_1)\exp(i\phi)\exp(i\mathbf{q}\cdot\mathbf{R}_1 - \mu t/\hbar)\cdots$$

$$\phi_N(\mathbf{r}_N)\exp(i\phi)\exp(i\mathbf{q}\cdot\mathbf{R}_N - \mu t/\hbar), \tag{6.3}$$

where $\mathbf{R}_i$ is the position vector of mass-center of i-th ^{4}He atom, $\mathbf{q}$ is the common wave vector of N ^{4}He atoms, ϕ is the common phase of N ^{4}He atoms in the superfluid component of static He II, μ is the chemical potential, and $\phi_i(\mathbf{r}_i)$ represents an internal eigenstate of i-th ^{4}He atom.

Chapter 7

Extended Abrikosov Pseudo-Fermion Method

7.1. Introduction

It is a very difficult task to deal with local spin operator because it satisfies neither the boson nor the fermion commutation rules. There are already many methods to treat local spin operator [340]– [349]. Abrikosov introduced a pseudo-fermion representation for local spin operator in Kondo problem [340]. It was however pointed out by Keiter that in Abrikosov representation the linked cluster expansion theorem of ordinary many-body field theory can not be used [350]. Moreover, as pointed out by Larsen [341], the infinite pseudo-fermion energy in Abrikosov method has no physical meaning and complicates calculations. Therefore, the Larsen's purpose is to show how one may use zero pseudo-fermion energy still preserving the usual linked cluster expansion theorem [341]. However, Larsen work is still confined in Kondo problem, and does not give an obvious expression for the transformation factor. The many other methods are discussed in section 7.6.

This chapter is to extend Abrikosov pseudo-Fermion method to deal with local spin lattice and coexistence of both Kondo Hamiltonian and Heisenberg Hamiltonian, and gives an obvious transformation factor, which is the advanced preparation in method for our theory on the superconductivity and the pseudogap in chapter 3.

7.2. Abrikosov Pseudo-Fermion Representation

The local spin operator $\widehat{\mathbf{S}}_i$ at site i is represented in Abrikosov representation as [340]

$$\widehat{\mathbf{S}}_i = \sum_{\alpha\beta} \mathbf{S}_{\alpha\beta} d_{i\alpha}^{+} d_{i\beta}, \tag{7.1}$$

where $\mathbf{S}$ is spin matrix vector, $d_{i\alpha}^{+}$ and $d_{i\alpha}$ are creation and annihilation operator of quasi-particle with spin α at site i. The quasiparticle is called pseudo-fermion. $d_{i\alpha}^{+}$ and $d_{i\alpha}$ satisfy the following anticommutation rules.

$$[d_{i\alpha}^{+}, d_{j\beta}]_{+} = \delta_{ij}\delta_{\alpha\beta}, \tag{7.2}$$

$$[d^+_{i\alpha}, d^+_{j\beta}]_+ = [d_{i\alpha}, d_{j\beta}]_+ = 0. \tag{7.3}$$

Let the local spin magnitude be 1/2. The extension to high local spin will be given in section 7.5. The pseudo-fermion state at site i can be written as

$$|n>_i = |n\uparrow n\downarrow>_i . \tag{7.4}$$

The pseudo-fermion number of $\uparrow$ spin $n\uparrow = 0$ or 1. $n\downarrow = 0$ or 1. The pseudo-fermion states $|01>_i$ and $|10>_i$ are called physical spin states. The pseudo-fermion states $|00>_i$ and $|11>_i$ are called false spin states because we deal with the spin states which always have one and only one local spin with magnitudes of 1/2 or $-1/2$ at site i. The eigenstates of the pseudo-fermion operators at local spin site i are $|00>_i$, $|01>_i$, $|10>_i$, and $|11>_i$.

We assume that the form of Hamiltonian containing local spin operators is

$$H = H_0 + H_{Kondo} + H_{Heisenberg}, \tag{7.5}$$

$$H_0 = \sum_{\mathbf{k}\alpha} \varepsilon_{\mathbf{k}} c^+_{\mathbf{k}\alpha} c_{\mathbf{k}\alpha}, \tag{7.6}$$

$$H_{Kondo} = \frac{J_K}{N_c} \sum_{i\mathbf{k}\mathbf{k}'\alpha\beta} \vec{\sigma}_{\alpha\beta} \cdot \widehat{\mathbf{S}}_i c^+_{\mathbf{k}\alpha} c_{\mathbf{k}'\beta} e^{i(\mathbf{k}'-\mathbf{k})\cdot\mathbf{R}_i}, \tag{7.7}$$

$$H_{Heisenberg} = J \sum_{i<j} \widehat{\mathbf{S}}_i \cdot \widehat{\mathbf{S}}_j, \tag{7.8}$$

where H_0 is the free carrier Hamiltonian, $\varepsilon_{\mathbf{k}}$ the dispersion relation, $\vec{\sigma}$ Pauli matrix vector, N_c the total cell number in a lattice. In Eq. (7.8) only the nearest-neighbor interaction is considered.

The statistical operators of non-interacting and interacting system are, respectively,

$$\rho_0 = e^{-\frac{(H_0-\mu N)}{T}}, \tag{7.9}$$

$$\rho = e^{-\frac{(H-\mu N)}{T}}, \tag{7.10}$$

where μ is chemical potential, N carrier number operator. The full average of an operator R on $|A>$ and both the physical and the false spin states is defined as

$$<R>^f = \frac{Tr^f(\rho R)}{Tr^f(\rho)}, \tag{7.11}$$

$$<R>^f_0 = \frac{Tr^f(\rho_0 R)}{Tr^f(\rho_0)}, \tag{7.12}$$

$$Tr^f(\cdots) = Tr_A Tr^{(1)(2)\cdots}(\cdots), \tag{7.13}$$

$$Tr_A(\cdots) = \sum_{\{A\}} <A|\cdots|A>, \tag{7.14}$$

$$Tr^{(1)(2)\cdots}(\cdots) = \sum_{\{n\}} <n|\cdots|n>, \tag{7.15}$$

where $|A>$ is the eigenstate of H_0, $|n>$ the eigenstate of the pseudo-fermion operators at all local spin sites enumerated by the numbers in the left side of Eq. (7.15). The superscripts (1), (2), ... represent all sites of the local spins. Note that $|n>$ contains both the physical and the false spin states.

$$|n>=|n>_1|n>_2\cdots, \tag{7.16}$$

$$\begin{aligned}Tr^{(1)}(\cdots) = &\ {}_1<00|\cdots|00>_1 + {}_1<11|\cdots|11>_1 \\ &+{}_1<10|\cdots|10>_1 + {}_1<01|\cdots|01>_1,\end{aligned} \tag{7.17}$$

$$Tr^{(1)'}(\cdots) = {}_1<01|\cdots|01>_1 + {}_1<10|\cdots|10>_1, \tag{7.18}$$

$$Tr^{(1)''}(\cdots) = {}_1<00|\cdots|00>_1 + {}_1<11|\cdots|11>_1, \tag{7.19}$$

where we use single prime to represent the physical spin states, and the double prime the false spin states. The physical average of R, $<R>$, should be taken on $|A>$ and the physical spin states.

$$<R>=\frac{Tr(\rho R)}{Tr(\rho)}, \tag{7.20}$$

$$<R>_0=\frac{Tr(\rho_0 R)}{Tr(\rho_0)}, \tag{7.21}$$

$$Tr(\cdots) = \sum_{\{A\}} <A|\sum_{\{n'\}} <n'|\cdots|n'>|A>. \tag{7.22}$$

Because the physical spin state $|n'>$ is only a subspace of the pseudo-fermion operator, we can not use the linked cluster expansion theorem in the calculation of $<R>$.

7.3. Transformation from Full to Physical Average

In this section we show

$$<R>-<R>_0= w(J/T)(<R>^f-<R>_0^f), \tag{7.23}$$

and obtain the expression of $w(J/T)$. Assuming R is independent of the local spin operator, we obtain from definitions that $<R>_0=<R>_0^f$. $<R>$ can be rewritten formally as

$$<R>=\frac{Tr(\rho R)}{Tr^f(\rho R)}\frac{Tr^f(\rho)}{Tr(\rho)}<R>^f. \tag{7.24}$$

All local spin sites are equivalent. At any local spin site there are the same scattering between carrier and local spin and the same coupling between local spins. Let us choose a concrete local spin site such as 1. We study how to eliminate the effect of the false local spin states on statistical average. Green function may also be regarded as statistical average of certain operator. From definitions we have

$$Tr^f(\rho) = Tr_A Tr^{(2)(3)\cdots}Tr^{(1)''}(\rho)+Tr_A Tr^{(2)(3)\cdots}Tr^{(1)'}(\rho). \tag{7.25}$$

Our aim is to use the pseudo-fermion representation for the local spin operator to study scattering between carrier and local spin in magnetic ordering state. In magnetic ordering state single spin flip is prohibited due to the energy minimum principle. Therefore, we should only consider the z-component of local spin operator in Hamiltonians of Eqs. (7.7) and (7.8). In generic cases considering the x- and y-components as well as the z-component, our Eqs. (7.41) and (7.66) are approximate, which will be explained at the end of this chapter. The operations of $\sum_{\alpha\beta} S^{\gamma}_{\alpha\beta} d^{+}_{i\alpha} d_{i\beta}$ with $\gamma = x, y, z$ on $|00>_i$ or $|11>_i$ are zero, which can be verified by direct calculations. From the above considerations, assuming $H_{Kondo} = J_K/N_c \sum_{\mathbf{kk}'\alpha\beta} S^z_{\beta\beta} \sigma^z_{\alpha\alpha} d^{+}_{1\beta} d_{1\beta} c^{+}_{\mathbf{k}\alpha} c_{\mathbf{k}'\alpha} \exp[i(\mathbf{k}' - \mathbf{k}) \cdot R_1]$, we can rewrite Eq. (7.25) as

$$Tr^f(\rho) = Tr_A(\rho_0) 2 f_1 + Tr_A(\rho_0') f_2, \tag{7.26}$$

$$f_1 = Tr^{(2)(3)\cdots}\left(e^{-JX}\right), \tag{7.27}$$

$$X = \frac{1}{T} \sum_{2 \le i < j \alpha\beta} S^z_{\alpha\alpha} S^z_{\beta\beta} d^{+}_{i\alpha} d_{i\alpha} d^{+}_{j\beta} d_{j\beta}, \tag{7.28}$$

where the double subscript means spin summation.

$$f_2 = Tr^{(2)(3)\cdots}\left(e^{-JY}\right), \tag{7.29}$$

$$Y = X + \frac{1}{2T} \sum_{j\alpha}{}' S^z_{\alpha\alpha} d^{+}_{j\alpha} d_{j\alpha}, \tag{7.30}$$

where the prime means that summation of j only over the nearest neighbors of site 1.

$$\rho_0' = e^{-\frac{H_0 - \mu N + E}{T}} + e^{-\frac{H_0 - \mu N - E}{T}}, \tag{7.31}$$

$$E = \frac{J_K}{2N_c} \sum_{\mathbf{kk}'\alpha} \sigma^z_{\alpha\alpha} c^{+}_{\mathbf{k}\alpha} c_{\mathbf{k}'\alpha} e^{-i(\mathbf{k} - \mathbf{k}') \cdot \mathbf{R}_1}, \tag{7.32}$$

$$Tr(\rho) = Tr_A(\rho_0') f_3, \tag{7.33}$$

$$f_3 = Tr^{(2)'(3)'\cdots}\left(e^{-JY}\right). \tag{7.34}$$

After some basic derivations and taking $N_c \to \infty$, we have

$$\frac{Tr_A(\rho_0')}{Tr_A(\rho_0)} = 2. \tag{7.35}$$

If we assume $H_{Kondo} = J_K/N_c \sum_{\mathbf{kk}'\alpha\beta j} S^z_{\beta\beta} \sigma^z_{\alpha\alpha} d^{+}_{j\beta} d_{j\beta} c^{+}_{\mathbf{k}\alpha} c_{\mathbf{k}'\alpha} \exp[i(\mathbf{k}' - \mathbf{k}) \cdot \mathbf{R}_j]$, where $\sum_j$ represents summation over finite number of local spins, then we can also obtain Eq. (7.35).

From Eqs. (7.26), (7.33), and (7.35) we have

$$\frac{Tr^f(\rho)}{Tr(\rho)} = \frac{f_1 + f_2}{f_3}. \tag{7.36}$$

Considering that operator R is independent of the local spin operators, by analogy with Eqs. (7.26) and (7.33) we obtain

$$Tr^f(\rho R) = Tr_A(\rho_0 R) 2 f_1 + Tr_A(\rho_0' R) f_2, \tag{7.37}$$

$$Tr(\rho R) = Tr_A(\rho_0' R) f_3. \tag{7.38}$$

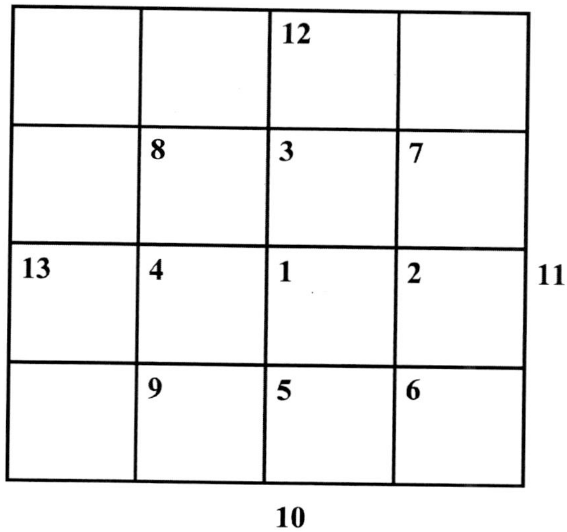

Figure 7.1. A numbered square lattice. The crossing points of two lines are the sites of the local spins.

From Eqs. (7.37), (7.26), and (7.35) we have

$$\frac{Tr_A(\rho_0' R)}{Tr_A(\rho_0)} = \frac{< R >^f (2f_1 + 2f_2) - < R >_0^f 2f_1}{f_2}. \tag{7.39}$$

From Eqs. (7.37) and (7.38) we have

$$\frac{Tr(\rho R)}{Tr^f(\rho R)} = \frac{f_3 \frac{Tr_A(\rho_0' R)}{Tr_A(\rho_0)}}{2f_1 < R >_0^f + f_2 \frac{Tr_A(\rho_0' R)}{Tr_A(\rho_0)}}. \tag{7.40}$$

Substituting Eqs. (7.40), (7.39), and (7.36) into Eq. (7.24), and considering $< R >_0^f = < R >_0$, we have

$$< R > - < R >_0 = \left(1 + \frac{f_1}{f_2}\right)(< R >^f - < R >_0^f). \tag{7.41}$$

f_1 and f_2 are dependent on the distribution of local spins. We take a square lattice as an example to calculate f_1 and f_2. Number the local spins as shown in Fig. 7.1. Using third neighbor approximation of site 1, we obtain

$$f_1 = \left\{ Tr^{(2)(6)(11)} \left[e^{-\frac{J}{T}\left[S^z_{\alpha\alpha} d^+_{2\alpha} d_{2\alpha} \left(S^z_{6\beta\beta} d^+_{6\beta} d_{6\beta} + S^z_{\beta\beta} d^+_{11\beta} d_{11\beta}\right)\right]} \right] \right\}^4, \tag{7.42}$$

$$f_2 = \left\{ Tr^{(2)(6)(11)} \left[e^{-\frac{J}{T}\left[S^z_{\alpha\alpha} d^+_{2\alpha} d_{2\alpha} \left(S^z_{\beta\beta} d^+_{6\beta} d_{6\beta} + S^z_{\beta\beta} d^+_{11\beta} d_{11\beta} + \frac{1}{2}\right)\right]} \right] \right\}^4. \tag{7.43}$$

Considering the definition of the trace for the pseudo-fermion operators in Eqs. (7.15), (7.16), and (7.17), and completing the trace operations in Eqs. (7.42) and (7.43), we have

$$f_1 = \left[32 + 2\left(2 + e^{-\frac{J}{4T}} + e^{\frac{J}{4T}}\right)^2\right]^4, \tag{7.44}$$

$$f_2 = \left[32 + \left(e^{-\frac{J}{4T}} + e^{\frac{J}{4T}}\right)\left(2 + e^{-\frac{J}{4T}} + e^{\frac{J}{4T}}\right)^2\right]^4. \tag{7.45}$$

Substituting Eqs. (7.44) and (7.45) into Eq. (7.41), we obtain $< R > - < R >_0$ from $< R >^f - < R >_0^f$.

It is easy to show that f_1 and f_2 in the first and second neighbor approximation are, respectively,

$$f_1 = 4^4, \tag{7.46}$$

$$f_2 = \left(2 + e^{-\frac{J}{4T}} + e^{\frac{J}{4T}}\right)^4, \tag{7.47}$$

and

$$f_1 = \left[8 + 2\left(2 + e^{-\frac{J}{4t}} + e^{\frac{J}{4T}}\right)\right]^4, \tag{7.48}$$

$$f_2 = \left[8 + \left(e^{-\frac{J}{4T}} + e^{\frac{J}{4T}}\right)\left(2 + e^{-\frac{J}{4T}} + e^{\frac{J}{4T}}\right)\right]^4. \tag{7.49}$$

The third neighbor approximation for the transformation factor in Eq. (7.41) is enough for application. The maximum variation range of f_1/f_2 is from 0 to 1. If there is no Heisenberg Hamiltonian, i.e., a pure Kondo problem, then $J = 0$ and $f_1/f_2 = 1$. That the transformation factor of Eq. (7.41) in pure Kondo problem is 2 was obtained only approximately by Cheung and Mattuck in 1968 [348]. The transformation factor of Eq. (7.41) is an open question before this book [89].

If the local spins constitute a 3-D tetragonal lattice, then under the third nearest-neighbor approximation the f_1 and f_2 become, respectively,

$$f_1' = \left[2 \times 4^5 + 2\left(2 + e^{-\frac{J}{4t}} + e^{\frac{J}{4T}}\right)^5\right]^6, \tag{7.50}$$

$$f_2' = \left[2 \times 4^5 + \left(e^{-\frac{J}{4t}} + e^{\frac{J}{4T}}\right)\left(2 + e^{-\frac{J}{4t}} + e^{\frac{J}{4T}}\right)^5\right]^6. \tag{7.51}$$

If the local spins constitute a chain, then under the third nearest-neighbor approximation the f_1 and f_2 become, respectively,

$$f_1'' = \left[2 \times 4 + 2\left(2 + e^{-\frac{J}{4t}} + e^{\frac{J}{4T}}\right)\right]^2, \tag{7.52}$$

$$f_2'' = \left[2 \times 4 + \left(e^{-\frac{J}{4t}} + e^{\frac{J}{4T}}\right)\left(2 + e^{-\frac{J}{4t}} + e^{\frac{J}{4T}}\right)\right]^2. \tag{7.53}$$

7.4. Transformation from Full to Physical Green Function

The Green function G is the average of the following operators [341].

$$D = T_\tau \left[e^{\tau(H-\mu N)} c_{\mathbf{k}\uparrow} e^{-\tau(H-\mu N)} c^+_{\mathbf{k}'\uparrow} \right], \tag{7.54}$$

$$D_0 = T_\tau \left[e^{\tau(H_0-\mu N)} c_{\mathbf{k}\uparrow} e^{-\tau(H_0-\mu N)} c^+_{\mathbf{k}'\uparrow} \right], \tag{7.55}$$

where T_τ is the imaginary-time-ordering operator. The D and D_0 contain three new symbols, i.e., T_τ, $\exp[\tau(H-\mu N)]$, and $\exp[\tau(H_0-\mu N)]$, in comparison with R. The physical Green functions are

$$G =< D >, \tag{7.56}$$

$$G_0 =< D_0 >_0 . \tag{7.57}$$

The physical Green function means the average of operators in Eqs. (7.54) and (7.55) on $|A>$ and physical spin states.

The full Green functions are

$$G^f =< D >^f, \tag{7.58}$$

$$G^f_0 =< D_0 >^f_0 . \tag{7.59}$$

The full Green function means the full average of operators in Eqs. (7.54) and (7.55) on $|A>$ and on both the physical spin states and the false spin states.

The final equation, Eq. (7.66), is not dependent on the sign of τ. Take $\tau > 0$. Considering that $c_{\mathbf{k}\uparrow}$ and the local spin operator in Eq. (7.54) are commutative, we obtain

$$Tr^f(\rho D) = Tr_A(\rho_0 D_0) 2f_1 + Tr_A\left(\rho_0''\right) f_2, \tag{7.60}$$

$$\rho_0'' = e^{-\frac{(H_0-\mu N+E)}{T}} e^{\tau(H_0-\mu N+E)} c_{\mathbf{k}\uparrow} e^{-\tau(H_0-\mu N+E)} c^+_{\mathbf{k}'\uparrow}$$
$$+ e^{-\frac{(H_0-\mu N-E)}{T}} e^{\tau(H_0-\mu N-E)} c_{\mathbf{k}\uparrow} e^{-\tau(H_0-\mu N-E)} c^+_{\mathbf{k}'\uparrow}, \tag{7.61}$$

$$Tr(\rho D) = Tr_A\left(\rho_0''\right) f_3. \tag{7.62}$$

From Eqs. (7.60) and (7.26) we have

$$\frac{Tr_A(\rho_0'')}{Tr_A(\rho_0)} = \frac{< D >^f 2(f_1+f_2) - < D_0 >_0 2f_1}{f_2}. \tag{7.63}$$

From Eqs. (7.62), (7.60), and (7.63) we have

$$\frac{Tr(\rho D)}{Tr^f(\rho D)} = \frac{\frac{f_3}{f_2}\left[< D >^f 2(f_1+f_2) - < D_0 >_0 2f_1\right]}{2 < D >^f (f_1+f_2)}. \tag{7.64}$$

From the definitions of $< D >$ and $< D >^f$, we have

$$< D >= \frac{Tr(\rho D)}{Tr^f(\rho D)} \frac{Tr^f(\rho)}{Tr(\rho)} < D >^f . \tag{7.65}$$

From Eqs. (7.64), (7.65), and (7.36) we have

$$G-G_0=\left(1+\frac{f_1}{f_2}\right)\left(G^f-G_0^f\right). \tag{7.66}$$

This is a very useful formula because one can easily obtain the full Green function G^f using the linked cluster expansion theorem, and many physical quantities can only be obtained in terms of G.

7.5. High Local Spin

Assume that the high local spin at site (1) is $j=S+1/2$, which has $(2j+1)$ z-components with $m=S+1/2,\cdots,-(S+1/2)$. We study at first the decomposition of high local spin at site (1) into two low local spins with $j_1=S$ and $j_2=1/2$, whose z-components are denoted by m_1 and m_2, respectively. $m_1=S,\cdots,-S$. $m_2=1/2,-1/2$. The state vector $|jm>$ can be expanded as

$$|jm>=\sum_{m_1m_2}<m_1m_2|jm>|m_1m_2>. \tag{7.67}$$

The trace of the operator D in Eq. (7.54), taking on the physical spin states of the local spin operator with spin $S+1/2$ at site (1), is

$$Tr^{(1)'}_{S+\frac{1}{2}}(D)=\sum_m<S+\frac{1}{2}m|D|S+\frac{1}{2}m>. \tag{7.68}$$

Substituting Eq. (7.67) into Eq. (7.68) gives

$$Tr^{(1)'}_{S+\frac{1}{2}}(D)=\sum_{m_1m_2m'_1m'_2}<m'_1m'_2|\sum_m|S+\frac{1}{2}m>$$

$$<S+\frac{1}{2}m|m_1m_2><m_1m_2|D|m'_1m'_2>. \tag{7.69}$$

The matrix element of D in $|m_1m_2>$ is diagonal if we consider only the z-component of the local spin operator, which is just the case discussed in our theory on the superconductivity and the pseudogap. Therefore, the Eq. (7.69) becomes

$$Tr^{(1)'}_{S+\frac{1}{2}}(D)=\sum_{m_1m_2}<m_1m_2|\sum_m|S+\frac{1}{2}m>$$

$$<S+\frac{1}{2}m|m_1m_2><m_1m_2|D|m_1m_2>. \tag{7.70}$$

It is easy to verify that for diagonal matrix element

$$\sum_m|S+\frac{1}{2}m><S+\frac{1}{2}m|=\left(\alpha_{S\frac{1}{2}}\widehat{I}+\beta_{S\frac{1}{2}}\widehat{S}^z_S\widehat{S}^z_{\frac{1}{2}}\right), \tag{7.71}$$

$$\alpha_{S\frac{1}{2}}=\frac{2S+2}{2(2S+1)}, \tag{7.72}$$

$$\beta_{S\frac{1}{2}} = \frac{4}{2(2S+1)}, \tag{7.73}$$

where $\widehat{I}$ is a unit operator. Substituting Eq. (7.71) into Eq. (7.70) gives

$$Tr^{(1)'}_{S+\frac{1}{2}}(D) = \alpha_{S\frac{1}{2}} Tr^{(1)'}_{(S\frac{1}{2})}(D) + \beta_{S\frac{1}{2}} Tr^{(1)'}_{(S\frac{1}{2})}(\widehat{S^z_S}\widehat{S^z_{\frac{1}{2}}}D), \tag{7.74}$$

where

$$Tr^{(1)'}_{(S\frac{1}{2})}(\cdots) = \sum_{m_1 m_2} < m_1 m_2|\ldots|m_1 m_2 > .$$

The statistical average of D on $|A>$ and all physical spin states $|S+1/2m>_{(1)}\ |S+1/2m>_{(2)}\cdots$ with $m = S+1/2,\cdots,-(S+1/2)$ is

$$G_{(S+\frac{1}{2})} = < D >_{S+\frac{1}{2}}$$

$$= \frac{Tr_A Tr^{(1)'(2)'\cdots}_{S+\frac{1}{2}}(\rho D)}{Tr_A Tr^{(1)'(2)'\cdots}_{S+\frac{1}{2}}(\rho)}. \tag{7.75}$$

The statistical average of D on $|A>$ and all $|m_1 m_2>_{(1)}\ |m_1 m_2)_{(2)}\cdots$ is defined as

$$G_{(S\frac{1}{2})} = < D >_{(S\frac{1}{2})}$$

$$= \frac{Tr_A Tr^{(1)'(2)'\cdots}_{(S\frac{1}{2})}(\rho D)}{Tr_A Tr^{(1)'(2)'\cdots}_{(S\frac{1}{2})}(\rho)}. \tag{7.76}$$

Substituting Eq. (7.74) into Eq. (7.75) leads to

$$G_{(S+\frac{1}{2})} = \frac{\alpha_{S\frac{1}{2}} < D >_{(S\frac{1}{2})} + \beta_{S\frac{1}{2}} < \widehat{S^z_S}\widehat{S^z_{\frac{1}{2}}}D >_{(S\frac{1}{2})}}{\alpha_{S\frac{1}{2}} + \beta_{S\frac{1}{2}} < \widehat{S^z_S}\widehat{S^z_{\frac{1}{2}}} >_{(S\frac{1}{2})}}. \tag{7.77}$$

It can be shown that

$$< \widehat{S^z_S}\widehat{S^z_{\frac{1}{2}}}D >_{(S\frac{1}{2})} = < \widehat{S^z_S}\widehat{S^z_{\frac{1}{2}}} >_{(S\frac{1}{2})} < D >_{(S\frac{1}{2})}. \tag{7.78}$$

Substituting Eq. (7.78) into Eq. (7.77) gives

$$G_{(S+\frac{1}{2})} = G_{(S\frac{1}{2})}. \tag{7.79}$$

Using Eq. (7.79) repeatedly leads to

$$G_{(S+\frac{1}{2})} = G_{(\frac{1}{2}\frac{1}{2}\cdots\frac{1}{2})}. \tag{7.80}$$

The physical meaning of Eq. (7.80) is that if we treat a local spin operators with spin $S+1/2$ and consider only their z-components in the Green function method, then the local spin operators with spin $S+1/2$ can be decomposed into summation of a series of local spin operators with spin 1/2, i.e., for $S+1/2 = n/2$

$$\widehat{S^z_i} = \widehat{S^z_{i_1}} + \widehat{S^z_{i_2}} + \cdots + \widehat{S^z_{i_n}}. \tag{7.81}$$

In Eq. (7.81) we use two subscripts to indicate the site of the high local spin and the number of the low local spins.

In the following we will show that

$$G_{(S+\frac{1}{2})}-G_{(S+\frac{1}{2})0}=G_{(\frac{1}{2}\cdots\frac{1}{2})}-G_{(\frac{1}{2}\cdots\frac{1}{2})0}=\left(1+\frac{F_1}{F_2}\right)\left[G^f_{(\frac{1}{2}\cdots\frac{1}{2})}-G^f_{(\frac{1}{2}\cdots\frac{1}{2})0}\right]. \tag{7.82}$$

Considering the decomposition in Eq. (7.81), Eq. (7.25) becomes

$$Tr^f(\rho)=Tr_A Tr^{(2)(3)\cdots}\left[Tr_1^{(1)''}+Tr_2^{(1)''}+\cdots+Tr_n^{(1)''}\right]\left(e^{-\frac{H_0-\mu N+H_{Kondo}+H_{Heisenberg}}{T}}\right)+Tr_A Tr^{(2)(3)\cdots}\left[Tr^{(1)'}\right]\left(e^{-\frac{H_0-\mu N+H_{Kondo}+H_{Heisenberg}}{T}}\right), \tag{7.83}$$

where $Tr_m^{(1)"}$ means that the trace is taken over the false spin states of m ($m \le n$) low local spin operators and over the physical spin states of (n-m) low local spin operators at site (1). Note that even only one eigenstate of one low local spin operator in $\widehat{S^z_{i_1}}+\widehat{S^z_{i_2}}+\cdots+\widehat{S^z_{i_n}}$ is the false spin state, the total eigenstate of all low local spin operators $\widehat{S^z_{i_1}}+\widehat{S^z_{i_2}}+\cdots+\widehat{S^z_{i_n}}$ is false spin state. Eq. (7.83) can be rewritten as

$$Tr^f(\rho)=2^n Tr_A(\rho_0)\left(F_1+F_2\right), \tag{7.84}$$

$$F_1=\sum_{m=1}^{n}\frac{n(n-1)\ldots[n-(m-1)]}{m!}Tr^{(2)(3)\cdots}e^{-\frac{J(n-m)}{2T}\sum_{j=2}^{5}\widehat{S^z_j}-\frac{J}{T}\sum_{2\le i<j}\widehat{S^z_i}\widehat{S^z_j}}, \tag{7.85}$$

$$F_2=Tr^{(2)(3)\cdots}e^{-\frac{Jn}{2T}\sum_{j=2}^{5}\widehat{S^z_j}-\frac{J}{T}\sum_{2\le i<j}\widehat{S^z_i}\widehat{S^z_j}}, \tag{7.86}$$

where $\{n(n-1)\ldots[n-(m-1)]\}/m!$ is the combination number of taking m low local spin operators from n low local spin operators, and $\sum_{j=2}^{5}$ comes from Fig. 7.1. We can also show that

$$Tr(\rho)=2^n Tr_A(\rho_0)F_3, \tag{7.87}$$

$$F_3=Tr^{(2)'(3)'\cdots}e^{-\frac{Jn}{2T}\sum_{j=2}^{5}\widehat{S^z_j}-\frac{J}{T}\sum_{2\le i<j}\widehat{S^z_i}\widehat{S^z_j}}. \tag{7.88}$$

From the similar derivations as for Eq. (7.60) we have

$$Tr^f(\rho D)=Tr_A(\rho_0 D_0)2^n F_1+Tr_A(\rho_0'')F_2. \tag{7.89}$$

From the similar derivations as for Eq. (7.62) we have

$$Tr(\rho D)=Tr_A(\rho_0'')F_3. \tag{7.90}$$

From Eqs. (7.89) and (7.84) we have

$$\frac{Tr_A(\rho_0'')}{Tr_A(\rho_0)}=\frac{1}{F_2}\left[<D>^f 2^n(F_1+F_2)-2^n<D_0>_0 F_1\right]. \tag{7.91}$$

Using Eq. (7.65), we have

$$< D > - < D_0 >_0 = \frac{Tr(\rho D)}{Tr^f(\rho D)} \frac{Tr^f(\rho)}{Tr(\rho)} < D >^f - < D_0 >_0 . \quad (7.92)$$

Substituting Eqs. (7.89), (7.84), (7.90), (7.91), and (7.87) into Eq. (7.92) gives

$$< D > - < D_0 >_0 = \left(1 + \frac{F_1}{F_2}\right) \left[< D >^f - < D_0 >_0^f\right] . \quad (7.93)$$

Because in all the derivations from Eq. (7.89) to Eq. (7.93) we assume that the local spin operators with spin $S+1/2$ have been decomposed into summation of a series of low local spin operators with spin 1/2 (See Eq. (7.81).), Eq. (7.93) is

$$< D >_{(\frac{1}{2}\cdots\frac{1}{2})} - < D_0 >_{(\frac{1}{2}\cdots\frac{1}{2})0} = \left(1 + \frac{F_1}{F_2}\right) \left[< D >^f_{(\frac{1}{2}\cdots\frac{1}{2})} - < D_0 >^f_{(\frac{1}{2}\cdots\frac{1}{2})0}\right] . \quad (7.94)$$

Eq. (7.94) is just the Eq. (7.82).

In the third nearest-neighbor approximation F_1 and F_2 are, respectively,

$$F_1 = \sum_{m=1}^{n} \frac{n(n-1)\ldots[n-(m-1)]}{m!} \left\{ \sum_{m'=1}^{n} 2^{m'} \frac{n(n-1)\ldots[n-(m'-1)]}{m'!} \right.$$
$$\left[Tr^{(6)} e^{-\frac{J(n-m')}{2T}\widehat{S}_6^z}\right]^2 \left(e^{-\frac{J(n-m)}{4T}} + e^{\frac{J(n-m)}{4T}}\right)^{n-m'}$$
$$\left. + \left[Tr^{(6)} e^{-\frac{Jn}{2T}\widehat{S}_6^z}\right]^2 \left(e^{-\frac{J(n-m)}{4T}} + e^{\frac{J(n-m)}{4T}}\right)^n \right\}^4 , \quad (7.95)$$

$$F_2 = \left\{ \sum_{m=1}^{n} 2^m \frac{n(n-1)\ldots[n-(m-1)]}{m!} \left[Tr^{(6)} e^{-\frac{J(n-m)}{2T}\widehat{S}_6^z}\right]^2 \left(e^{-\frac{Jn}{4T}} + e^{\frac{Jn}{4T}}\right)^{n-m} \right.$$
$$\left. + \left[Tr^{(6)} e^{-\frac{Jn}{2T}\widehat{S}_6^z}\right]^2 \left(e^{-\frac{Jn}{4T}} + e^{\frac{Jn}{4T}}\right)^n \right\}^4 , \quad (7.96)$$

$$Tr^{(6)} e^{-\frac{J(n-m')}{2T}\widehat{S}_6^z} = \sum_{m''=1}^{n} 2^{m''} \frac{n(n-1)\ldots[n-(m''-1)]}{m''!}$$
$$\left[e^{-\frac{J(n-m')}{4T}} + e^{\frac{J(n-m')}{4T}}\right]^{n-m''} + \left[e^{-\frac{J(n-m')}{4T}} + e^{\frac{J(n-m')}{4T}}\right]^n . \quad (7.97)$$

7.6. Methods Treating Local Spin Operators

Andrei and Coleman also studied the Hamiltonians in Eqs. (7.5) - (7.8) in terms of the transformation of Eq. (7.1) [346]. Their method to rule out the effect of the false spin states is to introduce Gutzwiller projection operator, and to exploit the path integral. The partition function in their treatment has complex structure, which is inconvenient in application.

Popov and Fedotov introduced a virtual chemical potential to rule out the effect of the false spin states in the partition function [344]. Xiao, Liu, and Zhang used Popov virtual chemical potential method to treat the many channel kondo problem [345]. We feel that it is still inconvenient for many authors because the virtual chemical potential method is connected with the path integral as well.

Keiter proposed a very general method to treat any operator satisfying neither fermion nor boson commutative rules [342, 343]. Keiter's method is a new diagrammatic perturbation technique completely different from the ordinary many body Green function.

Although the original Abrikosov pseudo-fermion method does not satisfy the linked cluster expansion theorem, it still can be used in some magnetic impurity problems [351]. However, the limit procedure from large pseudo-fermion energy in Ref. [340] brings a bit of inconveniences [351].

The old linearized Holstein-Primakoff transformation, which transforms the local spin operator into Boson operator, was also used as an approximate method in treating the t-J model of high-T_c cuprates [347].

The method proposed in this chapter also possesses a weakness. Eq. (7.66) is exact only under the condition that we consider the z-component of local spin operator. This condition is satisfied in our theory on the superconductivity and the pseudogap in chapter 3. However, in generic case considering the x- and y-components as well as the z-component of local spin operator, Eq. (7.66) is an approximate formula. To show this point, at first we note that the diagonal matrix elements of spin states for odd number power of $\widehat{S^x}$ or $\widehat{S^y}$ are zero. In expansion of ρ in Eq. (7.10) in terms of H_{Kondo} and $H_{Heisenberg}$, the contribution of $\widehat{S^z}$ to the diagonal matrix element of spin states is larger than that of $\widehat{S^x}$ and $\widehat{S^y}$ because the former appears from the first order of H_{Kondo} and $H_{Heisenberg}$, whereas the latter from the second order. The estimation for the approximation can be done according to specific circumstances.

Liu *et al.* showed that if Hamiltonian in Eq. (7.5) is $H_{Heisenberg}$ alone, then the transformation factor in Eqs. (7.41) and (7.66) is a simple number 1, i.e., $<R>=<R>^f$, which has been used in studying the dependence of spin wave of antiferromagnetism on temperature [349]. Using the method in this chapter, we can also obtain that the transformation factor is 1. If we take E=0, i.e., $J_K = 0$, in Eq. (7.31), then $\rho_0' = 2\rho_0, Tr^f(\rho R) = Tr_A(\rho_0)2(f_1+f_2), Tr(\rho R) = Tr_A(\rho_0 R)2f_3$. Substituting the above relations and Eq. (7.36) into Eqs. (7.40) and (7.24) gives $<R>=<R>^f$.

Appendix A

Abrikosov Pseudo-Fermion Representation

A.1. Introduction

In this note we shall describe in some detail the Larsen's procedure, which allows one to treat the case of a single local spin $\widehat{\mathbf{S}}$ of magnitude S=1/2 interacting with some external system or field in the framework of ordinary many-body perturbation theory at finite temperature [340,341]. Using the Larsen's procedure [341], we may overcome some difficulties associated with the previous Abrikosov pseudo-fermion representation [340].

The class of interaction Hamiltonian, considered in Ref. [341], may be written in the form

$$\widehat{V} = \mathbf{A} \cdot \widehat{\mathbf{S}}, \tag{A.1}$$

where $\mathbf{A}$ is a vector, operator or c-number, describing the coupling of local spin $\widehat{\mathbf{S}}$ to the external system or field. There are particularly two situations. In case 1

$$\mathbf{A} = -J_K \sum_{\alpha\beta} \vec{\sigma}_{\alpha\beta}\, \psi_\alpha^+ \psi_\beta, \tag{A.2}$$

where ψ_α is conduction electron field operator of spin α ($\uparrow$ or $\downarrow$) at the position of the local spin $\widehat{\mathbf{S}}$, $\vec{\sigma}_{\alpha\beta}$ is the Pauli matrix, and J_K is a coupling constant. In this case $\widehat{V}$ is the so-called Kondo Hamiltonian, which describes the interaction between conduction electron and the local impurity spin in dilute magnetic alloys.

In case 2

$$\mathbf{A} = -g\mu_B \mathbf{H}, \tag{A.3}$$

describing the interaction of magnetic moment $g\mu_B\mathbf{S}$ (g is the gyromagnetic factor, μ_B the Bohr magneton.) under a magnetic field $\mathbf{H}$.

A.2. Pseudo-Fermion Representation

In order to be able to use the standard diagram methods of quantum statistical physics in case 1 Abrikosov introduced a pseudo-fermion representation of the local spin operator

[340]

$$\widehat{\mathbf{S}} = \sum_{\beta\gamma} \mathbf{S}_{\beta\gamma} d^+_\beta d_\gamma, \tag{A.4}$$

where d^+_β and d_γ are creation and annihilation operators of the so-called pseudo-fermion obeying the usual fermion anticommutation rules

$$[d^+_\beta, d_\gamma]_+ = \delta_{\beta\gamma}, \tag{A.5}$$

$$[d^+_\beta, d^+_\gamma]_+ = [d_\beta, d_\gamma]_+ = 0, \tag{A.6}$$

where the subscript β corresponds to $-1/2$ ($\downarrow$) or $1/2$ ($\uparrow$), and $\mathbf{S}$ is the local spin matrix vector.

$$S^x = \begin{pmatrix} 0 & \frac{1}{2} \\ \frac{1}{2} & 0 \end{pmatrix}. \tag{A.7}$$

$$S^y = \begin{pmatrix} 0 & -\frac{i}{2} \\ \frac{i}{2} & 0 \end{pmatrix}. \tag{A.8}$$

$$S^z = \begin{pmatrix} \frac{1}{2} & 0 \\ 0 & -\frac{1}{2} \end{pmatrix}. \tag{A.9}$$

The definitions in Eqs. (A.4) - (A.9) make that the components of $\widehat{\mathbf{S}}$ satisfy the basic angular momentum commutation relations. For example, let us demonstrate

$$[\widehat{S}^x, \widehat{S}^y] = i\widehat{S}^z. \tag{A.10}$$

Substituting Eq. (A.4) into the left side of Eq. (A.10), we have

$$\left[\widehat{S}^x, \widehat{S}^y\right] = \left[\sum_{\beta\gamma} S^x_{\beta\gamma} d^+_\beta d_\gamma, \sum_{\beta'\gamma'} S^y_{\beta'\gamma'} d^+_{\beta'} d_{\gamma'}\right]. \tag{A.11}$$

Eq. (A.11) can be rewritten as

$$\left[\widehat{S}^x, \widehat{S}^y\right] = \frac{1}{2}\sum_{\beta\beta'} S^x_{\beta,-\beta} S^y_{\beta',-\beta'} [d^+_\beta d_{-\beta} d^+_{\beta'} d_{-\beta'} - d^+_{\beta'} d_{-\beta'} d^+_\beta d_{-\beta}]. \tag{A.12}$$

Substituting Eqs. (A.5) and (A.6) into Eq. (A.12), we obtain

$$\left[\widehat{S}^x, \widehat{S}^y\right] = \frac{1}{2} i \left(d^+_\uparrow d_\uparrow - d^+_\downarrow d_\downarrow\right) = i \sum_\delta S^z_{\delta\delta} d^+_\delta d_\delta = i\widehat{S}^z. \tag{A.13}$$

The Hamiltonian of non-interacting pseudo-fermion system may be written as

$$H_\lambda = \lambda \sum_\beta d^+_\beta d_\beta = \lambda \sum_\beta \widehat{n}_\beta = \lambda \widehat{n}, \tag{A.14}$$

where λ is the pseudo-fermion energy, $\widehat{n}_\beta$ is the pseudo-fermion number operator for spin β, and $\widehat{n}$ the total pseudo-fermion number operator (in this special case we use $\widehat{n}$ to distinguish the operators from their values n.). The corresponding set of many-particle eigenstates of the non-interacting pseudo-fermion system may be written in terms of pseudo-fermion number ($n_\beta = 0, 1$) in the form

$$|n_{\frac{1}{2}} n_{-\frac{1}{2}} >= |n> . \quad \text{(A.15)}$$

If the total pseudo-fermion number $n = 0$, then the eigenstate is $|00>$. If the total pseudo-fermion number $n = 1$, then the eigenstates are $|10>$ and $|01>$. If the total pseudo-fermion number $n = 2$, then the eigenstate is $|11>$.

From $\widehat{S^z}|10>= 1/2|10>$ and $\widehat{S^z}|01>= -1/2|01>$ we see that $|10>$ and $|01>$ correspond to the correct eigenvalues of local spin operator $\widehat{S^z}$. Therefore, $|10>$ and $|01>$ are called physical spin states. From $\widehat{S^z}|00>= 0$ and $\widehat{S^z}|11>= 0$ we see that $|00>$ and $|11>$ do not correspond the correct eigenvalues of local spin operator $\widehat{S^z}$. Therefore, $|00>$ and $|11>$ are called false spin states. However, both the physical spin states and the false spin states are the eigenstates of pseudo-fermion operators.

In order to "freeze out" the false spin states from statistical average Abrikosov sets the pseudo-fermion energy $\lambda \to \infty$ after taking the average and multiplication by $\exp(\lambda/T)/(2S+1)$ $(S = 1/2)$, where T is the temperature. However, it was pointed out by Keiter [350] that in this method one can not use the linked cluster theorem [47]. Larsen pointed out that in this method a disadvantage is the large pseudo-fermion energy, which has no physical meaning and complicates the calculation [341].

A.3. Statistical Average

In Larsen's method of pseudo-fermion representation $\lambda = 0$ [341]. When $\lambda = 0$ the pseudo-fermion operators will not enter the non-interacting Hamiltonian. The only contribution to this non-interacting Hamiltonian will eventually come from the external part of the system (A-part) interacting with the local spin operator through $\widehat{V}$. We call the A-part of Hamiltonian $H_A^{(\circ)}$ and its eigenstates $|A>$ for short. Then the basic states of the non-interacting system are $|n>|A>$, where the many-pseudo-fermion states are classified according to their value of n, the total number of pseudo-fermion. The total Hamiltonian is given by

$$H = H_A^{(\circ)} + \widehat{V}. \quad \text{(A.16)}$$

The statistical operators for the non-interacting and the interacting system are given by

$$\rho_\circ = e^{-\frac{H_A^{(\circ)}}{T}}, \quad \text{(A.17)}$$

$$\rho = e^{-\frac{H}{T}}, \quad \text{(A.18)}$$

where the chemical potentials are included in λ and $H_A^{(\circ)}$.

The average value of an operator R on all $|A>$ and all pseudo-fermion states $|n>$ is given by

$$< R >^f = \frac{Tr^f(\rho R)}{Tr^f(\rho)}, \quad \text{(A.19)}$$

where the trace can be expressed as

$$Tr^f(\cdots) = \sum_{(n)(A)} < A| < n| \cdots |n> |A>$$

$$= \sum_{(A)} < A| \sum_{n=0}^{2S+1} Tr_n(\cdots)|A >, \tag{A.20}$$

where $S = 1/2$, and Tr_n means trace over the subset of many-pseudo-fermion states containing n pseudo-fermions. Averages of this type can be calculated by the usual diagrammatic methods of field theory including the linked cluster theorem [47], because the averages are on the full spin space of pseudo-fermion operators. However, these averages are not exactly what we want, because they include the false spin state as well. The above average is called full average. The correct average including only the proper $2S+1$ ($S = 1/2$) states of local spin operator may be expressed as

$$< R >= \frac{Tr(\rho R)}{Tr(\rho)}, \tag{A.21}$$

where

$$Tr(\cdots) = \sum_{(A)} < A|Tr_1(\cdots)|A > . \tag{A.22}$$

The above average is called physical average. In order to be able to use the linked cluster theorem we want to express $< R >$ solely in terms of average $< R >^f$ and some related full averages.

To obtain $< R >$ we express Eq. (A.21) in the form

$$< R >=< R >^f AL, \tag{A.23}$$

$$A = \frac{Tr(\rho R)}{Tr^f(\rho R)}, \tag{A.24}$$

$$L = \frac{Tr^f(\rho)}{Tr(\rho)}. \tag{A.25}$$

Here A depends on the particular R, whereas L is the same for all R. We will also use the expression for the partition functions

$$Z_\circ = Tr^f(\rho_\circ), \tag{A.26}$$

$$Z = Tr^f(\rho), \tag{A.27}$$

and for the non-interacting average of the finite temperature "S-matrix" $\widehat{U} = \rho_\circ^{-1}\rho$.

$$< \widehat{U} >_\circ^f = \frac{Tr^f(\rho_\circ\rho_\circ^{-1}\rho)}{Tr^f(\rho_\circ)} = \frac{Z}{Z_\circ}. \tag{A.28}$$

Since

$$\widehat{S}^x|00 >= \widehat{S}^y|00 >= \widehat{S}^z|00 >= \widehat{S}^x|11 >= \widehat{S}^y|11 >= \widehat{S}^z|11 >= 0, \tag{A.29}$$

and since $\widehat{V}$ has the form Eq. (A.1), then by Eqs. (A.16) - (A.18), (A.20), and (A.22)

$$Tr^f(\rho) = \sum_{(A)} < A|[Tr_0(\rho_\circ) + Tr_1(\rho) + Tr_2(\rho_\circ)]|A >$$

$$= \frac{2}{4} Tr^f(\rho_\circ) + Tr(\rho). \tag{A.30}$$

This gives at once by Eqs. (A.25) - (A.28)

$$L = \frac{Tr^f(\rho)}{Tr^f(\rho) - \frac{1}{2} Tr^f(\rho_\circ)} = \frac{2 < \widehat{U} >_\circ^f}{2 < \widehat{U} >_\circ^f - 1}. \tag{A.31}$$

If the R contains pseudo-fermion operators only in such combinations, that it may be reduced to a sum of product-terms of the type $\widehat{S}^i \cdots \widehat{S}^j$, where $i, j, \cdots = x, y, z$, then one has

$$Tr^f(\rho R) = \sum_{(A)} < A | Tr_1(\rho R) | A > = Tr(\rho R), \tag{A.32}$$

and

$$A = 1. \tag{A.33}$$

Therefore by Eq. (A.23)

$$< R > = L < R >^f. \tag{A.34}$$

Eq. (A.34) allows us to obtain physical average $< R >$ through calculating the related full averages, which can be obtained by using the linked cluster theorem.

If on the other hand R does not contain pseudo-fermion operators at all, then we find by analogy with Eq. (A.30)

$$Tr^f(\rho R) = \frac{1}{2} Tr^f(\rho_\circ R) + Tr(\rho R). \tag{A.35}$$

Substituting Eq. (A.35) into Eq. (A.24) gives

$$A = 1 - \frac{1}{2} \frac{Tr^f(\rho_\circ R)}{Tr^f(\rho R)}. \tag{A.36}$$

Eq. (A.36) can be rewritten as

$$A - 1 - \frac{1}{2} \frac{\frac{Tr^f(\rho_\circ R)}{Tr^f(\rho_\circ)}}{\frac{Tr^f(\rho R)}{Tr^f(\rho)} \frac{Tr^f(\rho)}{Tr^f(\rho_\circ)}}$$

$$= 1 - \frac{1}{2} \frac{< R >_\circ^f}{< R >^f < \widehat{U} >_\circ^f}. \tag{A.37}$$

Since in this case simply $< R >_\circ = < R >_\circ^f$ we have

$$< R > - < R >_\circ = L \left(< R >^f - < R >_\circ^f \right). \tag{A.38}$$

Eq. (A.38) allows us to obtain physical average $< R >$ through calculating the related full averages $< R >^f$, which can be obtained by using the linked cluster theorem.

If $< R >$ has other structures more complicated situations may arise, for instance, the conduction electron Green function treated in the following. We restrict the discussions to $S = 1/2$ and the case 1 in section A.1. Consider the operators

$$D = T_\tau \left(\psi(\tau) \psi^+ \right) = T_\tau \left(e^{\tau H} \psi e^{-\tau H} \psi^+ \right), \tag{A.39}$$

$$D_\circ = T_\tau\left(e^{\tau H_A^{(\circ)}}\psi e^{-\tau H_A^{(\circ)}}\psi^+\right), \tag{A.40}$$

where T_τ is the imaginary-time-ordering operator and ψ's are the conduction electron field operators at the local spin site operating on the part $|A>$ (suppressing their spin index). The full Green functions for the interacting and non-interacting systems may be written respectively in the forms

$$G^f = < D >^f, \tag{A.41}$$

$$G^f_\circ = < D_\circ >^f_\circ. \tag{A.42}$$

From the corresponding quantities averaged only over the proper $2S+1$ states of the local spin we have physical Green functions G and $G_\circ$.

$$G = < D >. \tag{A.43}$$

Considering $D_\circ$ does not contain pseudo-fermion operators, we have

$$G_\circ = < D_\circ >_\circ = < D_\circ >^f_\circ = G^f_\circ. \tag{A.44}$$

Considering Eqs. (A.29) and (A.20), we have

$$Tr^f(\rho D) = \sum_{(A)} < A|Tr_0(\rho_\circ D_\circ) + Tr_1(\rho D) + Tr_2(\rho_\circ D_\circ)$$

$$= \frac{1}{2}Tr^f(\rho_\circ D_\circ) + Tr(\rho D). \tag{A.45}$$

Substituting Eq. (A.45) into Eq. (A.24) gives

$$A = 1 - \frac{1}{2}\frac{Tr^f(\rho_\circ D_\circ)}{Tr^f(\rho D)}. \tag{A.46}$$

Eq. (A.46) can be rewritten as

$$A = 1 - \frac{1}{2}\frac{\frac{Tr^f(\rho_\circ D_\circ)}{Tr^f(\rho_\circ)}}{\frac{Tr^f(\rho D)}{Tr^f(\rho)}\frac{Tr^f(\rho)}{Tr^f(\rho_\circ)}}$$

$$= 1 - \frac{1}{2}\frac{G^f_\circ}{G^f < \widehat{U} >^f_\circ}. \tag{A.47}$$

Considering Eqs. (A.44), (A.47), and (A.23), we have

$$G - G_\circ = G^f AL - G^f_\circ = G^f\left(1 - \frac{1}{2}\frac{G^f_\circ}{G^f < \widehat{U} >^f_\circ}\right)L - G^f_\circ$$

$$= G^f L - \frac{1}{2}\frac{G^f_\circ}{< \widehat{U} >^f_\circ}L - G^f_\circ. \tag{A.48}$$

Substituting Eq. (A.31) into Eq. (A.48) gives

$$G - G_\circ = L(G^f - G^f_\circ). \tag{A.49}$$

Eq. (A.49) allows us to obtain physical Green function G through calculating the full Green function G^f and L, which can be obtained by using the linked cluster theorem.

We demonstrate exactly that the L in Eqs. (A.49), (A.38), and (A.34) is equal to 2 in case 1 of section A.1. At first we demonstrate exactly $<\widehat{U}>^f_{\circ}=1$. Substituting Eqs. (A.1), (A.2), and (A.16) - (A.18) into Eq. (A.28) gives

$$<\widehat{U}>^f_{\circ}=\frac{Tr^f\left(\exp-\frac{H_A^{(\circ)}-J_K\sum_{\alpha\beta}\vec{\sigma}_{\alpha\beta}\psi^+_{\alpha}\psi_{\beta}\cdot\widehat{\mathbf{S}}}{T}\right)}{Tr^f\left(\exp-\frac{H_A^{(\circ)}}{T}\right)}. \tag{A.50}$$

$H_A^{(\circ)}$ is free electron Hamiltonian.

$$\widehat{H}_A^{(\circ)}=\sum_{\mathbf{k}\alpha}(\varepsilon_{\mathbf{k}}-\mu)c^+_{\mathbf{k}\alpha}c_{\mathbf{k}\alpha}. \tag{A.51}$$

$$\psi^+_{\alpha}=\frac{1}{\sqrt{N}}\sum_{\mathbf{k}}e^{-i\mathbf{k}\cdot\mathbf{R}}c^+_{\mathbf{k}\alpha}, \tag{A.52}$$

$$\psi_{\beta}=\frac{1}{\sqrt{N}}\sum_{\mathbf{k}'}e^{i\mathbf{k}'\cdot\mathbf{R}}c_{\mathbf{k}'\beta}, \tag{A.53}$$

where N is the total cell number of a lattice, and $\mathbf{R}$ is the site of local spin. Substituting Eqs. (A.51) - (A.53) into Eq. (A.50) and considering that the $c^+_{\mathbf{k}\alpha}c_{\mathbf{k}'\beta}$ appears in trace, thus $\mathbf{k}=\mathbf{k}'$ and $\alpha=\beta$, Eq. (A.50) becomes

$$<\widehat{U}>^f_{\circ}=\frac{Tr^f\left(\exp-\frac{\sum_{\mathbf{k}\alpha}[\varepsilon_{\mathbf{k}}-\mu-J_K/N\vec{\sigma}_{\alpha\alpha}\cdot\widehat{\mathbf{S}}]c^+_{\mathbf{k}\alpha}c_{\mathbf{k}\alpha}}{T}\right)}{Tr^f\left(\exp-\frac{\sum_{\mathbf{k}\alpha}[\varepsilon_{\mathbf{k}}-\mu]c^+_{\mathbf{k}\alpha}c_{\mathbf{k}\alpha}}{T}\right)}. \tag{A.54}$$

Noting $N\to\infty$, we have

$$<\widehat{U}>^f_{\circ}=1. \tag{A.55}$$

Substituting Eq. (A.55) into Eq. (A.31), we have

$$L=2. \tag{A.56}$$

The value of L did not be given in Ref. [341].

References

[1] Onnes, K. K., *Leiden Comm.* 1911, 120b, 122b, 124c.

[2] Bardeen, J., Cooper, L. N., Schrieffer, J. R., *Phys. Rev.* 1957, 108, 1175.

[3] Josephson, B. D., *Phys. Lett.* 1962, 1, 251.

[4] Ketterson J. B., Song, S. N. *Superconductivity*, Cambridge University Press, New York, 1999, p. 91.

[5] Bednorz, G., Müller, K. A., *Z. Phys.* 1986, B64, 189.

[6] Meissner, W., Ochsenfield, R., *Naturwissenschaften* 1933, 21, 787.

[7] Phillips, N. E., *Phys. Rev.* 1959, 114, 676.

[8] Shoenberg, D. *Superconductivity*, At the University Press, Cambridge, 1952, p. 61.

[9] Corak, W. S., Goodman, B. B., Satterthweite, C. B., Wexley, A., *Phys. Rev.* 1954, 96, 1442.

[10] Corak, W. S., Goodman, B. B., Satterthweite, C. B., Wexley, A., *Phys. Rev.* 1956, 102, 656.

[11] Maxwell, E., *Phys. Rev.* 1950, 78, 477.

[12] Anderson, P. W. In *Lecture on the Many-Body Problem* VII, Calianiello, E. R., Ed., Academic Press, New York, 1964, p. 113.

[13] Mahan, G. D. *Many-Particle Physics*, Plenum Press, New York, 1981, pp. 820, 814.

[14] Josephson, B. D. In *Superconductivity*, Parks, R. D., Ed., Marcel Dekker, Inc., New York, 1969, Vol. 1, p. 423.

[15] Tinkham, M. *Introduction to Superconductivity*, McGraw-Hill, Inc., New York, 1996, p. 197.

[16] Ekino, T., Sezaki, Y., Fujii, H., *Phys. Rev.* 1999, B60, 6916.

[17] Ambegaokar, V., Baratoff, A., *Phys. Rev. Lett.* 1963, 10, 486.

[18] Timusk, T., Statt, B., *Rep. Prog. Phys.* 1999, 62, 61.

[19] Loktev, V. M., Quick, R. M., Sharapov, S. G., *Phys. Rep.* 2001, 349, 1.

[20] Ding, H., Yokoya, T., Campuzano, J. C., Takahashi, T., Randeria, M., Norman, M. R., Mochiku, T., Kadowaki, K., Giapintzakis, J., *Nature (London)* 1996, 382, 51.

[21] London, F., London, H., *Proc. Roy. Soc. (London)* 1935, A149, 71.

[22] Tinkham, M. *Introduction to Superconductivity*, McGraw-Hill, Inc., New York, 1996, p. 5.

[23] London, F. *Superfluids* II, Wiley, New York, 1950.

[24] Tilley, D. R., Tilley, J. *Superfluidity and Superconductivity*, Adam Hilger Ltd, Bristol, 1986, pp. 40, 147.

[25] Kittel, C. *Introduction to Solid State Physics*, John Wiley and Sons, New York, 1976, pp. 154, 356.

[26] Pippard, A. B., *Proc. Roy. Soc. (London)* 1953, A216, 547.

[27] Faber, T. E., Pippard, A. B., *Proc. Roy. Soc. (London)* 1955, A231, 336.

[28] Ginzburg, V. L., Landau, L. D., *Zh. Eksperim. i Teor. Fiz.* 1950, *20*, 1064.

[29] Nettel, S., *Phys. Rev.* 1993, B48, 13709.

[30] Abrikosov, A. A., *Zh. Eksperim. i Teor. Fiz.* 1957, 32, 1442. [*Sov. Phys. - JETP* 1957, 5, 1174].

[31] Alexandrov, A. S. *Theory of Superconductivity*, Institute of Physics Publishing, Bristol, 2003, pp. 23, 26, 28.

[32] Deaver, B. S., Fairbank, W. M., *Phys. Rev. Lett.* 1961, 7, 43.

[33] Doll, R., Näbauer, M., *Phys. Rev. Lett.* 1961, 7, 51.

[34] Cooper, L. N., *Phys. Rev.* 1956, 104, 1189.

[35] Schrieffer, J. R. *Theory of Superconductivity*, Benjamin, New York, 1964, pp. 28, 35, 57.

[36] Rose-Innes, A. C., Rhoderick, E. H. *Introduction to Superconuctivity*, Pergamon, Oxford, 1978, p. 123.

[37] Tinkham, M. *Introduction to Superconductivity*, McGraw-Hill, New York, 1996, p. 52.

[38] Bogoliubov, N. N., *Nuovo Cimento* 1958, 7, 794.

[39] Valatin, J., *Nuovo Cimento* 1958, 7, 843.

[40] Gradshteyn, I. S., Ryhik, I. M. *Table of Integrals, Series, and Products*, Academic Press, New York, 1980, p. 958.

[41] Liu, Fu-sui, Chen, Wan-fang, Peng, Kuang-ding, *J. Phys.: Condens. Matter* 2001, 13, 2817.

[42] Inderhees, S. E., Salamon, M. B., Goldenfeld, N., Rice, J. P., Pazol, B. G., Ginsburg, D. M., *Phys. Rev. Lett.* 1988, 60, 1178.

[43] Dunlap, B. D., Nevitt, M. V., Slaski, M., Klippert, T. E., Sungaila, Z., McKale, A. G., Capone, D. W., Poeppel, R. B., Flandermeyer, B. K., *Phys. Rev.* 1988, B35, 7210.

[44] Liu, Fu-sui, Chen, Wan-fang, *J. Phys.: Condns. Matter* 2000, 12, 8475.

[45] Nakata, T., Momono, N., Takahashi, S., Oda, M., Ido, M., *Physica* 2001, C364-365, 430.

[46] Gorkov, L. P., *Zh. Eksperim. i Teor. Fiz.* 1959, 36, 1918. [*Soviet Phys. JETP.* 1959, 9, 1364.]

[47] Abrikosov, A. A., Gorkov, L. P., Dzyaloshinski, I. E. *Methods of Quantum Field Theory in Statistical Physics*, Prenticl-Hall, Englewood Cliffs, N. J., 1963.

[48] Alexandrov, A. S. *Theory of Superconductivity*, Institute of Physics, Bristol, 2003, p. 70.

[49] Cohen, M. H., Falikov, M., Philips, J. C., *Phys. Rev. Lett.* 1962, *8*, 316.

[50] Mahan, G. D. *Many-Particle Physics*, Plenum Press, New York, 1981. p. 798.

[51] Fetter, A. L., Walecka, J. D. *Quantum Theory of Many-Particle Systems*, McGraw-Hill, New York, 1971.

[52] Josephson, B. D. In *Superconductivity*, Parks, R. D., Ed., Marcel Dekker, New York, 1969, Vol. 1, p. 435.

[53] Tanaka, Y., Kashiwaya, S., *Phys. Rev.* 1997, B56, 909.

[54] Liu, Fu-sui, Chen, Wan-fang, *Inter. J. Mod. Phys. B*, 2006, accepted.

[55] Liu, Fu-sui, Hou, Yumin, Chen, Wan-fang, *Commun. Theor. Phys.* accepted.

[56] Josephson, B. D., thesis, unpublished, University of Cambridge, 1964.

[57] Rogovin, D., Scully, M., *Phys. Lett.* 1977, A64, 106.

[58] Greiner, W. *Quantum Mechanics*, Springer-Verlag, Berlin, 1994, p. 189.

[59] Josephson, B. D., *Adv. Phys.* 1965, 14, 419.

[60] Anderson, P. W., Dayem, A. H., *Phys. Rev. Lett.* 1964, 13, 195.

[61] Stephen, M. J. Suhl, H., *Phys. Rev. Lett.* 1964, 13, 797.

[62] Shapiro, S., *Phys. Rev. Lett.* 1963, 11, 80.

[63] Schrieffer, J. R. *Theory of Superconductivity*, Benjamin, New York, 1964.

[64] de Gennes, P. G. *Superconductivity in Metals and Alloys*, Benjamin, New York, 1966.

[65] Ketterson, J. B., Song, S. N. *Superconductivity*, Cambridge University Press, New York, 1999.

[66] Tinkham, M. *Introduction to Superconductivity*, McGraw-Hill, New York, 1996.

[67] Rickayzen, G. *Theory of Superconductivit*, Wiley, New York, 1965.

[68] Alexandrov, A. S. *Theory of Superconductivity*, Institute of Physics, Bristol, 2003.

[69] Buckel, W., Kleiner, R. K. *Superconductivity*, WILEY-VCH, Weinheim, 2004.

[70] Rose-Innes, A. C., Rhoderick, E. H. *Introduction to Superconuctivity*, Pergamon, Oxford, 1978.

[71] Mahan, G. D. *Many-Particle Physics*, Plenum Press, New York, 1981.

[72] Tilley, D. R., Tilley, J. *Superfluidity and Superconductivity*, Adam Hilger Ltd, Bristol, 1986.

[73] Anderson, P. W., *Science* 1987, 235, 1196.

[74] Liu, Fu-sui, *Phys. Lett.* 1997, A224, 185.

[75] Liu, Fu-sui, Chen, Wan-fang, *Phys. Rev. 1*998, B58, 8812.

[76] Wen, Xiao-Gang, Lee, P. A., *Phys. Rev. Lett.* 1996, 76, 503.

[77] Lee, P. A., Wen, Xiao-Gang, *Phys. Rev. Lett.* 1997, 78, 4111.

[78] Scalapino, D. J., *Phys. Rep.* 1990, 250, 239.

[79] Pines, D., *Physica* 1994, B199-200, 300.

[80] Zhang, Shou-cheng, *Science* 1997, 275, 1089.

[81] Lee, P. A., Nagaosa, N., Wen, Xiao-Gang, *Rev. Mod. Phys*. 2006, 78, 17.

[82] Basov, D. N., Timusk, T., *Rev. Mod. Phys.* 2005, 77, 721.

[83] Matsukawa, H., Fukuyama, H., *J. Phys. Soc. Jpn.* 1989, 58, 2845.

[84] Tokyama, T., *Physica* 1992, C191, 193.

[85] Hybertsen, M., Stechel, E. B., Schulter, M., Jennison, D. R., *Phys. Rev.* 1990, B42, 11068.

[86] Eskes, H., Sawatzky, G. A., *Phys. Rev.* 1991, B44, 9656.

[87] Grant, J. B., *Phys. Rev.* 1992, B46, 8440.

[88] Mahan, G. D. *Many-Particle Physics*, Plenum Press, New York, 1981, pp. 123, 819.

[89] Mattuck, R. D. *A Guide to Feynman Diagram in the Many-Body Problems*, McGraw-Hill International Book Company, New York, 1976, p. 310.

[90] Annett, J. F., Goldenfeld, N., Leggett, J. In *Physical Properties of High Temperature Superconductors*, Ginsberg, D. M., Ed., World Scientific, Singapore, 1996, Vol. 5, p. 396.

[91] Ren, Yong, Xu, Ji-hai, Ting, C. S., *Phys. Rev.* 1996, B53, 2249.

[92] Masaelian, K. A., Betouras, J., Chubukov, A. V., Joynt, R., *Phys. Rev.* 1996, B53, 3598.

[93] Oda, M., Ohguro, T., Matsuki, H., Yamada, N., Ido, M., *Phys. Rev.* B41, 1990, 2606.

[94] Nakano, T., Momono, N., Ido, M., *J. Phys. Soc. Jpn.* 1998, 67, 2622.

[95] Oda, M., Hoya, K., Kubota, R., Manabe, C., Momono, N., Nakano, T., Ido, M., *Physica* 1997, C281, 135.

[96] Doniach, S., Inui, M., *Phys. Rev.* 1990, B41, 6668.

[97] Rose-Innes, A. C., Rhoderick, E. H., *Introduction to Superconductivity*, second ed., Pergamon, Oxford, 1978, p. 167.

[98] Hebard, A. F. In *Inhomogeneous Superconductors-1979*, Francavilla, T. L., Gubsor, D. V., Leibiwitz, J. R., Wolf, S. A., Ed., *AIP Conference Proceedings* No. **58**, American Institute of Physics, New York, 1979, p. 129.

[99] Tanaka, Y., Kashiwaya, S., *Phys. Rev.* 1997, B56, 892.

[100] Anderson, P. W. In *Lecture on the Many-Body Problem* VII, Calianiello, E. R., Ed., Academic, New York, 1964, p. 127.

[101] Simanek, E. *Inhomogeneous Superconductors, Granular and Quantum Effects*, Oxford University Press, Oxford, 1994, p. 32.

[102] Doniach, S., *Phys. Rev.* 1981, B24, 5063.

[103] Ogg, Jr. R. A., *Phys. Rev.* 1946, 69, 243.

[104] Schafroth, M. A., *Phys. Rev.* 1955, 100, 463.

[105] Blatt, J. M., Butler, S. T., *Phys. Rev.* 1955, 100, 476.

[106] Rickayzen, G. In *Superconductivity*, Parks, R. D., Ed., Marcel Dekker, New York, 1969, Vol. 1, p. 81.

[107] Alexandrov, A. S. *Theory of Superconductivity*, Institute of Physics Publishing, Bristol, 2003, p. 11.

[108] Enz, C. P. *A Course on Many-Body Theory Applied to Solid-State Physics*, World Scientific Publishing, Singapore, 1992, p. 180.

[109] Monthoux, P., Balatsky, A. V., Pines, D., *Phys. Rev.* 1992, B46, 14803.

[110] Chakraverty, S., Sudbo, A., Anderson, P. W., Strong, S., *Science* 1993, 261, 337.

[111] Bickers, N. E., Scalapino, D. J., White, S. R., *Phys. Rev. Lett.* 1989, 62, 961.

[112] Moriya, T., Takahashi, Y., Ueda, K., J. *Phys. Soc. Jpn.* 1990, 59, 2905.

[113] Monthoux, P., Pines, D., *Phys. Rev. Lett.* 1991, 67, 3448.

[114] Schrieffer, J. R., Wen, X. -G., Zhang, S. -C., *Phys. Rev. Lett.* 1988, 60, 944.

[115] Lee, P. A., *Phys. Rev. Lett.* 1993, 71, 1887.

[116] Sigrist, M., Rice, T. M., J. *Phys. Soc. Jpn.* 1992, 61, 4283.

[117] Shen, Z. -X., Dessau, D. S., Wells, B. O., King, D. M., Spicer, W. E., Arko, A. J., Marshall, D., Lombardo, L. W., Kapitulnik, A., Dickinson, P., Doniach, S., DiCarlo, J., Loeser, A. G., Park, C. H., *Phys. Rev. Lett.* 1993, 70, 1553.

[118] Dessau, D. S., Shen, Z. -X., Marshall, D. M., *Phys. Rev. Lett.* 1993, 71, 4278.

[119] Liu, Fu-sui, Chen, Wan-fang, Peng, Kuang-ding, *Commun. Theor. Phys. (China)* 2003, 40, 739.

[120] Williams, G. V. M., Tallon, J. L., Haines, E. M., Michalak, R., Dupree, R., *Phys. Rev. Lett.* 1997, 78, 721.

[121] Shen, Z. -X., Dessau, D. S., *Phys. Rep.* 1995, 253, 142.

[122] Dessau, D. S., Shen, Z. -X., King, D. M., Marshall, D. S., Lombardo, L. W., Dickinson, P. H., Loeser, A. G., DiCarlo, J., Park, C. -H., A. Kapitulnik, A., Spicer, W. E., *Phys. Rev. Lett.* 1993, 71, 2781.

[123] Chen, Han-dung, Vafec, O., Zhang, Shou-Cheng, *Phys. Rev. Lett.* 2004, 93, 187002.

[124] Norman, M. R., Pines, D.,Kallin, C., *Adv. Phys.* 2005, 54, 715.

[125] Fauqui, B., Sidis, Y., Hinkov, V., Pailhes, S., Lin, C. T., Chaud, X., Bourges, P., *Phys. Rev. Lett.* 2006, 96, 197001.

[126] Barzykin, V., Pines, D., *Phys. Rev. Lett.* 2006, 96, 247002.

[127] Macridin, A., Jarrell, M., Maier, T., Kent, P. R. C., D'Azevedo, E., *Phys. Rev. Lett.* 2006, 97, 036401.

[128] Ohkawa, F. J., *Phys. Rev.* 2006, B74, 134500.

[129] Passov, C. A. C., Orlando, M. T. D., Passamai, Jr. J. L., de Mello, E. V. L., Corraa, M. P. S., *Phys. Rev.* 2006, B74, 094514.

[130] Liu, Fu-sui, Hou, Yumin, Chen, Wan-fang, *Commun. Theor. Phys.* accepted.

[131] Liu, Fu-sui, Chen, Wan-fang, *Inter. J. Modern Phys. B.* accepted.

[132] Feng, D. L., Lu, D. H., Shen, K. M., Kim, C., Eisaki, H., Damascalli, A., Yoshizaki, R., Shimoyama, J. -i., Kisho, K., Gu, G. D., Oh, S., Andrus, A., Donnell, J. O., Eckstein, J. N., Shen, Z. -X., *Science* 2000, 289, 277.

[133] Corson, J., Mallozzi, R., Orenstein, J., Eckstein, J. N., Bozovic, I., *Nature (London)* 1999, 398, 221.

[134] Buckel, W., Kleiner, R. *Superconductivity*, WILEY-VCH, Weinheim, 2004, p. 309.

[135] Presland, M. R., *Physica* 1991 C165, 391.

[136] Cooper, S. L., Gray, K. E. In *Physical Properties of High Temperature Superconductors*, Ginsberg, D. M., Ed., World Scientific, Singapore, 1994, Vol. 4, p. 74.

[137] Pan, S. H., O'Neal, J. P., Badzey, R. L., Chamon, C., Ding, H., Engelbrecht, J. R., Elsaki, Z., Uchida, S., Gupta, A. K., Ng, K. -W., Hudson, E. W., Lang, K. M., Davis, J. C., *Nature (London)* 2001, 413, 282.

[138] Liu, Jin-Xiang, Goldman, J. C., *Phys. Rev. Lett.* 1991, 67, 2195.

[139] Wang, Lan-ping, He, Jian, Wang, Guowen, *Phys. Rev.* 1989, B40, 10594.

[140] Adkins, E. A., Chandlev, C. J., *J. Phys.* 1987, C20, L1009.

[141] Matsuda, A., Fijii, T., Watanabe, T., *Physica* 2003, C388 - 389, 207.

[142] Corson, J., Mallozi, R., Orenstein, J. N., Bozovic, I., *Nature* 1999, 398, 221.

[143] Ranninger, J., Tripodi, L., *Phys. Rev.* 2003, B67, 174521.

[144] Wang, Yayu, Li, Lu, Naughton, M. J., Gu, G. D., Uchida, S., Ong, N. P., *Phys. Rev. Lett.* 2005, 95, 247002.

[145] Iguchi, I., Yamaguchi, T., Sugimoto, A., *Nature (London)* 2001, *412*, 420.

[146] Xu, Z. A., Ong, N. P., Wang, Y., Kakeshita, T., Uchida, S., *Nature (London)* 2000, 406, 486.

[147] Howard, C., Fournier, P., Kapitulnik, A., *Phys. Rev.* 2001, B64, 100504.

[148] Lang, K. M., Madhavan, V., Hoffman, J. E., Hudson, E. W., Eisaki, H., Uchida, S., Davis, J. C. *Nature (London)* 2002, 415, 412.

[149] Kato, T., Morimoto, H., Katagiri, A., Okitsu, S., Sakata, H., *Physica* 2003, C392 - 396, 221.

[150] Wang, Ziqiang, Engelbrecht, J. R., Wang, Shancai, Ding, H., Pan, S. H., Cond-mat/0107004, 2001.

[151] Joglekar, Y. N., Castro Neto, A. H., Balatsky, A. V., *Phys. Rev. Lett.* 2004, 92, 037004.

[152] Martin, I., Balatsky, A. V., *Physica* 2001, C357-360, 46.

[153] Birgeneau, P. J., Gabbe, D. R., Jensen, H. P., Kastner, M. A., Picone, P. J., Thurston, T. R., Shirane, G., Endoh, Y., Sato, M., Yamada, K., Hidaka, Y., Oda, M., Enomoto, Y., Suzuki, M., Murakumi, T., *Phys. Rev.* 1988, B38, 6614.

[154] Hasekawa, T.,Ikuta, H., Kitazawa, K. In *Physical Properties of High Temperature Superconductors*, Ginsberg, D. M., Ed., World Scientific, Singapore, 1992, Vol. 3, p. 533.

[155] Liu, Fu-sui, Hou, Yumin In *Progress in Superconductivity Research*, Nova, New York, accepted.

[156] Liu, Fu-sui, Chen, Wan-fang, *Inter. J. Modern Phys.* 2001, B27, 3513.

[157] Suzuki, M., Watanabe, T., Matsuda, A., *Phys. Rev. Lett.* 1999, 82, 5361.

[158] Suzuki, M., Watanabe, T., *Phys. Rev. Lett.* 2000, 85, 4787.

[159] Krasnov, V. M., Yurgens, A., Winkler, D., Delsing, P., Cleason, T., *Phys Rev. Lett.* 2000, 84, 5860.

[160] Renner, Ch., Revaz, B., Genoud, J. -Y., Kadowaki, K., Fisher, O., *Phys. Rev. Lett.* 1998, 80, 149.

[161] Suzuki, M., Watanabe, T., Matsuda, A., *Phys. Rev. Lett.* 1998, 81, 4248.

[162] Suzuki, M., Watababe, T., Matsuda, A., *IEEE Trans. Appl. Supercond.* 1999, 9, 4511.

[163] Gaifullin, M. B., Matsuda, Y. N., Chikmoto, N., Shimoyama, J., Kishio, K., Yoshizaki, R., *Phys. Rev. Lett.* 1999, 83, 3928.

[164] Mahan, G. D. *Many-Particle Physics*, Plenum Press, New York, 1981, p. 819.

[165] Deutscher, G., *Nature (London)* 1999, 397, 410.

[166] Clem, J. R., *Physica* 1989, C162 - 164, 1137.

[167] Bernhard, C., J. L. Tallon, J. L., Blasius, Th., Gornik, A., Niedermayer, Ch., *Phys. Rev. Lett.* 2001, 86, 1614.

[168] Emery, V. J., Kivelson, S. A., *Nature (London)* 1995, 374, 434.

[169] Chakraverty, B. K., Ramakrishnan, T. V., *Physica C* 1997, 282-287, 290.

[170] Szotek, Z., Gyorffy, B. L., Temmerman, W. N., *Phys. Rev.* 2000, B62, 3997.

[171] Bernhard, C., Niedermayer, Ch., Binninger, U., Hofer, A., Wenger, Ch., Tallon, J. L., Williams, G. V. M., Ansaldo, E. J., Budnik, I., Stronach, C. E., Noakes, D. R., Blankson-Mills, M. A., *Phys. Rev.* 1995, B52, 10488.

[172] Salamon, M. B. In *Physical Properties of High Temperature Superconductors*, Ginsberg, D. M., Ed., World Scientific Publishing, Singapore, 1989, Vol. 1, p. 48.

[173] Liu, Fu-sui, Chen, Wan-fang, *Physica* 2002, C377, 531.

[174] Zuev, Y., Kim, M. S., Lemberger, T. R., *Phys. Rev. Lett.* 2005, 95, 137002.

[175] Liang, R., Bonn, D. A., Hardy, W. N., Broun, D., *Phys. Rev. Lett.* 2005, 94, 117001.

[176] Williams, G. V. M., Tallon, J. C., Quilty, J. W., Trodahl, H. J., Flower, N. E., *Phys. Rev. Lett.* 1998, 80, 377.

[177] Fisher, D. S., Millis, A. J., Shraiman, B., Bhatt, R. N., *Phys. Rev. Lett.* 1988, 61, 482.

[178] Hafliger, P. S., Podlesnyak, A., Conder, K., Pomjakushina, E., Furrel, A., *Phys. Rev.* 2006, B74, 184520.

[179] Smilde, H. J. H., Golubov, A. A., Ariando, Rijnders, G., Dekkers, J. M., Harkema, S., Blank, D. H. A., Rogalla, H., Hilgenkamp, H., *Phys. Rev. Lett.* 2005, 95, 257001.

[180] Tanner, D. B., Timusk, T. In *Physical Properties of High Temperature Superconductors*, Ginsberg, D. M., Ed., World Scientific Publishing, Singapore, 1992, Vol. 3, p. 416.

[181] Khurana, A., *Physics Today* 1989, 42, 17.

[182] Paget, K. M., Cuha, S., Cieplak, M. Z., Trofimov, I. F., Turneaure, S. J., Lemberger, R., *Phys. Rev.* 1999, B59, 641.

[183] Fisher, R. A., Phillips, N. E., Schilling, A., Buffeteau, B., Calemczuk, R., Hargreaves, T. E., Marcena, C., Dennis, K. W., McCallum, R. W., Marcenat, C., *Phys. Rev.* 2000, B61, 1473.

[184] Keimer, B., Belk, N., Birgeneau, R. J., Cassanho, A., Chen, C. Y., Greven, M., Kastner, M. A., Aharony, A., Endoh, Y., Erwin, R. W., Shirane, G., *Phys. Rev.* 1992, B46, 14034.

[185] Liu, Fu-sui, Hou, Yumin, Chen, Wan-fang, *Inter. J. Modern Phys.* B, will be submitted.

[186] Norman, N. M., Ding, H., Randeria, M., Campuzano, J. C., Yokoya, T., Takeuchi, T., Takahashi, T., Mochiku, T., Kadowaki, K., Hinks, D. G., *Nature (London)* 1998, 392, 157.

[187] Harris, J. M., Shen, Z. -X., White, P. J., Marshal, D. S., Schebel, M. C., *Phys. Rev.* 1996, B54, R15665.

[188] Harris, J. M., White, P. J., Shen, Z. -X., Ikeda, H., Yoshizaki, R., Eisaki, H., Uchida, S., Si, W. D., Xiong, J. W., Shen, Z. -X., Dessau, D. S., *Phys. Rev. Lett.* 1997, 79, 143.

[189] Mesot, J., Norman, M. R., Ding, H., Randeria, M., Campuzano, J. C., Paramekanti, A., Fretwell, H. M., Kaminski, A., Takeuchi, T., Yokoya, T., Sato, T., Takahashi, T., Mochiku, T., Kadowaki, K., *Phys. Rev. Lett.* 1999, 83, 840.

[190] Mannella, N., Yang, W. L., Zhou, X. J., Zheng, H., Mitchell, J. F., Zaanen, J., Devereaux, Nagaosa, N., Hussain, Z., Shen, Z. -X., *Nature (London)* 2005, 438, 474.

[191] Damascelli, A., Shen, Z. -X., Hussain, Z., *Rev. Mod. Phys.* 2003, 75, 473.

[192] Callaway, J. *Quantum Theory of the Solid State*, Academic Press, London, 1976, p. 219.

[193] Greiner, W. *Quantum Mechanics*, Springer-Verlag, Berlin, 1994, p. 204.

[194] Liu, Fu-sui, Hou, Yumin, Chen, Wan-fang, *Phys. Rev. B*, submitted.

[195] Zheng, Guo-qing, Clark, W. G., Kitaoka, Y., Asayama, K., Kodama, Y., Kuhns, P., Moulton, W. G., *Phys. Rev.* 1999, B60, 9947.

[196] Gorny, K., Vyaselev, O. M., Martindale, J. M., Nander, V. A., Rennington, C. H., Hammel, P. C., Hults, W. L., Smith, J. L., Kuhns, P. L., Reyes, A. P., Moulton, W. G., *Phys. Rev. Lett.* 1999, 82, 177.

[197] Krasnov, V. M., Kovalev, A. E., Yurgens, A., Winkler, D., *Phys. Rev. Lett.* 2001, 86, 2657.

[198] Chen, Qijin, Levin, K., Kosztin, I., *Phys. Rev.* 2001, B63, 184519.

[199] Renner, Ch., Revaz, B., Kadovaki, K., Maggio-Aprile, I., Fischer, O., *Phys. Rev. Lett.* 1998, 80, 3606.

[200] Gradshteyn, I. S., Ryzhik, I. M. *Tables of Integrals, Series, and Products*, Academic Press, New York, 1980, p. 705.

[201] Larsen, Ulf., *Phys. Lett.* 1981, A185, 471.

[202] Kim, Jae Gil, Lee, Eok Kyun, Lee, Soonchil, *Phys. Rev.* 1995, B51, 670.

[203] Darby, M. I., *Am. J. Phys.* 1969, 37, 354.

[204] Kitaoka, Y., Ohsugi, S., Ishida, K., Asayama, K., *Physica* 1990, C170, 189.

[205] Han Rushan, Chew, C. W., Phua, K. K., Gan, Z. Z., *J. Phys.: Condens. Matter* 1991, 3, 8059.

[206] Batlogg, B., *Solid State Comm.* 1998, 107, 639.

[207] Lyons, K. B., Fleury, P. A., Schneemeyer, L. F., Waszczak, J. V., *Phys. Rev. Lett.* 1988, 60, 732.

[208] Timusk, T., Tanner, D. B. In *Physical Properties of High Temperature Superconductors*, Ginsberg, D. M., Ed., World Scientific, Singapore, 1992, Vol. 3, p. 363.

[209] Schruter, M. A., *Mater. Sci. Eng.* 1993, B19, 129.

[210] Abrikosov, A. A., L. P. Gorkov, L. P., Dzyaloshinski, I. E. *Methods of Quantum Field Theory in Statistical Physics*, Prenticl-Hall, Englewood Cliffs, N. J., 1963, p. 398.

[211] Golovashkin, A. I., Kraiskaya, K. V., Shelakov, A. L., *Sov. Phys. Solid State* 1990, 32, 98.

[212] Iye, Y. In *Physical Properties of High Temperature Superconductors*, Ginsberg, D. M., Ed., World Scientific, Singapore, 1992, Vol. 3, p. 303.

[213] Buckel, W., Kleiner, R. K. *Superconductivity*, WILEY-VCH, Weinheim, 2004, p. 95.

[214] Jiang, H., Huang, Y., How, H., Zhang, S., Vittoria, C., Widom, A., Chrisey, D. B., Horwitz, J. S., Lee, R., *Phys. Rev. Lett.* 1991, 66, 1785.

[215] Naito, M., Sato, H., Yamamoto, H., *Physica* 1997, C293, 36.

[216] Kashiwaya, S., Ito, T., Oka, K., Ueno, S., Takashima, H., Koyanagi, M., Tanaka, Y., Kajimura, K., *Phys. Rev.* 1998, B57, 8680

[217] Andreone, A., Cassinese, A., Chiara, A. D., Vaglio, R., Gupta, A., Sarnelli, E., *Phys. Rev.* 1994, B49, 6392.

[218] Wu, D. H., Mao, J., Mao, S. N., Peng, J. L., Xi, X. X., Venkatesan, T., Greene, R. L., Anlage, S. M., *Phys. Rev. Lett.* 1993, 70, 85.

[219] Cooper, J. R., *Phys. Rev.* 1996, B54, R3753.

[220] Luke, G. M., Fudamoto, Y., Kojima, K., Larkin, M., Merrin, J., Nachumi, B., Uemura, Y. J., Sonier, J. E., Ito, T., Oka, K., de Andrade, M., Maple, M. B., Uchida, S., *Physica* 1997, C282 - 287, 1465.

[221] Alff, L., Meyer, S., Kleefish, S., Schoop, U., Marx, A., Sato, H., Naito, M., Gross, R., *Phys. Rev. Lett.* 1999, 83, 2644.

[222] Tsuei, C. C., Kirtley, J. R., *Phys. Rev. Lett.* 2000, 85, 182.

[223] Luo, H. G., Xiang, T., *Phys. Rev. Lett.* 2005, 94, 027001.

[224] Tsuei, C. C., Private commun. June 2000.

[225] Aprili, M., Corington, M., Paraoanu, E., Neidermeier, B., Greene, L. H., *Phys. Rev.* 1998, B57, 8139.

[226] Wu, H., Zhao, L., Yuan, J., Cao, L. X., Zhong, J. P., Gao, I. J., Xu, B., Dai, P. C., Zhu, B. Y., Qiu, X. G., Zhao, B. R., *Phys. Rev.* 2006, B73, 104512.

[227] Krotkov, P., Chubukov, A. V., *Phys. Rev. Lett.* 2006, 96, 107002.

[228] Higging, J. S., Dagan, Y., Barr, M. C., Weaver, B. D., Greene, R. L.,*Phys. Rev.* 2006, B73, 104510.

[229] Koitzsch, A., Blumberg, G., Gozar, A., Dennis, B. S., Foumier, P., Greene, R. L., *Phys. Rev.* 2003, B67, 184522.

[230] Onose, Y., Taguchi, Y., Ishizaka, K. Tokura, Y.,*Phys. Rev. Lett.* 2001, 87, 217001.

[231] Mackenzie, A. P., Julian, S. R., Diver, A. J., McMullan, G. J., M. P. Ray, M. P., Lonzarick, G. G., Maeno, Y., Nishizaki, S., Fujita, T., *Phys. Rev. Lett.* 1996, 76, 3786.

[232] Bergmann, C., Julian, S. R., Mackenzie, A. P., Nishizaki, S., Maeno, Y., *Phys. Rev. Lett.* 2000, 84, 2662.

[233] Hill, S., Brooks, J. S., Mao, Z. Q., Maeno, Y., *Phys. Rev. Lett.* 2000, 84, 3374.

[234] Singh, D., *Phys. Rev.* 1995, B52, 1385.

[235] Riseman, T. M., Kealey, P. G., Forgan, E. M., Mackenzie, A. P., Galvin, L. M., Tyley, A. M., Lee, S. L., Ager, C., D. Paul, D. McK., Aegerter, C. M., Cubitt, R., Mao, Z. Q., Akima, T., Maeno, Y., *Nature (London)* 1998, 396, 242.

[236] Riseman, T. M., Kealey, P. G., Forgan, E. M., Mackenzie, A. P., Galvin, L. M., Tyley, A. M., Lee, S. L., Ager, C., D. Paul, D. McK., Aegerter, C. M., Cubitt, R., Mao, Z. Q., Akima, T., Maeno, Y., *Nature (London)*2000, 404, 629.

[237] Imai, T., Hunt, A. W., Thurber, K. R., Chou, F. C., *Phys. Rev. Lett.* 1998, 81, 3006.

[238] Rice, T. M., Sigrist, M., *J. Phys.: Condens. Matter* 1995, 7, 643.

[239] Ishita, K., Mukuda, H., Kitaoka, Y., Asayama, K., Mao, Z. Q., Mori, Y., Maeno, Y., *Nature (London)* 1998, 396, 658.

[240] Laube, F., Goll, G., Lohneysen, H. V., Fogelstrom, M., Lichtenberg, F., *Phys. Rev. Lett.* 2000, 84, 1595.

[241] Zutic, I., Mazin, I., *Phys. Rev. Lett.* 2005, 95, 217004.

[242] Luke, G. M., Fudamoto, Y., Kojima, K. M., Larkin, M., Nachumi, B., Uemura, Y. J., Maeno, Y., Mao, Z. Q., Mori, Y., Nakamura, H., Sigrist, M., *Nature (London)* 1998, 394, 558.

[243] Sigrist, M., Agterberg, D., Furusaki, A., Honerkamp, C., Ng, K. K., Rice, T. M., Zhitomirsky, M. E., *Physica* 1999, C317-318, 134.

[244] Mazin, I. I., Singh, D. J., *Phys. Rev. Lett.* 1997, 79, 733.

[245] Maeno, Y., *Nature (London)* 1997, 372, 532.

[246] Maeno, Y., *Physica* 1997, C282-287, 206.

[247] Maeno, Y., Nishizaki, S., Yoshida, K., Ikeda, S., Fujita, T., *J. Low Temp. Phys.* 1997, 105, 1577.

[248] Ishida, K., Kitaoka, Y., Asayama, K., Ikeda, S., Nishizaki, S., Maeno, Y., Yoshida, K., Fujita, T., *Phys. Rev.* 1997, B56, 505.

[249] Miyake, K., Nariyako, O., *Phys. Rev. Lett.* 1999, 83, 1423.

[250] Bonalde, I., Yanoff, B. D., Salamon, M. B., Van Harlingen, D. J., Chia, E. M. E., *Phys. Rev. Lett.* 2000, 85, 4775.

[251] Nishizaki, S., Maeno, Y., Farner, S., Ikeda, S., Fujita, T., *Physica* 1997, C282 - 287, 1413.

[252] Nishizaki, S., Maeno, Y., Farner, S., Ikeda, S., Fujita, T., *J. Phys. Soc. Jpn.* 1998, 67, 560.

[253] Nishizaki, S., Maeno, Y., Mao, Z. Q. unpublished.

[254] Ishida, K., Mukuda, H., Kitaoka, Y., Mao, Z. Q., Mort, Y., Maeno, Y., *Phys. Rev. Lett.* **84** (2000) 5387.

[255] Kuwabara, T., Ogata, M., *Phys. Rev. Lett.* 2000, 85, 4586.

[256] Liu, Fu-sui, Chen, Wan-fang, Peng, Kuang-ding, *Physica* 2001, C361, 210.

[257] Basov, D. N., Liang, R., Bonn, D. A., Hardy, W. N., Dabrowski, B., Quijada, M., Tanner, D. B., Rice, J., Ginsberg, D. M., Timusk, T., *Phys. Rev. Lett.* 1995, 74, 598.

[258] Edwards, H. L., Derro, D. J., Barr, A. L., Markert, J. T., de Lozanne, A. L., *Phys. Rev. Lett.* 1995, 75, 1387.

[259] Gagnon, R., Pu, S., Ellman, B., Taillefer, L., *Phys. Rev. Lett.* 1997, 78, 1976.

[260] Sridhar, S., Srikanth, H., Zhai, Z., Willemsen, B. A., Jacobs, T., Erb, A., Walker, E., Srikiger, R., *Physica* 1997, C282-287, 256.

[261] Steglich, F. In *Superconductivity Springer Series in Solid State Sciences*, Bednortz, J. G., Müller, K. A., Ed., Springer-Verlag, Berlin, 1990, p. 315.

[262] Donovan, S., Schwartz, A., Gruner, G., *Phys. Rev. Lett.* 1997, 79, 1401.

[263] Tou, H., Kitaoka, Y., Asayama, K., Kimura, N., Onuki, Y., Yamamoto, E., Maezawa, K., *Phys. Rev. Lett.* 1996, 77, 1374.

[264] Schottl, S., Schuberth, E. A., Flachbart, K., Kycia, J. B., Hong, J. I., Seidman, D. N., Halperin, W. P., Hufnagl, J., Bucher, E., *Phys. Rev. Lett.* 1999, 82, 2378.

[265] Amann, A., Mota, A. C., Maple, M. B., Lohneysen, H. V., *Phys. Rev.* 1998, B57, 3640.

[266] Schuberth, E. A., Struckler, B., Andres, K., *Phys. Rev. Lett.* 1992, 68, 117.

[267] Suderrow, H., Brison, J. P., Huxley, A., Flouquet, J., *J. Low Temp. Phys.* 1997, 108, 11.

[268] Luke, G. M., Keren, A., Le, L. P., Wu, W. D., Uemura, Y. J., Bonn, D. A., Taillefer, L., Garrett, J. D., *Phys. Rev. Lett.* 1993, 71, 1466.

[269] Jourdan, M., Huth, M., Adrian, H., *Nature (London)* 1999, 398,47.

[270] Kimura, N., Ito, K., Saitoh, K., Umeda, Y., Aoki, H., *Phys. Rev. Lett.* 2005, *95*, 247004.

[271] Edwards, D. M., *J. Phys.* 1982, F12, 1789.

[272] Edwards, D. M., *Physics* 1991, B169, 271.

[273] Lebed, A. G., *Phys. Rev.* 1999, B59, 729.

[274] Schmalian, J., *Phys. Rev. Lett.* 1998, 81, 4232.

[275] Jerome, J. In *Advances in Superconductivity*, Deaver, B., Ruvalds, J., Ed., Plenum Press, New York, 1983, p. 129.

[276] Abrikosov, A. A., *J. Low Temp. Phys.* 1983, 53, 539.

[277] Gorkov, L. P., Jerome, D., *J. Phys. (Paris)* 1985, 46, 64.

[278] Hasegawa, Y., Fukuyama, H., *J. Phys. Soc. Jpn.* 1987, 56, 877.

[279] Hawley, M. E., Gray, K. E., Tarris, B. D., Wang, H. H., Carlson, K. D., Williams, J. M., *Phys. Rev. Lett.* 1986, 57, 629.

[280] Toyota, N., Sasaki, T., *Solid State Comm.* 1990, 74, 361.

[281] Lee, I.J., Brown, S. E., Yu, W., Naughton, M. J., Chaikin, P. M., *Phys. Rev. Lett.* 2005, 94, 197001.

[282] Uehara, M., Nakata, T., Akimitsu, J., Takahashi, H., Mori, N., Kinoshita, K., J. *Phys. Soc. Jpn.* 1996, 65, 2764.

[283] Osafune, T., Motoyama, N., Eisaki, H., Uchida, S., Tajima, S., *Phys. Rev. Lett.* 1999, 82, 1313.

[284] Katano, S., Nagata, T., Akimitsu, J., Nishi, M., Kakurai, K., *Phys. Rev. Lett.* 1999, 82, 636.

[285] Dagotto, E., Rice, T. M., *Science* 1996, 271, 618.

[286] Azuma, M., Hiroi, Z., Takano, Z., Bando, M., Takeda, Y., *Nature (London)* 1992, 36, 775.

[287] Zou, Z., Kawanaka, H., Minawa, T., Bando, H., Nishihara, Y. In *Advances in Superconductivity*, Koshizuka, N., Tajima, S., Ed., Springer-Verlag, Tokyo, 1999, p. 379.

[288] Wu, Hua, Zheng, Qing-qi, Gong, Xin-gao, Lin, H. Q., *J. Phys.: Condens. Matter* 1999, 11, 4637.

[289] Petrakovski, G. A., Losawa, G. V., Kuselev, N. I., Ovchenikov, S. G., Ivanova, N. B., Chernov, V. K., *Pis'ma Zh. Eksp. Fiz.* 1991, 54, 224.

[290] Vaglio, R., Terris, B. D., Zasadzinski, J. F., Gray, K. E., *Phys. Rev. Lett.* 1984, 53, 149.

[291] Blagoev, K. B., Engelbrecht, J. R., Bedell, K. S., *Phys. Rev. Lett.* 1999, 82, 133.

[292] Lide, D. R. In *Handbook of Chemistry and Physics*, 73rd edition 1992-1993, CRC Press, London, 1993, p. 12-106.

[293] Allen, P. B. In *Physical Properties of High Temperature Superconductors*, Ginsberg, D. M. Ed., World Scientific Publication Co. Singapore, 1989, Vol. 1, p. 242.

[294] Zheng, X. G., Tsumi, N., Tanaka, S., Sujuki, M., Xu, C. N., *Physica* 1999, C321, 67.

[295] Igbal, Z., Baughman, R. H., Ramakrishnan, B. L., Khare, S., Murthy, N. S., Barnemann, H. J., Morris, D. E., *Science* 1991, 254, 826.

[296] Ramirez, A. P., Kortan, A. R., Rosseinsky, M. J., Duclos, S. J., Mujsce, A. M., Haddon, R. C., Murphy, D. W., Makhija, A. V., Zahurak, S. M., Lyons, K. B., *Phys. Rev Lett.* 1992, 68, 1058.

[297] Fuhrer, M. S., Cherrey, K., Zettl, A., Cohen, M. L., *Phys. Rev. Lett.* 1999, 83, 404.

[298] Zhang, Z., Chen, Chia-Chun, Kelty, S. P., Dai, Hongjie, Lieber, C. M., *Nature (London)* 1991, 353, 333.

[299] Palstra, T. T. M., Haddon, R. C., Hebard, A. F., Zaanen, J., *Phys. Rev. Lett.* 1992, 68, 1054.

[300] Li, C. Y., Zhao, W. B., private communication.

[301] Simovic, B., Jerome, D., Rachdi, F., Baumgartner, G., Forro, L., *Phys. Rev. Lett.* 1999, 82, 2298.

[302] Moore, D. E., Beasley, M. R., Rowell, K., *J. Phys. (Paris)* 1978, 39, C6-1390.

[303] Totto, I., Ruvalds, J., *Phys. Rev.* 1979, B20, 952.

[304] Vieland, L. J., Wicklund, A. W., *Phys. Rev.* 1968, 166, 424.

[305] Kihlstrom, K. E. In *Novel Superconductivity*, Wolf, S. A., Kresin, V. Z. Ed., Plenum Press, New York, 1987, p. 95.

[306] Bozovic, I., Eckstein, J. N. In *Physical Properties of High Temperature Superconductors*, Ginsberg, D. M. Ed., World Scientific Publishing Co., Singapore, 1996, Vol. 5, p. 101.

[307] Bozovic, I., Eckstein, J. N., Klausmeiex, M. E., Virshup, G. F., *J. Supercond.* 1992, 5, 19.

[308] Bozovic, I., J. N. Eckstein, J. N., Virshup, G. F., *Physica* 1994, C235-240, 178.

[309] Li, X., Kawai, T., Kawai, S., *Jpn. J. Appl. Phys. Lett.* 1994, 33, 18.

[310] Norton, D. P., Chakoumakos, B. C., Budai, J. D., Lowndes, D. H., B. C. Sales, B. C., Thomson, J. R., Christen, D. K., *Science* 1994, 265, 2074.

[311] Anderson, P. W. *The Theory of Superconductivity in the High-T_c Cuprate Superconductors*, Princeton University Press, Preceton, 1997.

[312] Takagi, H., Nohare, M., Cava, R. J., *Physica* 1997, B237-238, 292.

[313] Tomy, C. V., Chang, L. J., Paul, D. McK., Ritter, P. C., *Physica* 1997, C230-232, 872.

[314] Iwamoto, Y., Oda, T., Ueda, K., Kohara, T., *Physica* 1997, B230-232, 886.

[315] Terashina, T., Haworth, C., Takeya, H., Uji, S., Aoki, H., *Phys. Rev.* 1997, B56, 5120.

[316] Jacobs, T., Willemsen, B. A., Sridhar, S., Nagarajam, R., Gupta, L. C., Hossain, Z., Mazumbar, C., Canfield, P. C., Cho, B. K., *Phys. Rev.* 1995, *B52*, 7022.

[317] Bommli, F., Degiorgi, L., Wachter, P., Cho, B. K., Canfield, P. C., Chau, R., Maple, M. B., *Physica* 1997, B230-232, 879.

[318] Kishimoto, Y., Shibata, N., Ohno, T., Kitaoka, Y., Asayama, K., Amaya, K., Kanashino, T., *J. Phys. Soc. Jpn.* 1992, 61, 696.

[319] Annett, J. F., *Physica* 1999, C317-318, 1.

[320] Pfleiderer, C., Uhlarz, M., Hayden, S. M., Vollmer, R., Loheysen, H. v., Bernhoeft, N. R., Lonzarich, G. G., *Nature (London)* 2001, 412, 58.

[321] Takabataka, T., Iga, F., Yoshino, T., Echizen, Y., Katoh, K., Kobayashi, K., Higa, M., Shimizu, N., Bando, Y., Nakamoto, G., Fujii, H., Izawa, K., Suzuki, T., Fujita, T., Sera, M., Hiroi, M., Maezawa, K., Mock, S., Lohneysen, H. V., Bruckl, A., Neumaiex, K., Andres, K. J., *Magn. Magn. Mater.* 1998, 177-181, 277.

[322] Hou, Yumin, Yan, Shou-sheng, *Physica* 2000, C341-348, 883.

[323] Kagan, Y., Kikoin, K. A., Mishchenko, A. S., *Phys. Rev.* 1997, B55, 12348.

[324] Chahara, Ken-ichi, Ohno, T., Kassai, M., Kozono, Y., *Appl. Phys. Lett.* 1993, 63, 1990.

[325] Von Helmolt, R., Wecker, J., Holzapfel, B., Schultz, L., Samwer, K., *Phys. Rev. Lett.* 1993, 71, 2331.

[326] Dessau, D. S., Saitoh, T., C. -H. Park, C. -H., Shen, Z. -X., Moritomo, Y., Tokura, Y., *Inter. J. Modern Phys.* 1998, 12, 3389.

[327] Dessau, D. S., Saitoh, T., C. -H. Park, C. -H., Shen, Z. -X., Villella, P., Hamada, N., Moritomo, Y., Tokura, Y., *Phys. Rev. Lett.* 1999, 81, 192.

[328] Dardel, B., Malterre, D., Grioni, M., Weibel, P., Baer, Y., Levy, F., *Phys. Rev. Lett.* 1991, 67, 3144.

[329] Terrasi, A., Mersi, M., Berger, H., Margaritondo, G., Kelley, R. J., Onellion, M., *Phys. Rev.* 1995, B52, 5592.

[330] Joynt, R., *Science* 1999, 284, 277.

[331] Saxena, S. S., Agarvar, P., Ahilan, K., Grosche, F. M., Naselulmmer, R. K. W., Steiner, M. J., Pugh, E., Walker, I. R., Jullan, S. R., Monthoux, P., Lonzarich, G. G., Huxley, A., Shelkin, I., Bralthwaite, D., Flouquet, J., *Nature (London)* 2000, 406, 588.

[332] Huxley, A., Sheikin, I., Ressouche, E., Kernavanois, N., Braithwaite, D., Calemczuk, R., 2001, *Phys. Rev.* B63, 144519.

[333] Anderson, P. W., Morel, P., *Phys. Rev.* 1961, 123, 1911.

[334] Balian, R., Werthamer, N. R., *Phys. Rev.* 1963, 131, 1553.

[335] Tilley, D. R., Tilley, J. *Superfluidity and Superconductivity*, Adam Hilger Ltd, Bristol, 1986, pp. 27, 336, 348.

[336] Legget, A., *Rev. Mod. Phys.* 1975, 47, 331, 343.

[337] Li, Qi, Belitz, D., Kirpattick, T. R., *Phys. Rev.* 2006, B74, 134503.

[338] Tilley, D. R., Tilley, J. *Superfluidity and Superconductivity*, Adam Hilger Ltd, Bristol, 1986, pp. 1, 171, 31, 4.

[339] Mayers, J., *Phys. Rev.* 2006, B74, 014516.

[340] Abrikosov, A. A., *Physica* 1965, 2, 5.

[341] Larsen, U., *Z. Phys.* 1972, 256, 65.

[342] Keiter, H., Morandi, G., *Phys. Rep.* 1984, 109, 277.

[343] Brandt, U. B., Keiter, H., Liu, Fu-sui, *Z. Phys.* 1985, 58, 267.

[344] Popov, V. N., Fedotor, S. A., *Sov. Phys. JETP.* 1988, 67, 535.

[345] Xiao, Ming, Liu, Fu-sui, Zhang, Xin-wei, *Phys. Rev.* 1996, B54, 4052.

[346] Andrei, N., Coleman, P., *Phys. Rev. Lett.* 1989, 62, 595.

[347] Nazarenko, A., Dagotto, E., *Phys. Rev.* 1996, B54, 13158.

[348] Cheung, C. Y., Mattuck, R. D., *Phys. Rev.* 1968, 2, 466.

[349] Liu, Fu-sui, Wang, Zhong-cheng, Chen, Wan-fang, Yuan, Xiao-jian, *Phys. Rev.* 1995, B51, 12491.

[350] Keiter, H., *Z. Phys.* 1968, 213, 466.

[351] Liu, Fu-sui, Roshen, W. A., Ruvalds, J., *Phys. Rev.* 1987, B36, 492.

Index